# Neutron Stars and Their Birth Events

# NATO ASI Series

**Advanced Science Institutes Series**

*A Series presenting the results of activities sponsored by the NATO Science Committee, which aims at the dissemination of advanced scientific and technological knowledge, with a view to strengthening links between scientific communities.*

The Series is published by an international board of publishers in conjunction with the NATO Scientific Affairs Division

| | | |
|---|---|---|
| **A** | **Life Sciences** | Plenum Publishing Corporation |
| **B** | **Physics** | London and New York |
| | | |
| **C** | **Mathematical** | Kluwer Academic Publishers |
| | **and Physical Sciences** | Dordrecht, Boston and London |
| **D** | **Behavioural and Social Sciences** | |
| **E** | **Applied Sciences** | |
| | | |
| **F** | **Computer and Systems Sciences** | Springer-Verlag |
| **G** | **Ecological Sciences** | Berlin, Heidelberg, New York, London, |
| **H** | **Cell Biology** | Paris and Tokyo |

**Series C: Mathematical and Physical Sciences - Vol. 300**

# Neutron Stars and Their Birth Events

edited by

## Wolfgang Kundt

Institut für Astrophysik,
Universität Bonn,
Bonn, F.R.G.

Springer-Science+Business Media, B.V.

Proceedings of the NATO Advanced Study Institute on
Neutron Stars: Their Birth, Evolution, Radiation and Winds
Erice, Sicily, Italy
September 5–17, 1988

**Library of Congress Cataloging in Publication Data**

NATO Advanced Study Institute on Neutron Stars, Their Birth Evolution,
   Radiation, and Winds (1988 : Erice, Italy)
     Neutron stars and their birth events : proceedings of the NATO
   Advanced Study Institute on Neutron Stars, Their Birth, Evolution,
   Radiation, and Winds, Erice, Sicily, Italy, 5-17 September 1988 /
   edited by Wolfgang Kundt.
        p.    cm. -- (NATO ASI series. Series C, Mathematical and
   physical sciences ; v. 300)
     "Published in cooperation with NATO Scientific Affairs Division."

     1. Neutron stars--Congresses.  2. Pulsars--Congresses.
   3. Supernova remnants--Congresses.   I. Kundt, Wolfgang.  II. North
   Atlantic Treaty Organization.  Scientific Affairs Division.
   III. Title.  IV. Series.
   QB843.N4N37  1988
   523.8'874--dc20                                            89-26695
                                                                  CIP

ISBN 978-94-010-6718-8        ISBN 978-94-009-0515-3 (eBook)
DOI 10.1007/978-94-009-0515-3

# TABLE OF CONTENTS

# PREFACE

This volume is the documentation of the second Course on 'Neutron Stars, Active Galactic Nuclei and Jets', held at Erice in September 1988.

This second Course was devoted to our knowledge about neutron-star sources. The poster spoke of: pulsars, accreting X-ray sources and jet engines, perhaps also UHE pulsars, $\gamma$-ray bursters and black-hole candidates. Neutron stars have even been proposed as the primary cosmic-ray boosters. Most of their properties are still controversial, such as their birth mechanism (neutrino versus magnetic piston), internal structure (neutrons, quarks, strange particles), magnetic, thermal and spin histories, wind generation (hydrogen versus pair plasma, radiation versus centrifugal pressure), magnetospheric structure and accretion modes (along field lines versus quasi-Keplerian).

The listed controversies have largely survived through the Course and entered into the proceedings. Several lecturers speak of 'magnetic-field decay' in neutron stars, of the 'recycling' of old pulsars, and of 'accretion-induced collapse' of white dwarfs as though such processes were textbook knowledge. Terms and abbreviations like RPSR (=recycled pulsar), spinup line, AIC, and ADC (=accretion disk corona) help to foster the assumptions. It is not clear to me at this time whether any of these notions has an application to reality.

Feynman had confidence in quantum electrodynamics when he could (correctly) predict Dirac's number, the magnetic moment $\mu$ of the electron in natural units, to within ten significant figures: $\mu 2mc/e\hbar$ = 1.001159652193. Astrophysics is not equally confirmed. Even our planet Earth holds plenty of secrets: Every educated geophysicist 'knows' that oil and coal are of biogenic origin, i.e. consist of former plants left to themselves at high pressure and temperature, and that salt deposits are evaporated seas; yet there are a handful of respectable sceptics, for some hundred years, who are impressed by the order, quantity, and cleanliness of the deposits and who are convinced that most of it ($\gtrsim$ 90% by mass) is the result of outgassing, carbon enrichment, dehydration, and diffusive separation.

Our experimental knowledge decreases when we leave Earth and take interest in the other planets and the moons. It decreases when we study the Sun, and again when we study more distant stars. It decreases when we leave the Galaxy for other galaxies and galaxy clusters, and reaches its minimum –

if such exists - when we beg the question of the origin of the Universe. With increasing distance, we lose information and replace it by hypotheses, in order to make predictions.

There is nothing wrong with such a procedure. The scientific approach consists of replacing gaps in our knowledge by plausible hypotheses and exploring their implications to the point where they might lead to inconsistencies. But even at the speed of modern research, such inconsistencies can take decades to grow obvious. By that time, generations of publications have been printed and have used the term "general agreement" or "fact" for what earlier approaches called an assumption. This long timescale of falsification makes astrophysics more vulnerable to tacit assumptions than laboratory physics and causes scientific evolution to happen in the form of revolutions, in the sense highlighted by Thomas Kuhn (MIT, 1970).

Tacit assumptions help keeping a presentation simple, yet they can hamper our insight when they turn out to be wrong. Examples of what I have in mind are listed on pages 51/2 below. In his book 'Surely you're joking, Mr. Feynman', the author states on page 341: "Details that could throw doubt on your interpretation must be given, if you know them". The whole chapter 'Cargo Cult Science' explains what he means by it.

So what can be said about this book? The reader will find different explanations of the same facts by the different authors. He or she should not deplore but rather appreciate such a situation. Research need not be unanimous. It benefits from controversies. When Rahel Dewey calls her review of radio pulsars (in NATO ASI C 262) "the standard three-quarter-truths", she is aware of the situation.

An example may clarify my point: how old are the ms-pulsars? Are they as old as their spindown ages $P/2\dot{P}$ ($=10^{8.5\pm1}$ yr) indicate, or are they as young as their kinematic ages $(z-z_s)/\dot{z}$ ($=10^{6.5\pm1.5}$ yr) suggest? Are their companions as old as their colour temperature seems to tell us? Has their magnetic moment (half-way) decayed or not? Have they been recycled or born fast? In a situation as confused as this, I don't think we are in need of a quick decision but rather of an open mind. All inferences about their birthrates, histories and futures strongly depend on their age.

Another astrophysical hypothesis of importance is whether or not diffusive shock acceleration works, i.e. whether or not particles can be efficiently boosted to relativistic energies 'in situ', in strong interstellar or extragalactic shocks. The question is of relevance both to the non-thermal radiation of supernova shells, to the generation of the cosmic rays, and to the supersonic jets of the bipolar-flow sources. On page 303, Samuel Falle argues that diffusive shock acceleration cannot be very efficient.

Before this preface comes to an end, I should like to cordially thank Antonino Zichichi and his team at the CCSEM for their strong support; further NATO for the ASI award and the European Physical Scoiety and National Science Foundation for additional help. I should also like to thank Christine Tilly-Schröder and Kathy Schrüfer from the Bonn astrophysics Institute for years of reliable service, and the photographic laboratory of our neighbouring Max-Planck Institute for ready help. Finally, my thoughts go to the participants - both lecturers, students and company - who made the Course a big event in my scientific life.

Bonn, September 1989

Wolfgang Kundt
Institut für Astrophysik
der Universität

# LIST OF PARTICIPANTS

| | |
|---|---|
| Ames, Susan | I.f.R.A., Bonn, F.R.G. |
| Banhatti, Dilip | Madurai-Kamaraj Univ., India |
| Baykal, Altan | Middle East Techn. Univ.,Ankara, Turkey |
| Bhattacharya, Dipankar | R.R.I., Bangalore, India |
| Branch, David | Norman, Oklahoma, U.S.A. |
| Camenzind, Max | Landessternwarte, Heidelberg, F.R.G. |
| Carioli, Maurizio | M.P.I.f.K., Heidelberg, F.R.G. |
| Chopra, Namrata | Delhi University, India |
| Coté, Jacqueline | Amsterdam Univ., Netherlands |
| Dal Fiume, Daniele | C.N.R., Bologna, Italy |
| Falle, Samuel | Univ. of Leeds, UK |
| Gök, Fatma | Middle East Techn. Univ., Ankara, Turkey |
| Gürbüz, Gülsen | M.S.U., Istanbul, Turkey |
| Kallrath, Josef | I.f.A., Bonn, F.R.G. |
| Kulkarni, Shri | Caltech, Pasadena, CA, U.S.A. |
| Kundt, Wolfgang | I.f.A., Bonn, F.R.G. |
| Kunte, Prabhakar | T.I.F.R., Bombay, India |
| Lewin, Walter | M.I.T., Cambridge, Mass., U.S.A. |
| Lundquist, Peter | Lund Observatory, Sweden |
| Pols, Onno | Amsterdam Univ., Netherlands |
| Salinas, Ener | Stockholm Univ., Sweden |
| Sang, Yeming | Louisiana State Univ., Baton Rouge, U.S.A. |
| Schaaf, Reinhold | I.f.A., Bonn, F.R.G. |
| Sorrell, Wilfred | Univ. of Wisconsin, Madison, U.S.A. |
| Srinivasan, Gianesan | R.R.I., Bangalore, India |
| Strom, Richard | Radio Obs., Dwingeloo, Netherlands |
| Van Paradijs, Jan | Amsterdam Univ., Netherlands |
| Verbunt, Frank | Utrecht Univ., Netherlands |
| Zapata, Arauco, Juan | San Andres Univ., La Paz, Bolivia |

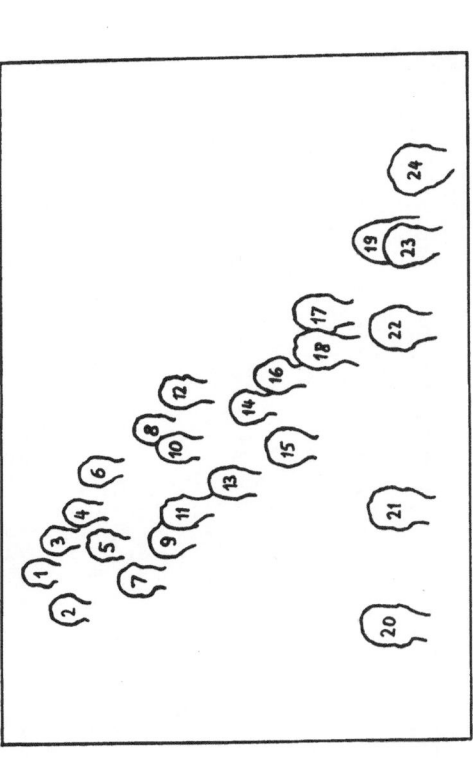

1. Dal Fiume, Daniele
2. Kallrath, Josef
3. Banhatti, Dilip
4. Sang, Yeming
5. Kulkarni, Shri
6. Srinivasan, Gianesan
7. Bhattacharya, Dipankar
8. Ames, Susan
9. Branch, David
10. Kunte, Prabhakar
11. Van Paradijs, Jan
12. Schaaf, Reinhold
13. Verbunt, Frank
14. Baykal, Altan
15. Gök, Fatma
16. Chopra, Namrata
17. Salinas, Ener
18. Lundquist, Peter
19. Coté, Jacqueline
20. Pols, Onno
21. Falle, Samuel
22. Carioli, Maurizio
23. Kundt, Wolfgang
24. Lewin, Walter

# 1. OBSERVED NEUTRON-STAR PROPERTIES

W. Kundt
Institut f. Astrophysik
der Universität Bonn
Auf dem Hügel 71
D 5300 Bonn 1, FRG

ABSTRACT. The Galaxy contains at least $10^{9\pm1}$ neutron stars which are thought to be born in supernova explosions of massive stars, and which we observe as radio pulsars if isolated and as X-ray binaries if attached to a close companion. Their beams are likely to have fan shape. Young radio pulsars are also observed as $\gamma$-ray pulsars, perhaps even as ultrahard $\gamma$-ray pulsars, and occasionally as infrared, optical, and X-ray pulsars. Pulsars die after $10^{6.7\pm0.3}$ years, probably due to flux alignment (with the spin axis).

Binary neutron stars are probably spun down by the wind of their companion - generating cosmic rays - until an accretion disk forms which confines the corotating magnetosphere and reverses the spindown, though hardly to the msec regime. One then sees an X-ray source which may flicker (down to several msec intervals, perhaps because of inverse-Compton losses of relativistic pair plasma) and pulse if a significant amount of matter accretes onto the polar caps. Non-pulsing X-ray sources can emit X-ray bursts, explained by nuclear-chemical explosions (type 1) or spasmodic accretion (type 2), and show enhanced variability in a preferred frequency interval (=QPO). Even $\gamma$-ray bursts may be emitted by (old) neutron stars.

The jet sources Sco X-1, SS 433, and Cyg X-3 and the black-hole candidates - headed by Cyg X-1 and A 0620-00 - may likewise be binary neutron stars.

All observed neutron stars appear to be strongly magnetized, with dipole moments $BR^3 := \mu = 10^{30.8\pm0.5}$ G cm³ except for the (eight) msec pulsars whose moments may be $10^{3\pm1}$ times weaker. Their surface temperatures are $\leq 10^{5.8}$ K after a few $10^2$ yr, cooler than expected for conductive plus radiative cooling.

Supernova shells are initially almost spherical, or barrel-shaped; when older than $10^4$ years they can look exotic, like CTB 80, IC 433, the flying duck, and others.

1

*W. Kundt (ed.), Neutron Stars and Their Birth Events, 1–58.*
© *1990 Kluwer Academic Publishers.*

Perhaps all of them contain a neutron star near their center.


## 1.1 THE (RADIO) PULSARS

More than 440 pulsars in the Galaxy are presently known, with repetition periods between 1.5 (or even $\lesssim 1$) ms and 4.3 s, average radio luminosities (near 0.4 GHz) between $10^{25}$ and $10^{31}$ erg s$^{-1}$, low-frequency cutoffs mostly near $10^2$ MHz. Several of them - the youngest and/or nearest - have also been reported to emit optical and $\gamma$-ray pulses and even ultrahard $\gamma$-ray pulses, with photon energies between $10^{11}$ and $10^{17}$ eV, and luminosities a small but non-negligible fraction of the spindown power $I\Omega\dot{\Omega}$ . Some of the properties of the fastest known pulsars are listed in table 1, ordered w.r.t. increasing pulse period. For references on pulsar data see Lyne et al (1985), Taylor & Stinebring (1986), Kundt (1985b, 1988a), the UHE report by Schwarzschild (1988), and the lectures by Kulkarni in this volume.

The pulsars form a disk population of the Galaxy, with a scale height of some 350 pc and a possible clustering in a ringlike domain of radius $\leqslant 5$ Kpc. One therefore observes an isotropic distribution around the Sun out to some 200 pc beyond which it changes into a Milky-Way distribution; but completeness is lost beyond distances of 0.5 Kpc. The proper motions of some 25 to 50 well-observed pulsars have an inferred 3-d mean value of $\langle \vec{v}^2 \rangle^{1/2} = 10^{2.3}$ Km s$^{-1}$, much higher than that of any other stellar population but consistent with the large scale height and a statistical age of $10^{6.7}$ years. Unless SN explosions are strongly asymmetric - which is not expected on theoretical grounds - the only known mechanism for boosting the pulsars is recoil from a binary (multiple) system during the explosion of their progenitor stars.

One infers that pulsars derive from multiple stars, consistent with the fact that massive stars are either multiple or runaway stars, the latter with space velocities of order (50±25) Km s$^{-1}$ (Stone, 1979). In this interpretation, the runaway stars receive their kick during the first supernova explosion and have a neutron-star companion. (This companion is easily hidden by their windzone, cf. Indulekha et al (1988)). Pulsars are normally formed during the second (last) SN explosion in the system, already at a large scale height; but there is the possibility that even the first-born neutron star in a binary system spins fast enough, at the time of the second explosion, to become a (slow and long-lived) pulsar. If a fraction $\varepsilon$ of all first-born neutron stars turn eventually into (slow) pulsars of long lifetime $\xi t$ (where t is the lifetime of last-born pulsars), the percentage $\varepsilon\xi$ of slow pulsars in the sky may be comparable to that of fast pulsars.

Pulsars of (spin) period $\lesssim 0.1s$ are called ms-PSRs (Dewey et al, 1986; Stokes et al, 1986). According to table 1, we presently know eight of them, all within 20 pc of the middle of the Galactic disk or near the center of a globular cluster. They have spindown ages of $10^{8.5\pm1}$ years during which they could reach the scaleheight of the 'ordinary' pulsars at a (tiny) peculiar velocity of 1 Km s$^{-1}$. There is a widespread conviction that this very different distribution is a pure selection effect, so that we shall soon hear of dozens of ms-PSRs at large Galactic heights. Maybe. For the time being, the ms-PSRs differ from the ordinary pulsars by a 30 times smaller scale height – hence at least 30 times smaller age or peculiar velocity – by a $10^{3\pm1}$ times higher spindown power, and by a $10^{3\pm1}$ times smaller transverse magnetic dipole moment $\mu_\perp$ evaluated according to the vacuum dipole radiation formula

$$-I\Omega\dot\Omega = 2\ddot\mu_\perp^2/3c^3 = (2\Omega^4/3c^3)\mu_\perp^2 \tag{1}$$

in which I = moment of inertia, $\Omega$ = angular velocity; see figure 1. Are they born fast, or are they 'recycled', i.e. spun up by accretion?

Spinup by accretion is observed for the $\gtrsim 30$ pulsing X-ray sources listed in table 2. It is inefficient: the secular spinup rates are $\varepsilon := 10^{-1\pm0.5}$ times slower than maximal, i.e. slower by $\varepsilon$ than if the X-ray power were supplied with the specific angular momentum of the corotation radius $r_{cor} = (GM/\Omega^2)^{1/3}$ (Kundt, 1985b). (A slightly different, and more pessimistic view of spinup is held by Henrichs (1983)). Consequently, the accreting pulsators would take

$$t \gtrsim (3/4\varepsilon)\ (\Omega^2/GM)^{2/3}\ I/\dot M = 10^{7.7}\ yr\ \Omega_{3.6}^{4/3}\ I_{45}/\dot M_{18}\ \varepsilon_{-1} \tag{2}$$

to reach a period of P = 1.56 ms (corresponding to $\Omega = 10^{3.6}$ s$^{-1}$) at the (large) Eddington mass rate of $\dot M = 10^{-8}$ $M_\odot$/yr for a constant I and (high) spinup efficiency $\varepsilon = 10\%$. In reality, Cowsik et al (1983) have shown that I grows by more than a factor of 3 when $\Omega$ approaches its critical value, of order $\Omega_{crit} \approx 10^4$ s$^{-1}$, both because of centrifugal growth and a reduction of gravitational binding energy. At the same time, $\varepsilon$ shrinks due to enhanced wind losses. A 1.56 ms pulsar would therefore take more than $10^{8.5}$ yr of spinup at the Eddington rate, during which time it would radiate some $10^{38}$ erg s$^{-1}$ of X-rays and accrete more than a solar mass; but from whom? Most X-ray binaries radiate less than $10^{37}$ erg s$^{-1}$, implying minimal spinup times of $\gtrsim 10^{9.5}$ yr, comparable to, or perhaps much longer than, the lifetime of a ms-PSR. This estimate conflicts with the reported (small number of) $10^{2.2}$ X-ray sources of luminosity $L_x \gtrsim 10^{35}$ erg s$^{-1}$ (between 1 and 10 KeV).

4

Table 1: Some data for fast and/or young and/or irregular (radio) pulsars, in order of increasing pulse period, improved over table 1 of ref. 82. $\tau := P/2\dot{P}$, $B_\perp$ is defined in fig. caption 1, $n := \Omega\ddot{\Omega}/\dot{\Omega}^2$, $\Delta\Omega$ and $\Delta\dot{\Omega}$ are observed jumps in $\Omega$ and $\dot{\Omega}$, and $\Delta t$ is their repetition time. The column 'spectrum' lists detections at non-radio frequencies, and $P_{orb}$ is the orbital period of binary pulsars. The table is complete for the detected fast PSRs ($P < 0.1s$), young PSRs ($\tau < 10^{4.5}$ yr), strongly jumping PSRs ($\Delta\Omega/\Omega > 10^{-8}$), binary PSRs, and hard-spectrum PSRs (detected at $\geq$ IR frequencies). Note that only 3% of all PSRs are binary. New data on ms-PSRs are from refs. 30 and 144, the largest jump from ref. 103 . The first entry interprets the recent discovery by Kristian et al (Nature 338, 234, 1989) as that of a PSR with an almost identical interpulse. See also Kulkarni's lecture.

| NAME | POSITION | P/s | log(τ/yr) | log(B⊥/G) | n | log(ΔΩΩ) | log(ΔΩ̇/Ω̇) | Δt/yr | SPECTRUM | P_orb/d |
|---|---|---|---|---|---|---|---|---|---|---|
| SN 1987 A | 0536-6930 | 0.00102 | | | | | | | 0     ? | 0.3 |
| | 1937+214 | 0.00156 | 8.4 | 8.6 | $10^{3.3}$ | | | | VHE | |
| eclipsing | 1957+20 | 0.00161 | 9.2 | 8.2 | | | | | | 0.38 |
| M 28 | 1821-24 | 0.00305 | 7.5 | 9.3 | | | | | | |
| 47 Tuc | 0021-72 A | 0.00448 | | | | | | | | 0.0223 |
| | 1855+097 | 0.00536 | 9.6 | 8.5 | | | | | | 12.327 |
| 47 Tuc | 0021-72 B | 0.00613 | | | | | | | | 51 |
| | 1953+29 | 0.00613 | 9.5 | 8.6 | | | | | VHE | 117.35 |
| M 13 | 1639+36 | 0.0104 | | | | | | | | |
| M 4 | 1620-26 | 0.0111 | 8.3 | 9.5 | | | | | | 191.44 |
| M 15 | 2127+11 C | 0.0305 | | | | | | | | 0.335 |
| Crab | 0531+21 | 0.0331 | 3.1 | 12.6 | 2.52 | -7.7 | -3.7 | 6 | IR,O,X,γ,UHE | |
| CTB 80 | 1951+32 | 0.0395 | 5.0 | 11.7 | | | | | | |
| LMC | 0540-693 | 0.0503 | 3.2 | 12.7 | 2.01 | | | | noR,O,X | |
| M 15 | 2127+11 B | 0.0561 | | | | | | | | |
| 2 n** | 1913+16 | 0.0590 | 8.1 | 10.3 | | | | | | 0.32 |
| Vela | 0833-45 | 0.0892 | 4.0 | 12.5 | $10^{1.8}$ | -5.7 | -1.7±0.5 | 2.5 | 0   γ | |
| | 1823-13 | 0.101 | 4.3 | 12.4 | | | | | | |
| M 15 | 2127+11 A | 0.111 | ? | | | | | | | ? |
| | 1802+23 | 0.112 | | | | | | | VHE | |
| | 1800-21 | 0.134 | 4.2 | 12.6 | | | | | X | |
| MSH 15-52 | 1509-58 | 0.150 | 3.2 | 13.2 | 2.83 | | | | | |
| | 0355+54 | 0.156 | 5.8 | 11.9 | $10^{3.3}$ | -5.4 | -1 | >10 | | |
| | 0655+64 | 0.196 | 9.7 | 10.0 | | | | | | 1.03 |
| | 0950+08 | 0.253 | 7.2 | 11.4 | | | | | VHE | |
| W 44 | 1853+01 | 0.267 | 4.3 | 12.9 | | | | | | |
| | 1820-11 | 0.280 | 6.5 | 11.8 | | | | | | 357.76 |
| G 201+8.2 | 0656+14 | 0.385 | 5.0 | 12.7 | | | | | X    ? | |
| | 1641-45 | 0.455 | 5.6 | 12.5 | | -6.7 | -2.7 | > 2 | | |
| | 1831-00 | 0.521 | 8.8 | 10.9 | | | | | | 1.81 |
| | 1325-43 | 0.533 | 6.5 | 12.1 | | -7 | | | | |
| | 1737-30 | 0.607 | 4.3 | 12.3 | | | | | | |
| | 2224+65 | 0.683 | 6.0 | 12.4 | | -5.8 | <-3.4 | | | |
| | 1508+55 | 0.740 | 6.4 | 12.3 | $10^{3.5}$ | -9.7 | -2.4 | | | |
| | 0820+02 | 0.865 | 8.1 | 11.5 | | | | | | 1232.40 |
| | 2303+46 | 1.066 | 7.6 | 11.8 | | | | | | 12.340 |

Table 2: The 33 (strong) pulsing binary X-ray sources, in order of increasing spin period P, improved over table 2 of ref. 82. $\langle \dot{P} \rangle$ is the secular period derivative , whose magnitude can be smaller, by a factor of $\lesssim 10^2$, than short-term values of $|\dot{P}|$. The superorbital period P₃ is taken from ref. 126. L stands for the X-ray power. The first entry is from ref. 19. The source 1 E 2259+586 is in a globular cluster, hence its 'companion' is unknown. The distances of 4U 1627-67 and 1E 0630+178 were assumed 10 and 1 Kpc, respectively.

| NAME | POSITION | P/s | log(P/⟨-Ṗ⟩yr) | P$_{orb}$/d | P$_3$/d | log(L$_{37}$) | γ | COMPANION |
|------|----------|-----|---------------|---------|--------|---------|---|-----------|
| 1E | 1024-5732 | 0.061 | | | | ≲-1 | | |
| A | 0538-66 | 0.069 | P/Ṗ = 4yr | 16.66 | | 1.9 tr | | B 2 III e |
| SMC X-1 | 0115-73 | 0.714 | 3.1 | 3.892 | 60 | 1.8 | | B 0 I |
| Her X-1 | 1656+35 | 1.24 | 5.5 | 1.700 | 35 | 0 | VHE | A9-B |
| H | 0850-42 | 1.78 | | | | 0 | | |
| 4U | 0115+63 | 3.61 | 4.5 | 24.31 | | 0.5 tr | VHE | B e |
| V | 0332+53 | 4.38 | > 4 | 34.25 | | ≲ 0.3 | | B e |
| Cen X-3 | 1119-60 | 4.84 | 3.5 | 2.087 | ≈130 | 0.7 | | O 7 III-V e |
| 1E | 1048-5937 | 6.44 | | | | ≲-1 | | |
| 1E | 2259+586 | 6.98 | | 0.03 | | -1.7 | | ε G 109.1-1.0 |
| 4U | 1627-67 | 7.68 | 3.7 | 0.0288 | | -0.2 | | UV |
| 2S | 1553-54 | 9.26 | | 30.6 | | 0      ? | | B e |
| LMC X-4 | 0533-66 | 13.5 | 3.6 | 1.408 | 30.4 | 1.5 | UHE | O 7 III-V e |
| 2S | 1417-62 | 17.6 | | >15 | | ≤ 0.6 | | B e |
| GPS | 1840+01 | 29.5 | | | | | | |
| OAO | 1653-40 | 38.2 | > 2.3 | 7.8  ? | | -1.4 | | B 0.5 I ae |
| EXO | 2030+375 | 41.2 | | 38 | | ≲ 1.6 | | B e |
| 1E | 0630+178 | 60 | < 3 | | | -5 | | |
| Cep X-4 | 1734-275 | 66.25 | | | | -1 | | B e |
| 4U | 1700-37 | 67.4 | | 3.4 | | -0.5 | | O 6.5 f |
| A | 0535+26 | 104 | 3.0 | 111.0 | | ≤ 0.3 tr | | O 9.7 III e |
| GX 1+4 | 1728-247 | 122 | 1.7 | >15 | | 0.6 | | M 6 III |
| 4U | 1230-61 | 191 | | | | | | |
| GX 304-1 | 1258-61 | 272 | | 132.5 | | -1.3 | | B 2 V ne |
| Vela X-1 | 0900-40 | 283 | ≳ 3.5 | 8.965 | 93.3 | -0.8 | UHE | B 0.5 I b |
| 4U | 1145-619 | 292 | ≳ 3 | 187.5 | | -1.5 | | B 0-1 V e |
| 1E | 1145-6141 | 297 | ≳ 2.5 | >12 | | -0.5 | | B 1 I |
| A | 1118-61 | 405 | | | | -0.3 tr | | O 9.5 III-V e |
| GPS | 1722-366 | 414 | | | | | | |
| 4U | 1907+097 | 437 | > 2 | 8.38 | 41.6 | ≲ 0.6 | | OB e |
| 4U | 1538-52 | 529 | ≳ 2.7 | 3.730 | | -0.4 | | B 0 I b |
| GX 301-2 | 1223-624 | 696 | ≳ 2 | 41.52 | | 0 | | B 1.5 I ae |
| X Per | 0352+309 | 835 | ≳ 3.1 | 581 | | -3.4 | | O 9.5 III-V e |

6

I therefore prefer to think that most - if not all - ms-pulsars are born fast, thanks to a lower-than-average magnetic coupling to the blown-off supernova shell, and that recycling is inefficient. Their small scale height hints at a very massive progenitor star or rather at a young age.

Single ms-PSRs are sometimes thought to have 'ablated' their former donor star. But the only known system likely to ablate, the occulting PSR 1957+20, is hardly ablating at all (Michel, 1989).

Pulsars die after $10^{6.7 \pm 0.3}$ yr, though not from magnetic flux decay (Kundt, 1981b , 1988a; Michel, 1987). This statement is inferred from (i) fig. 1 which shows a de-

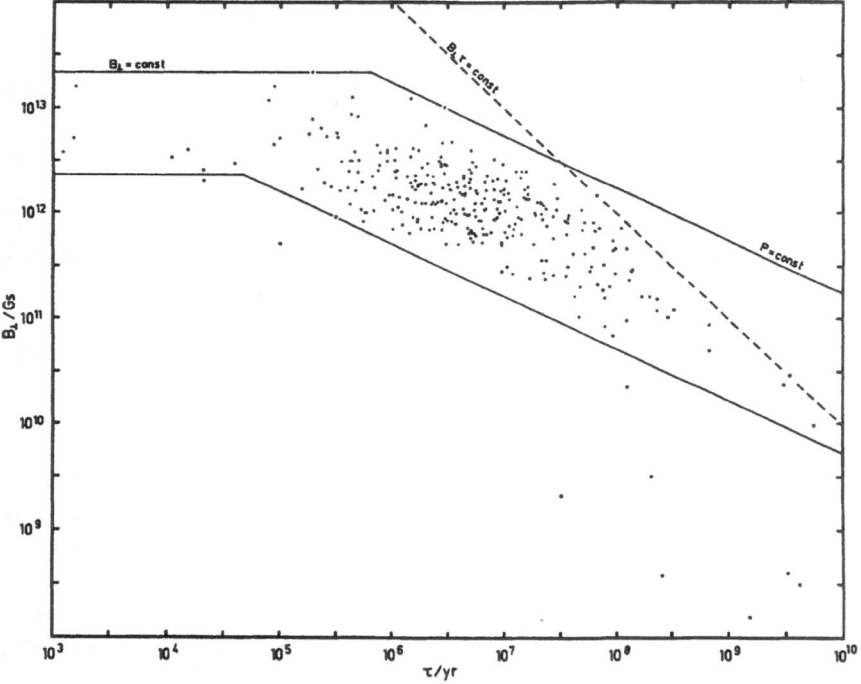

<u>Figure 1</u>. Values of $(P\dot{P}/10^{-15}\,\mathrm{s})^{1/2}$ =: $B_\perp/10^{12}$ G $(I/R^6)_9^{1/2}$ versus spin-down age $P/2\dot{P}$ =: $\tau$ for some 300 pulsars, taken from ref. 82 with a few recent points added. The broken line $B_\perp \tau = 10^{20}$ Gs is a suggestive detectability limit. The plot supports a PSR evolution starting with $B_\perp$ = const and ending with $P$ = const, whereby PSRs die before reaching the detectability limit. Only a small number of (mostly short-period) PSRs fall beyond the drawn strip; their projected magnetic dipole moments are weaker by a factor of $\lesssim 10^2$.

creasing number density of pulsars of spindown age $\tau := P/2\dot{P} \gtrsim$ $10^{6.7}$ yr (in the $B_\perp$ versus $\tau$-plane, $B_\perp := \mu_\perp/R^3$) - well above the detection limit $B_\perp \tau \lesssim 10^{20}$ G s - further from (ii) fig. 2 which shows that their kinetic age $t_{kin} := (z-z_\ast)/\dot{z}$ saturates near $10^7$ yr, and from (iii) the histograms dN/dQ and histories Q(t) of additional measured quantities Q such as $\ddot{P}$, the pulse width W, and the phase dependence of linear polarization $\theta(\varphi)$; see also Candy & Blair (1983,1986) and Kuz'min et al (1984).

Pulsars probably die by dipole alignment (with the spin). The alignment time scale $\tau_\mu$, defined by $\mu_\perp \sim \exp(-t/2\tau_\mu)$, is of order $10^{5.8}$ yr and corresponds to a true age t given by

$$t/\tau_\mu = \ln\left[1+(\tau-\tau_\ast)/(\tau_\mu+\tau_\ast)\right] , \qquad (3)$$

where

$$\tau_\ast = 3 \; Ic^3/\mu_\perp^2 \; \Omega_\ast^3 = 10^9 s \quad I_{45}/\mu_{31}^2 \; \Omega_2^3 \qquad (4)$$

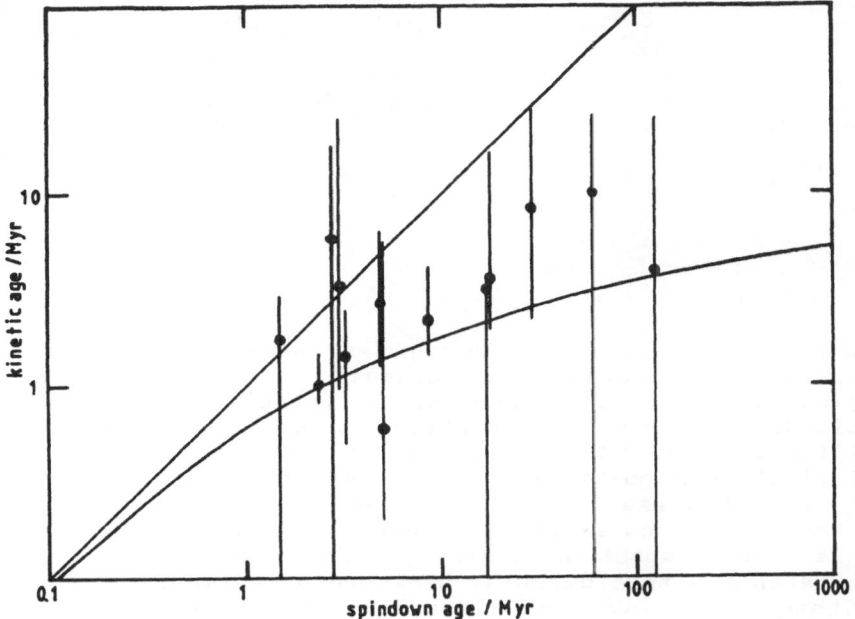

<u>Figure 2</u>. Kinetic age $t_{kin} := (z-z_\ast)/\dot{z}$ versus spindown age $P/2\dot{P} =: \tau$ for the 13 PSRs measured by Lyne, Anderson & Salter (M.N. <u>201</u>, 503, 1982), with error bars taking care of an uncertain galactic height at birth and a warp of the Milky Way. A reasonable fit is obtained by the curve $t_{kin}/\tau_\mu = \ln(1+\tau/\tau_\mu)$ with $\tau_\mu = 10^{5.8}$ yr.

is the spindown timescale at birth. The further correlation $|n| := |\Omega \ddot{\Omega} / \dot{\Omega}^2| \lesssim \tau/\tau_c$ with $\tau_c \approx 10^{2.5}$ yr suggests that dipole alignment takes place in an oscillatory manner, with an oscillation (quasi-) period of $10^{1\pm0.5} \tau_c$ (Kundt, 1985b).

Some doubt has been recently cast on the foregoing interpretation by Lyne & Manchester (1988). They repeat the Narayan & Vivekanand (1983) analysis of the swing of the linear polarization $\theta(\varphi)$ through the pulse window on a larger and less noisy sample and conclude that pulsars may well have pencil beams, i.e. that figure 3 may be mistaken, and that alignment is not the cause of death. They also derive (in the text) a mean beaming fraction of 0.14, inconsistently low with the supernova rate.

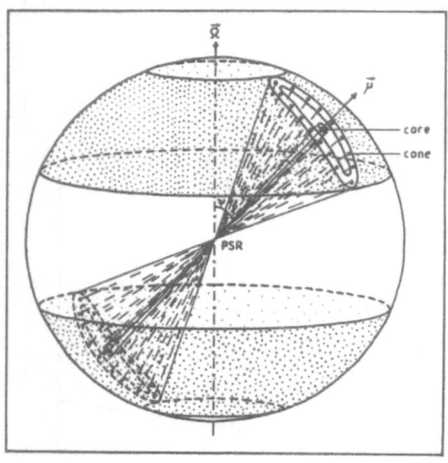

Figure 3. Preferred beaming pattern of a PSR into two antipodal banana-shaped spherical domains which corotate rigidly with it, taken from ref. 89. Retardation (spiraling) is suppressed. $\vec{\Omega}$ = spin axis, $\vec{\mu}$ = dipole moment, $\Upsilon$ = obliqueness angle, and "core" and "cone" designate the inner and outer beam components which have different spectral properties (ref. 129).

How can we find out whether pulsar beam cross sections are circular, elliptical or banana? Earth finds herself periodically in the beam of a pulsar at fixed pulsar latitude, but no prescription is known to measure this latitude. Lyne & Manchester assume that the linear polarization $\theta$ points always towards the center of the beam. They can then use the function $\theta(\varphi)$, or just its derivative at the center and range, to measure the offset from the beam center (or 'impact parameter') $\beta$. In this way, they find consistency with a circular beam cross section.

There is, however, an alternative assumption that one can make: One can propose that pulsar radiation has its preferred polarization parallel to, or at right angles to the curvature plane of the radiating particles' orbit. The local curvature planes of particles escaping from the pulsar along a trumpet-shaped surface will intersect a cross section more or less at right angles. It is therefore reasonable to assume that the linear polarization direction $\theta$ is at right angles to the beam circumference (rather than pointing towards the center), or anywhere in between (because the cross section will change). When the circumference is approximated by an ellipse and $\theta(\varphi)$ – at distance x from the center – is chosen as $\theta = $ arc cos $(x/\alpha(\varphi))$, this function $\theta(\varphi)$ is indistinguishable from that of a circumscribed circle of radius $\alpha(\varphi)$; see figure 4. In other words: without a detailed model for the geometry of the polarized radio emission, we cannot distinguish between fan beams and pencil beams. The safest 'handle' remains the beaming factor (with its relation to the birthrate).

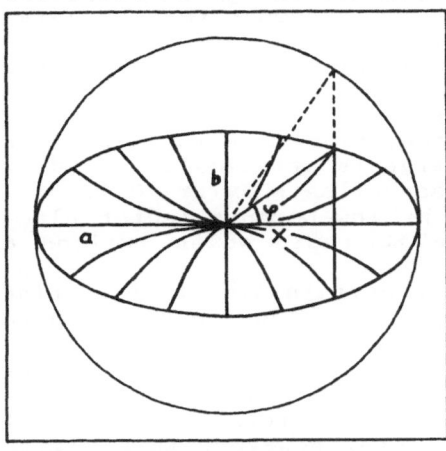

Figure 4. Cross section through an elliptical beam coming from one polar cap of a PSR. If the curved radial lines (through the center) give the direction of linear polarization, an observer finds the same polarization swing as for a spherical beam.

Connected with the beam shape is the quest for the emission mechanism and distance from the neutron star's surface at which the pulses at the various frequencies are radiated. Recent opinions on this problem can e.g. be found in Kundt (1985b), Cheng et al (1986), and Rankin & Gil (1989). The spectrum alone suggests that the pulses at infrared and higher frequencies are emitted incoherently whereas at radio frequencies they are emitted coherently. Consistently with this interpretation, the radio pulses are much narrower and more variable. The pulse windows open up towards 180°, though, when the dynamic range is increased beyond $10^2:1$. Even for pencil beams, therefore, we may see every pulsar at sufficient sensitivity.

The emission sites at the various frequencies cannot be very near the stellar surface if we assume that radiation is emitted almost parallel to the magnetic field lines, because the emission cones would be too narrow, and there would be no phase offsets at different frequencies. Even at (extreme) radio frequencies, non-dispersive phase delays have been reported that increase with wavelength (Shitov & Malofeev, 1985). The pulsar magnetosphere is transparent at all frequencies below the pair production threshold, (Heintzmann

et al., 1975; Melrose, 1980). A natural distance is that of
the speed-of-light cylinder (=SLC) where most of the post-
acceleration is expected to happen, i.e. where the
corotating magnetosphere is forced open by the inertia of
its plasma load. This is in accord with Pacini's (modified)
prediction for the pulsed optical and X-ray power: $L \sim \mu_\perp^4 \, \Omega^{11}$
(Kundt, 1985b).

At distances much beyond the speed-of-light cylinder, the
charges will $\vec{E} \times \vec{B}$-drift with the outgoing electromagnetic
(Poynting) flux, both wave and steady, so that the only
radiation losses are inverse Compton on the ambient photon
bath. Such inverse Compton losses can account for the X-ray
pulses (of the Crab: Kundt 1980).

It is often said that the (coherent) radio pulses are
emitted not far above the polar caps. Against the claim
speak the following considerations:
(1)   The pulse wings (at low intensities) are too wide, and
      the beaming factor is too large (statistically): $f \gtrsim 0.5$.
(2)   Interpulses tend to have phase offsets $\Delta\psi/2\pi \lesssim 0.1$.
(3)   Induced Compton absorption by low-Lorentz-factor elec-
      trons makes the SLC opaque to radio waves (Wilson &
      Rees, 1978).
(4)   Coherent radiation requires a certain amount of pumping
      which is not easily provided in the inner magnetosphere
      (Melrose, 1980).
(5)   The size of the emission region of PSR 1237+25 was
      found - via refractive scintillation - to equal
      $\ell = 10^{-1.5 \pm 0.5}$ c$/\Omega$ , i.e. several percent of the SLC,
      (Wolszczan & Cordes, 1987). Note, however, that a much
      larger size $\ell\Omega/c = 10^{3.5}$ has been found by Gwinn et al
      (1988) for PSR 1933+16, whose interpretation is not
      clear.

How many pulsars are there in the Galaxy? The birth
interval $\Delta t$ is related to the mean age t, beaming factor f,
and number (of pulsars) N via

$$\Delta t(PSR) = tf/N = 10^{1.5 \pm 0.3} \text{ yr } f_{-0.3} , \qquad (5)$$

i.e. one pulsar every 16 to 60 years in the Galaxy for a
beaming factor f near 0.5, (cf. Narayan, 1987). The large
uncertainty enters through both t and f. This high rate,
when evaluated for a 0.5 Kpc neighbourhood of the solar
system, requires that all stars more massive than 6 $M_\odot$ (or
even somewhat lower) end up as pulsars (Blaauw, 1985).

If pulsars are the younger brothers of binary neutron
stars - as suggested both by their high space velocities and
independently by their respective distributions (Blair &
Candy, 1985) - we have two neutron stars for every pulsar,
hence a neutron-star birth interval in the Galaxy of

$$\Delta t(n\star) = 10^{1.2 \pm 0.3} \text{ yr} \qquad (6)$$

which must not be shorter than the birth interval of both supernovae and supernova shells, because it is highly unlikely that the liberation of a neutron star's binding energy, $E_{bind} = 10^{53.2\pm0.5}$ erg, does not give rise to a SN, or at least a SN shell (when a certain fraction of the mechanical energy of the ejecta is thermalized and radiated). At last year's Bad Honnef workshop, $\Delta t(n*)$ was evaluated near 10 yr (cf. Kundt, 1988b).

Blaauw's estimate of progenitor masses of neutron stars is thereby lowered towards $M \gtrsim 5\ M_\odot$. This estimate conflicts with common wisdom that neutron stars are the remnants of type II SN explosions which derive from stars of mass $\gtrsim 8\ M_\odot$ at birth, (Woosley & Weaver, 1986). It also makes SN 1987A a highly unlikely event if its progenitor's mass exceeded $15\ M_\odot$ (Dopita, 1988). The significance of this statement can be gleaned from fig. 5 which shows the stellar luminosity function for the Galaxy, both linearly and logarithmically. Luminosities can be converted to masses by the (coarse and evolving) power-law dependence $L \sim M^{3.8\pm1}$ for $0.5 \lesssim M/M_\odot \lesssim 20$.

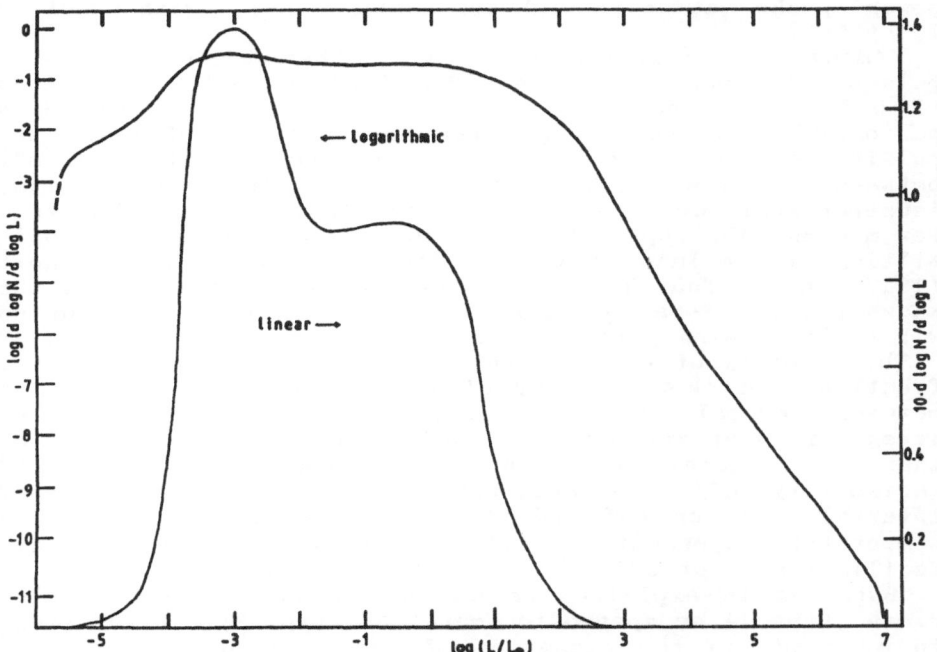

Figure 5. The luminosity function of the stars in our Galaxy, both linearly (scale to the right) and logarithmically (scale to the left), taken from ref. 88. Note that bright (massive) stars are very rare. See also M. Mateo, Astrophys. J. 391, 261 (1988).

What evidence do we have of pulsar winds? In the cases of
the Crab and Vela PSR, the winds are probably best traced at
X-rays, see next lecture and Pelling et al (1987: Crab),
Harnden et al (1985: Vela), and Kundt (1981b). In the case
of the binary pulsar 1913+16, an optical detection of the
windzone is discussed in (Kundt, 1980b). For the eclipsing
ms-PSR 1957+20, Kulkarni & Hester (1988) succeeded in
mapping it at Hα, as a Balmer nebula.

## 1.2 THE X-RAY BINARIES

Some 150 (bright) Galactic stellar X-ray sources have X-ray
luminosities between $10^{35}$ and $10^{39}$ erg s$^{-1}$, most of them
between $10^{36}$ and $10^{38}$ erg s$^{-1}$. Most or all of them are
neutron-star binaries. A ten times larger number of fainter
X-ray sources, between $10^{32}$ and $10^{35}$ erg s$^{-1}$, are thought to
be predominantly white-dwarf binaries; though the latter can
still be much fainter ($\gtrsim 10^{29}$ erg s$^{-1}$). Blair et al (1989)
find an overall luminosity function of the Galaxy of
$N_L dL \sim L^{-1/2}$ dL between $\lesssim 10^{33}$ and $10^{39}$ erg s$^{-1}$, (so that the
power peaks at the high-luminosity end), with $\int N_L dL =$
$10^{3.4 \pm 0.4}$.
    Among the bright sources, more than 30 show regular
pulses, with periods between 60 and $10^3$ s, pulsed fractions
of order 20%. The pulse period is explained as the spin
period of their accreting polar caps. On top of this, the
pulsing sources tend to show a second, orbital period,
between less than an hour and several months, and a third,
'superorbital' or 'precession' period, of duration one to a
few months. The superorbital period may be due to a periodic
tilting of the inner accretion disk, cf. Priedhorsky & Holt
(1987) and my fourth lecture. Some of the properties of the
known pulsing X-ray sources are listed in table 2, ordered
w.r.t. increasing pulse period.
    The majority of (bright stellar) X-ray sources have pulsed
fractions $\Delta I/I < 1\%$. Many of them flicker, on timescales
between several ms and years, many of them burst, at
irregular intervals between hours and days and duration less
than a minute, and many of them show preferred
variability ($\Delta I/I$ = several upto 30%) in a certain frequency
interval (between 0.5 and $10^2$ s$^{-1}$) which is intensity- and
temperature-dependent and called 'quasi-periodic
oscillations', or QPO.
    Bursting is explained as due to nuclear-chemical explo-
sions (type 1) or spasmodic accretion (type 2). Flickering,
in my mind, is the signature of inverse-Compton losses of
relativistic pair plasma, generated by the spinning,
strongly magnetized neutron star in vacuum discharges
(Kundt, 1984a, 1985b, 1987c). Alternatively, flickering
may result from unsteady accretion and/or screening by
the corotating magnetosphere. Finally, a hump in the

fluctuation power spectrum (QPOing) carries the signature of a more or less corotating magnetosphere which screens the central neutron star (Kundt et al, 1987, Kundt & Fischer, 1989). Reasons for the latter interpretation are (i) the occasional strength of the modulation, which shows that essentially the whole accretion power can be modulated, (ii) the expectation of the presence of a non-transparent plasma-loaded magnetosphere which performs differential torsional oscillations (fourth lecture), and (iii) the frequency range which is reminiscent of (somewhat lowered) neutron-star spin rates. We would thus deal with neutron stars in the sub-sec to msec range.

Besides the dichotomy of pulsing and non-pulsing sources, there is the dichotomy of high-mass ($M \gtrsim 10 \, M_{\odot}$) and low-mass ($M \lesssim M_{\odot}$) companions. The former systems are short-lived ($\lesssim 10^{5 \pm 1}$ yr), the latter long-lived. For comparable observed numbers, this means that the low-mass systems form rarely. They are thus unimportant for statistics.

Remarkably, a superorbital period is detected in all four classes of sources (Priedhorsky & Holt, 1987). Another frequently observed property is intensity-dipping, i.e. an occasional lowering of the intensity, preferentially - though not exclusively, or strictly - at preferred orbital phases. These properties have been often attributed to the outer accretion disk ('accretion disk corona'), in spite of the dynamical difficulties of keeping enough matter at large heights above the disk. They may find a simpler explanation by the (variable structure of the) inner disk, by the plasma-loaded magnetosphere, and/or by clumped wind matter of the companion star that crosses the line-of-sight.

Among the non-periodic X-ray binaries are the (radio- and X-ray-) jet sources Sco X-1, SS 433, and Cyg X-3 which have found various interpretations in the literature. My own conviction is that their jets are driven by supersonically expanding pair plasma, generated in vacuum discharges near the surface of the neutron star. Their estimated ages are $10^{3.5}$ yr (Sco X-1: Kundt & Gopal-Krishna, 1984) and $\gtrsim 10^4$ yr (SS 433: Kundt, 1985a, 1987b; Cyg X-3: Strom et al, 1989) respectively. They are thus among the youngest X-ray binaries. Gneration of (relativistic) pair plasma requires strong magnetic fields (Blome & Kundt, 1988). Similarly, the detection of ultrahigh-energy (UHE) $\gamma$-ray pulses from Cyg X-3, upto and beyond $10^{16}$ eV, is a signature of strong magnetic fields (Kundt et al, 1987). For $\gamma$-ray detection at very-high-energies (= VHE: $10^{11} \div 10^{13}$ eV) and UHE-energies ($\gtrsim 10^{14}$ eV) see Hillas (1984), Dowthwaite et al (1984), Protheroe & Clay (1985), Ramana Murthy & Wolfendale (1986), and Schwarzschild (1988). The (non-pulsing) jet sources and UHE-sources - and probably all the (other) flickering sources as well - are therefore likely to have strong magnetic fields, equally strong as those of the pulsars.

The class of flickering X-ray binaries contains the black hole candidates, identified by the large inferred mass ($\gtrsim 6$ $M_\odot$) of the quasi-structureless X-ray source. The best candidates are presently the massive system Cyg X-1 and the X-ray nova A0620-00, (Ilovaisky, 1987). As will be discussed more in detail in my third lecture, mass determinations of systems with strong winds are unreliable because the windzone can be opaque in the lines. This reasoning does not apply to A0620-00 which has, however, several characteristics of a 1.4 $M_\odot$ neutron star inside a more or less massive accretion disk (Kundt & Fischer, 1989). I am not convinced that we have identified any black hole yet.

X-ray binaries are much more frequent in (the cores of) globular clusters than in the disk of the Galaxy, by a factor of several $10^2$, perhaps due to tidal capture. The disk binaries are hardly formed by tidal captures, because of insufficient space densities of progenitors. Once we agree on their formation mode, we may have to rethink the case of the globular-cluster binaries. It is worth noting that the tightest among them, the burst source 1820-303, has an orbital period of only 11 minutes.

A complete list of classes of Galactic neutron stars should not omit the $\gamma$-ray bursters whose true nature is still controversial. Whereas e.g. Lipunov et al (1982) argue strongly in favour of spasmodically accreting old neutron stars, Boer et al (1989) revive the white-dwarf interpretation because of the high required space density, $\gtrsim 10^4$ Kpc$^{-3}$, at distances $\lesssim 10^2$ pc from the Sun; see also Murakami et al (1988), Paczyński (1989). A difficulty of the white-dwarf interpretation is their hard spectra, between 30 KeV and $10^2$ MeV; though the 'repeaters', of which so far only three are known, have somewhat softer spectra. Even the most famous repeater, known since the 5 March 1979 event, had a harder spectrum during its first outburst.

$\gamma$-ray bursts have durations of $\lesssim 0.1s$, 1s, or $\gtrsim 10s$, with occasional repetitions, and with an isotropic distribution over the sky. For distances $\lesssim 10^2$ pc, their powers never exceed the Eddington luminosity of a 1.4 $M_\odot$ star except for a possible initial radiation spike. Features in the spectrum are reminiscent of magnetic fields above $10^{12}$ G and of pair annihilation in a strong gravitational field. Occasional periods between 4s and 10s suggest that we deal with defunct, slowly spinning pulsars, of which at least $10^9$ in the Galaxy are expected, hence a density of more than $10^6$ Kpc$^{-3}$. Bursts could be the result of spasmodic, clumped accretion (Kundt et al, 1987). If this interpretation is correct, we learn that even the oldest known neutron stars can have strong magnetic fields.

Only one class of (binary) neutron stars has been surmised not to have conserved their (strong) magnetic fields since birth: the non-pulsing galactic-bulge X-ray binaries. The

absence of pulsations has been blamed on a weakness of the
funneling magnetosphere, and the softness of their X-ray
spectra taken as a confirmation that accretion occurs onto a
significant fraction of the neutron star's surface, not just
onto its polar caps. Whereas the latter interpretation may
well be correct, it is not clear at all that strongly
magnetized rotators have to pulse. I consider the jet
sources and the VHE sources as counter examples. As will be
discussed in the fourth lecture, all that is required is a
smaller fraction of evaporated disk material that can
accrete onto the polar caps.

If all neutron stars are born with moderately strong
magnetic fields - the ms-PSRs being at the low end - and if
SN explosions are (predominantly) the birth sites of neutron
stars, why don't we see a pulsar in every SN shell? I don't
think one can put the burden on a small beaming factor, as
is occasionally done: The beaming factor is unlikely to be
small, and the pulsar windzone will be an even more
isotropic radiator (Kundt, 1981b). Rather, as Srinivasan
argues convincingly, many pulsars will simply be too weak.
Yet there is the class of slowly spinning X-ray pulsators;
how did those neutron stars lose their huge rotation energy?
According to Blair & Candy (1985), they are at least equal
in number to the pulsars. If their spin energy was given to
the braking wind of their companion star, this wind material
must acquire relativistic velocities. I have argued
repeatedly that the young binary neutron stars act as
relativistic grindstones and are the long-sought cosmic-ray
boosters (Holloway et al, 1978; Kundt, 1983a, 1984c). The
conventional cosmic-ray explanation via shock acceleration
(e.g. Drury, 1983) is likely to fail when high efficiencies
($\gtrsim 1\%$) are required, see Kundt (1984b), Sarris & Krimigis
(1985), Falle & Giddings (1987), and Falle's (third) lecture.
In this way we understand (i) why so many supernova shells
look empty, (ii) how the pulsing (and other) X-ray binaries
lost their spin energy, and (iii) why cosmic-ray boosters
appear to correlate with (certain) SN shells.

It has been argued above that the pulsars receive their
high space velocities from the orbital momenta of their
progenitor binaries. We would thus have at least as many
young binary neutron stars as single ones. In this
conclusion, it is assumed that the first supernova explosion
in a binary is unlikely to disrupt the system whereas the
second one generally does. This assumption is based on the
expression for the (numerical) eccentricity $\epsilon$ of a binary

$$\epsilon = (M_1^- - M_1)/(M_1 + M_2) \tag{7}$$

after an instantaneous, isotropic and interaction-free
removal of some mass $M_1^- - M_1$ from the first star, whereby $M_1$
and $M_2$ are the final masses, and the initial orbit has been

assumed circular. A system stays bound as long as $\iota < 1$, i.e. as long as less mass is lost during the explosion than remains. In a massive star system, this condition is expected to be satisfied for the first explosion but not for the second, so that the first-born neutron star tends to be multiple whereas the last-born one tends to be single, hence a pulsar.

From all the knowledge we have about neutron stars, both binary and single, we can draw conclusions on their mass function, radii, moments of inertia, magnetic dipole moments, cooling history, and internal structure. Our present knowledge is consistent with uniform masses of $(1.4 \pm 0.2)$ $M_\odot$ (cf. Taylor & Weisberg, 1989), radii $R = 10^{1 \pm 0.3}$ Km, moments of inertia $I = 10^{45 \pm 0.5}$ g cm$^2$, magnetic moments $\mu = 10^{30.8 \pm 0.5}$ G cm$^3$ except for the ms-PSRs, and surface temperatures $T_s \lesssim 10^{5.8}$ K after a few $10^2$ yr (Brinkmann & Ögelman, 1987). General-relativistic corrections make it difficult to determine $R$ any better than stated - rather worse - even when we can measure both the temperature and bolometric luminosity of X-rays from their surfaces (Goldman, 1979, and second lecture by Lewin). Much has been said already about uncertainties in determining $\mu$, cf. Kundt et al (1987). Throughout the years, calculations find a ratio $T_s/T_i = 10^{-2 \pm 0.2}$ between surface and interior temperature for various combinations of conductive and radiative neutron-star cooling which predicts slower cooling than observed, see Romani (1987). Quite likely, convective cooling - via volcanos - ought to be considered. Finally, the equation-of-state in the deep interior is constrained by a neutron star's maximal spin rate, of order 0.6 ms; cf. Krotscheck & Kundt (1978), Cowsik et al (1983), and Alcock (1989).

## 1.3 SUPERNOVA SHELLS

The birth of a neutron star liberates its binding energy

$$E_{bind} = E_{grav} - (E_{int} + E_{rot} + E_{turb} + E_{mag}) = 10^{53 \pm 0.5} \text{ erg}, \tag{8}$$

where $E_{grav} = (0.2 \pm 0.1) M c^2 \approx 10^{53.8}$ erg$(M/1.4 \, M_\odot)$ has to be obtained within General Relativity and can be measured as a redshift, $E_{rot} = I\Omega^2/2 \lesssim 10^{52.7}$ erg couples most easily to the outside world, and where the internal or elastic energy $E_{int}$ satisfies $E_{int}/E_{grav} = \{ {}^0_1; 5 \}$ for cold, $\left\{ \begin{array}{l} \text{non-relativistically} \\ \text{relativistically} \end{array} \right\}$ degenerate matter within Newtonian approximation. Consequently, we have $E_{rot} \lesssim E_{bind} \lesssim E_{grav}/2$, as already used in evaluating expression (8). Most of this energy, so we think, is emitted in the form of neutrinos, throughout the first $\lesssim 10$ sec of a neutron star's formation.

It is possible that the extreme spin of a forming neutron star is the cause for a supernova explosion (Kundt, 1976b, 1988b). In any case, the liberation of $E_{bind}$ within seconds is unlikely to go unnoticed by modern sky monitoring, either immediately as a bright flash, or centuries later as a glowing shell of decelerating ejecta. In other words: no neutron star birth without a big event.

The converse is not equally straight-forward because supernova shells have mechanical energies of $10^{51 \pm 1}$ erg only, which could be supplied by a big nuclear explosion (of a white dwarf). In other words: we could have supernova explosions without neutron-star formation. This possibility is likewise suggested by a large number of SN shells without an obvious compact remnant. Yet the statistics are against it: we are aware of at least as many neutron stars as SN explosions and as supersonically expanding SN shells, viz one in 10 yr (cf.eq.(6)). Only a small fraction of all SN explosions can afford not to give birth to a neutron star. Besides, Blaauw's (1985) result (mentioned above) warns against dismissing certain classes of SN explosions as the birth events of neutron stars.

This statistical conclusion is not in direct conflict with Branch's assertion that SN Ia events are explosions of white dwarfs, because he estimates them to be rare ($\gtrsim 20\%$). He has not convinced me, though: On the one hand, there is a large scatter among lightcurves of type Ia events. Second, the shapes of the lightcurves are indistinguishable from type Ib events, whose progenitor masses are estimated distinctly higher. Third, white dwarfs are often thought to lose mass secularly throughout their evolution, rather than gain (Ritter, 1986). Also, if many of them ($\gtrsim 1\%$) would accrete a significant fraction of a solar mass, their integrated UV-output would exceed that of ordinary stars ($L_{UV}$(galaxy) = $10^{41 \pm 0.5}$ erg s$^{-1}$). The corresponding HII regions are not seen. Fifth, the scale height of Galactic SN shells of type Ia, $\leq 180$ pc, equals that of the (massive) runaway stars, and smaller than that of (old) white dwarfs. Sixth, there is the problem of iron-overenrichment of the ISM. Finally, the filamentary morphology of all SN shells is reminiscent of magnetic Rayleigh-Taylor instabilities during the acceleration, different from the bloblike morphology of chemical explosions. Filamentary shells have different photon-storing properties from 1-component ones, so that one gets different predictions for their evolutions, in particular of their effective temperature and apparent size. In my last lecture, I shall argue that all SN explosions are spin-driven, and that their lightcurves can be understood as the result of delayed line-photon escape from their opaque shells.

In the literature, there is considerable confusion about the true size, shape, and association of SNe and SN shells with their stellar remnants, involving the Vela PSR, CTB 80, IC 443, MSH 15-52, and others. If SNe are almost isotropic explosions, at initial bulk velocities of $10^{3.8\pm0.3}$ Km s$^{-1}$, their ejecta should form rings in the sky unless the surrounding medium happened to be very different on two sides of the explosion center. (A better approximation to the true mass distribution of a SN shell may be a barrel: Manchester, 1987, Caswell, 1988). Instead, many shells look like crescents, others like mushrooms, yet others like a rabbit, or a flying duck (bird). Apparently, the visible shape differs from the true shape; the mass-to-light ratio of the ejecta may fluctuate rapidly as a function of time.

At deeper glance, SN shells are indeed (broken) rings. Here are a few examples, taken from Kundt (1988b): (1) The Vela SN shell looks oval, with an off-center position of 'its' PSR. At deeper look, both the optical and radio shape approach a (larger) circle, with the PSR at its center. (2) The PSR in CTB 80 has a spindown age of $10^5$ yr and a large proper motion towards the west. The radio shape of CTB 80 looks like a rabbit. The rabbit is surrounded by a ring of both optical and IR emission, centered on the likely birth site of the PSR. Apparently, a $10^5$ yr-old shell cannot easily be recognized, and the radio source CTB 80 is solely generated by the PSR. (3) The true diameter of the Crab and of Cas A are likely to be three times bigger than on typical maps. Even larger is the discrepancy for the Crab's 'double' in the LMC, around PSR 0540-693, whose extended optical radiation is patchy and large, as large as its radio size (radius $\gtrsim 0.7'$, corresponding to $\gtrsim 10^{19.8}$ cm: Mathewson et al, 1983, Mills et al, 1984). See also Milne (1987) for possibly extended linear polarization. (4) As a further example of incompleteness, fig. 6 shows IC 443 seen at various frequencies: the usual optical continuum photograph occupies one quarter in diameter of a larger shell which is suggested by emission-line photographs. (5) The flying duck has been found to be one half of a 'barrel' by Caswell et al (1987).

Young SN shells can be distinguished from H II regions by their larger radio/IR and [S II]/H$\alpha$ emission ratio: Fürst et al (1987) find $\nu S_\nu$ (11 cm)/$\nu S_\nu$ (60$\mu$) > $10^{-4.4}$ for SNSs but <$10^{-6}$ for H II regions. Arendt (1989) warns that this energy flux ratio varies strongly from shell to shell, through more than a factor of $10^4$, and that it can overlap for old SNSs with that of H II regions. Old shells are strong IR emitters. SNS catalogs are probably incomplete for old shells.- Young shells can also be strong X-ray emitters (Seward, 1985).

How reliable are age and size determinations? A simple estimate of age is offered by dividing the apparent angular radius of a shell by its (angular) velocity. It can be checked on the 'historical' shells, of age $\lesssim 10^3$ yr, and leads in general to satisfactory agreement. (A difficulty is

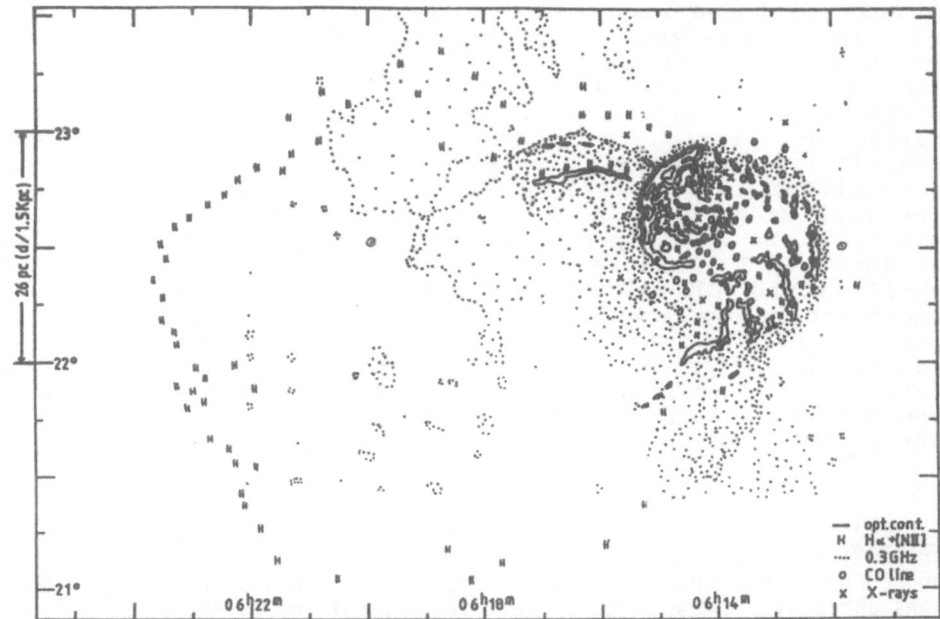

Figure 6. Sketch of the SN shell IC 443 as seen at various frequencies, based on the atlas by Parker et al (1979). The true (linear) size may be five times larger than what is mapped on optical continuum photographs.

posed by the existence of multiple velocity fields). Unfortunately, old shells expand slowly, and angular velocities are difficult to measure.

Instead, Doppler velocities can be determined spectroscopically and used as speed indicators. If they were reliable estimators of radial velocity, they would simultaneously yield values for the size and distance (under the assumption of isotropic expansion). But the Doppler broadening of a spectral line can be interpreted in two different ways: as the result of a projected radial motion, or of a locally isotropic velocity field created by shock excitation. The latter (wave) interpretation has been recently applied to SN 1006 by Kirshner et al (1987) and Long et al (1988), with systematically lower results for $v_r$ than in the former (shrapnel) interpretation. Their distance and size estimates may therefore be systematically low. Reliable size and velocity estimates are important for testing SN models on SN shells.- For shell expansions, the splinter model predicts

$$r/v_0 t_0 = \left\{ \begin{matrix} t/t_0 \\ [1+(t/t_0-1)/3]^{2/5} \end{matrix} \right\} \text{ for } t \left\{ \begin{matrix} \ll \\ \gg \end{matrix} \right\} t_0 \approx 10^{3.5} \text{yr}/\varrho_{-25}^{1/2} . \tag{9}$$

What is known about SNS-n∗ associations? To begin with, all (four) PSRs with spindown ages ≤ 10⁴ yr are inside their birth shell, in a central position, whereas only three of the older PSRs have been found surrounded by a shell. In particular, an unpublished search by W. Reich in 1986 has failed to detect radio shells around the PSRs 1800-21 and 1823-13 - whose spindown ages are $10^{4.2}$ and $10^{4.3}$ yr - with a sensitivity of $10^{-12}$ W/m² Hz ster at 0.4 GHz. These facts are consistent with a statistical SNS age of (only) 10⁴ years, which is independently indicated by the number ratio of 10⁴- over 10³-yr-old shells and also by the fact that the shell around CTB 80 has essentially disappeared from the radio and optical sky. Old shells can at best be found on IR maps. Similarly, only two binary X-ray sources are known inside SN shells - viz SS 433 and the X-ray pulsator 1 E 2259+586 - consistent with the estimate that most binary neutron stars do not turn on at X-rays (accrete) until after 10⁴ years. (They are in the grindstone stage before).

Table 3: Some SN shells (SNSs) of various ages t and their central pulsar (PSR) or binary neutron star, whenever known. PSR ages are assumed of order P/2Ṗ. The ages of the jet sources are discussed in the text. SNS ages agree with those given in ref. 117, except for IC 443 whose morphology and molecular lines tell that it is special , and W 44 which has been recently detected to contain the pulsar PSR 1859+01 .

| log(t/yr) | PSR | Binary n∗ | SNS (year of explosion) |
|---|---|---|---|
| 0.5 | 0535.8-6930 | - | all supernovae |
| 2 | - | - | SN 1885; extended, variable radio sources in M 82 |
| 2.5 | - | - | Cas A (1680), Kepler(1604), Tycho (1572) |
| 3 | Crab,1509-58,0540-69.3 | | Crab (1054), SN 1006, MSH 15-52, LMC 0540-69, 3C 58 |
| 3.5 | | Sco X-1 | Puppis A |
| 4 | Vela, 1800-21, 1823-13 1859+01 | SS 433,Cyg X-3, 1E 2259+586 | Vela, Cygnus Loop, W 50, G 109.1-1.0 (= CTB 109),W44 |
| 5 | 1951+32, 0656+14, ... | ? | CTB 80,IC 443,G 201.2+8.2 |
| 6.5 | most | ? | - |

Table 3 summarizes my understanding of SNS-n∗ associations, ordered w.r.t their age since the SN explosion. The first line contains the reported ms-PSR in SN 1987A, whose reality remains to be confirmed. The second line contains the $10^2$ yr-old SNS 1885 in the Andromeda galaxy which has been recently mapped in Fe-line absorption by R. Fesen, of diameter 0.3 arcsec. It also contains the (20) variable radio sources in M82 whose largest ones are resolved. They have been interpreted as SNSs because of their (small) size, declining radio flux, radio spectral index, and measured expansion (Bartel, 1988). In the then-following lines, some of the age estimates are tentative, such as those of Sco X-1 (Kundt & Gopal-Krishna, 1984) and of the shells W50, G109.1-1.0, and IC 443. Even though the table is far from complete - among others because of an insufficient knowledge of ages - it correctly conveys the impression that the majority of well-studied SNSs have ages below $10^{3.5}$ yr: SNSs fade after $\lesssim 10^4$ yr, except possibly at IR frequencies.

In table 3, I have not followed the suggestion by Winkler et al (1989) that Puppis A could be two SNSs in one: Apart from the low probability of two SNSs projecting onto each other in the sky - SNSs have a surface-covering factor smaller than $10^{-2}$ - Winkler et al face the problem of only blue-shifted emission at (comparatively) modest velocities. Puppis A is just a marvelous example for how heterogeneous SNSs can be.

## 2. THE CRAB NEBULA AND ITS PULSAR

ABSTRACT.    The  Crab  looks  (presently)  atypically  slow
because we only see its bright interior. The latter consists
of  (i)  a  thermal  component,  of    temperature  T  $\gtrsim$  $10^4$  K,
density n = $10^{3.5\pm0.5}$ cm$^{-3}$, mass $\gtrsim$ $1M_\odot$ , and volume-filling
factor  f  $\sim$  $10^{-3}$,  composed  of  $\gtrsim$  $10^4$  filaments.  The volume is
filled  by  (ii)  relativistic  pair  plasma,  of  energy
distribution $N_\gamma\, d\gamma \sim \gamma^{-2.2}\, d\gamma$ between $\gamma = 10^{5.5}$ and $10^{8.8}$,
which  is  continually  injected  by  the  pulsar;  it  is
essentially weightless. Moreover, (iii) there is a bath of
multiply reflected 30 Hz waves, strong enough to have post-
accelerated the thermal filaments by 8%.
    If  the  Crab's  progenitor  was  a  blue  giant  with  a  fast,
low-density wind, most of the ejected matter finds itself
presently in the low-density CSM where it is too thin to be
(easily) detected, both in emission and absorption.

## 2.1 MOTIVATION

Once  upon  a  time,  every  other  astrophysical  paper  was
devoted to the Crab nebula. I therefore attempted to enter
astrophysics via the Crab and chose Rees & Gunn (1974) as
the boarding gate. Many connections became clear, but a few
loose  ends  remained.  For  instance,  what  radial  force  is
strong  enough  to  explain  the  (then  15%,  now)  8%  post-
acceleration of the thermal filaments?
    This question led Eckhard Krotscheck and myself to a new
description of the Crab nebula, communicated in sections 2a
through 2z of a paper which finally appeared in A&A in 1980,
thanks to the goodwill of the fifth referee. A summary of it
is  contained  in  Kundt  (1980a).  Neither  of  them  conquered the
market.  Much  more  popular  became  a  simpler,  slightly
different model by Kennel & Coroniti (1984), presented also
by Coroniti & Kennel (1985) at the Fairfax workshop, even
though  it  left  important  features  unexplained  -  like  the
post-acceleration of the filaments and the soft part (radio
to IR) of the spectrum. Is the Crab worth all these efforts?
    The  Crab  is  one  out  of  only  four  cases  where  we  see  a
(young) PSR inside a SN shell. But the shell is atypical:
its expansion velocity is $v_r$ < $10^{8.3}$ cm s$^{-1}$, instead of the
typical    $10^{8.7}$  cm  s$^{-1}$.    Correspondingly,    its    kinetic
(expansion) energy equals $E_{kin} = 10^{49.2\pm0.2}$  erg (M/M$_\odot$) (d/2
Kpc)$^2$,  at  least  a  factor  of  10  below  the  $10^{51}$  erg  of  a
typical SN explosion (for a mass M between 1 and 3 M$_\odot$ which
has  not  been  slowed  down  yet,  rather  post-accelerated!).
Does  most  of  the  mechanical  energy  reside  inside  an
invisible outer shell - the low-density part of the CSM - at
3  times  larger  radii?  As  already  mentioned  in  my  first

lecture, this possibility is supported by both deep photoelectric and spectral measurements. Besides, the similar PSR 0540-693 in LMC is surrounded by a shell of similar spectrum and total power whose outer (radio) edge corresponds, however, to a time-averaged velocity of $\gtrsim 10^{9.1}$ cm $s^{-1}$.

Let us, therefore, not worry right now why the Crab is atypically slow; what can we learn from it? To me, the Crab has suggested that (i) strong magnetic waves can drive an extremely relativistic wind, of Lorentz factor $\gamma$ between $10^{5.5}$ and $10^8$; (ii) the wind of a pulsar consists predominantly of electron-positron pair plasma; (iii) in the absence of 'obstacles', a (supersonic) relativistic wind can be almost loss-free; and (iv) the shell of the Crab supernova consists of $\gtrsim 10^4$ separate filaments. In subsequent years, I convinced myself that every sufficiently magnetized fast rotator can drive an extremely relativistic pair-plasma wind and that such winds, if confined by thermal plasma, can alternatively give rise to the twin-jet phenomenon of active galactic nuclei and young stellar objects (Kundt, 1987c). I also convinced myself that (all?) supernova explosions produce filamentary shells (Kundt, 1985c, 1987a, 1988b).

## 2.2 THE THREE COMPONENTS

The Crab is an optical source, both line and continuum. Its thermal component has a temperature of T $\gtrsim 10^4$ K, hydrogen number density $n_{th} = 10^{3.5\pm0.5}$ cm$^{-3}$, and (visible) mass $M_{th}$ $\gtrsim 1$ $M_\odot$; see Davidson & Fesen (1985). The thermal pressure thus equals

$$p_{th} = 2\ n_{th}\ kT = 10^{-8\pm0.5}\ \text{dyn cm}^{-2},\qquad (10)$$

some $10^4$ times higher than (unperturbed) galactic. For an equivalent radius of R = $10^{18.7\pm0.1}$ cm (d/2 Kpc), hence volume of $V = (4\pi/3)\ R^3 = 10^{56.8}$ cm$^3$ $d_{21.8}{}^3$ , the volume-filling factor of the thermal component

$$f_{th} := M_{th}/\rho V = 10^{-3.2}\ M_{(0)}\qquad (11)$$

is of order $10^{-3}$, $(M_{(0)} := M/10^0\ M_\odot)$ ; most of the volume must be filled by the relativistic component.

The largest thermal filaments have diameters of $10^{16.5}$ cm. If the filaments are to cover the central source completely, with at least 6 of them on average aligned radially, their typical diameters must be $\leq 10^{14.5}$ cm $M_{(0)}$, and total number $\gtrsim 10^{4.3}/M_{(0)}$ (Kundt, 1980a). This estimate is consistent with the findings that both the dispersion and rotation measure of the Crab PSR have shown almost linear excursions

on the timescale of 2 years, totalling $\Delta\,DM := \Delta\int n_e\,ds = 0.05$
pc cm$^{-3}$ and $\langle B_{\shortparallel}\rangle = 10^{-5.9}$ G (DM/pc cm$^{-3}$) (rad m$^{-2}$/RM) $=$
$10^{-3.8}$ G respectively (Rankin et al, 1988). (A dispersion
measure change of $10^{-1.3}$ pc cm$^{-3}$ corresponds to a plasma
column length of $\Delta s = 10^{13.7}$ cm $n_{3.5}{}^{-1}$.) For the Vela PSR,
corresponding changes of $-0.6$ pc cm$^{-3}$ and $\langle B_{\shortparallel}\rangle = 10^{-4.6}$ G
throughout 15 yr have been reported.

The optical and UV continuum of the Crab is explained as
synchrotron radiation by relativistic electrons (and
positrons) in the local fields, whose strength must average
$\langle B^2\rangle^{1/2} = 10^{-3.3\pm0.3}$ G for pressure equilibrium with the
thermal component. Their typical Lorentz factor $\gamma$ follows
from the peak synchrotron frequency

$$\nu_{max} = e\,B_{\perp}\,\gamma^2/\pi\,m_e\,c = 10^{0.75}\ \text{MHz}\ \gamma^2\ B_0 \qquad (12)$$

in a transverse field $B_{\perp}$ , $B_0 := B_{\perp}/10^0$ G. For $B_{\perp} = 10^{-3.3}$ G
and $\nu_{max} \approx 10^{15}$ Hz, one finds a typical Lorentz factor of $\gamma =$
$10^{5.8}\ \nu_{15}{}^{1/2}$. Such electrons have a synchrotron lifetime

$$\gamma\,/\dot{\gamma}\ =:\ t_s = 8\ \text{yr}/\gamma_6\ B_{-3}{}^2 \qquad (13)$$

which is much shorter than the age of the nebula. Unless
most of them got multiply reaccelerated, we learn that the
pulsar injects relativistic electrons (and positrons) at a
rate

$$\dot{N}_{e^+} = L_{neb}/\langle\gamma\rangle\,m_e\,c^2 = 10^{38.5}\ s^{-1} = 10^{4.1}\ \dot{N}_{GJ}, \qquad (14)$$

some $10^4$ times the Goldreich-Julian rate $\dot{N}_{GJ} = \mu\,\omega^2/ec =$
$10^{34.4}\ s^{-1}\ \mu_{31}$ which would obtain if the (shorting out)
Goldreich-Julian charge density were extracted along the
open field lines at the (maximum) speed of light. This, to
me, is the best evidence that the Crab pulsar produces
electrons and - by charge conservation - positrons in the
vacuum between its polar caps and the (inner edge of the)
nebula. Its relativistic wind is created in vacuum
discharges.

From the spectrum of the Crab - figure 7 - and eq (12) it
follows that the nebular radiation requires electron Lorentz
factors $\gamma$ between $10^{5.5}$ and $10^{8.8}$. (Beyond a few MeV , it
depends on one's choice of the 'baseline' whether or not the
power is 100% pulsed.) The lower limit, $\gamma_{min} = 10^{5.5}$, just
guarantees that the charges don't lose phase w.r.t the
pulsar's outgoing 30 Hz wave, on their way to the inner edge
of the nebula, at r = 1 lyr. The upper limit, $\gamma_{max} = 10^{8.8}$,
is somewhat larger than Gunn & Ostriker's $f^{2/3} \lesssim 10^8$ near
the speed-of-light cylinder, where

$$f := e\,B/m_e\,c\,\omega = 10^{11.7}/r_8 \qquad (15)$$

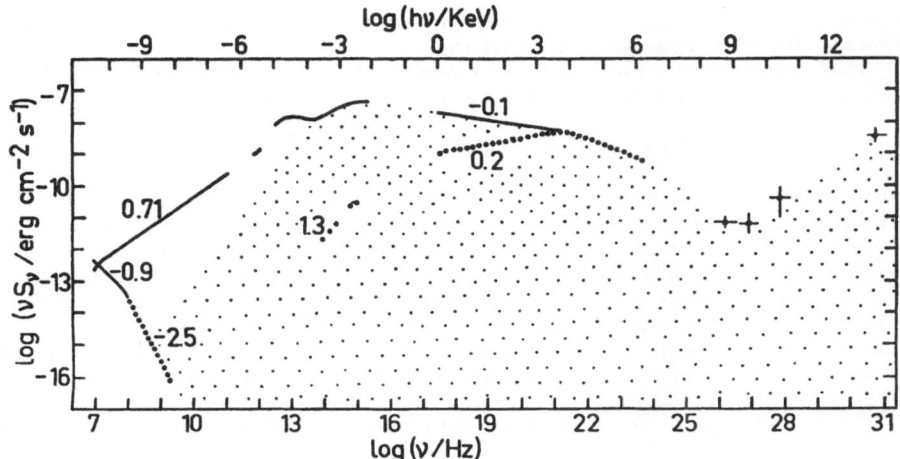

<u>Figure 7</u>. Spectrum of the Crab nebula and its pulsar (dotted), taken from ref. 82. Note the wide range of emitted frequencies (photon ener- gies), from $\nu < 10$ MHz to $h\nu > 10^{16}$ eV. The highest energy data are un- certain; if real, they mean a remarkable efficiency for radiation at cosmic-ray energies. The stippled portion of the spectrum is interpreted as due to present-day injection.

is the strength parameter of the outgoing dipole wave. Quite likely, the injected charges are somewhat post-accelerated at the inner edge of the nebula, by an ordered (convected) electric voltage ; which is independently suggested by the hard-X-ray appearance (Pelling et al, 1987). The injected $e\pm$-energy spectrum required by the radiated spectrum reads:

$$\dot{N}_\gamma \, d\gamma \sim \gamma^{-2.2} \, d\gamma \quad \text{for } 10^{5.5} \lesssim \gamma \lesssim 10^{8.8} . \quad (16)$$

This injected electron spectrum does not account for the large radio excess (and infrared bump); it is harder. The radio excess is explained by Kundt & Krotscheck (1980) p. 17 as the pile-up of degraded $e\pm$ since the SN explosion, whereby the electromagnetic fields were considerably stronger at an earlier epoch: An aging electron population evolves towards a $\delta$-distribution (in energy); successive generations of such aging spikes superpose to the observed hard- power-law distribution. As a check on the model, one finds that the time-integrated injection rate $\dot{N}_{e\pm}$ , eq (14), is almost equal to the total present number of soft electrons in the nebula: $N_{soft} = 10^{49.5}/B_{-3} \approx \int \dot{N}_{e\pm} \, dt$. Some post-acceleration may have taken place in reconnecting magnetic fields, in the wakes of the filaments, with the effect that every radio-emitting $e\pm$ is involved twice (on average). Alternatively, there may still be a pair-plasma excess due to the SN explosion.

The density contrast between the relativistic and the thermal component - in pressure equilibrium - amounts to

$$\varrho_{rel}/\varrho_{th} = 6 \; kT/m_p c^2 = 10^{-8 \cdot 2} \; T_4 . \tag{17}$$

Consequently, there is practically no dynamic interaction between the two. The observed 8% post-acceleration of the thermal component must therefore be due to a third component: the electromagnetic flux. But there is a problem: The spindown power of the pulsar falls short, by a factor of $\gtrsim 10^{1 \cdot 5}$, of the momentum required for the post-acceleration. The corresponding momentum-per-area balance, $\int p \; dt = \int \varrho_{th} \Delta v$ dr with $p = L/4\pi r^2 c$ and $\int \varrho \, dr = M/4\pi r^2$, implies

$$\Delta v/v = \int L dt/Mvc \lesssim I\omega^2/2Mvc = 10^{-2 \cdot 7} \; I_{45}/M_{(0)} . \tag{18}$$

Even if post-acceleration had only been effective for the large, identified filaments, of total mass $\lesssim 1 \; M_\odot$, there still remains a missing factor of $\approx 10^{1 \cdot 5}$ which led Kundt & Krotscheck to the conclusion that the 30 Hz waves from the pulsar are trapped inside the nebular shell and perform some 30 bounces on average before being absorbed. No other way is known to me which could solve the radial momentum budget.

The inferred 30 Hz wave bath inside the Crab solves a few further problems: It explains the high pressure required to (i) confine the thermal filaments, and (ii) shape the synchrotron spectrum, which exceeds $L/4\pi r^2 c$ by a factor of 30. (iii) It also explains the absence of circular polarization - at a level of 1% - expected for synchro-Compton radiation in a wave field of strength parameter $f \gtrsim 10$ (diluted by projection).

## 2.3 STRUCTURE OF THE CRAB NEBULA

Figure 8 sketches the various peculiarities of the Crab's appearance. All authors agree that there should be an inner shock front, where the supersonic pulsar wind is stalled, and an outer shock front, where the CSM is pushed out. They do not agree, however, on many of the details.

To begin with, some authors locate the inner shock front at $r_i = 10^{17 \cdot 5}$ cm, somewhere near the brightest wisp, (in order to explain the high pressure in the nebula). But there are several wisps in succession, which occasionally move at relativistic speeds; which one to choose? More importantly, figure 9 shows that a small inner radius would imply too strong a compressed magnetic field in the nebula. I therefore prefer $r_i = 10^{18}$ cm for the inner edge, the wisps being a laser phenomenon in the shock front.

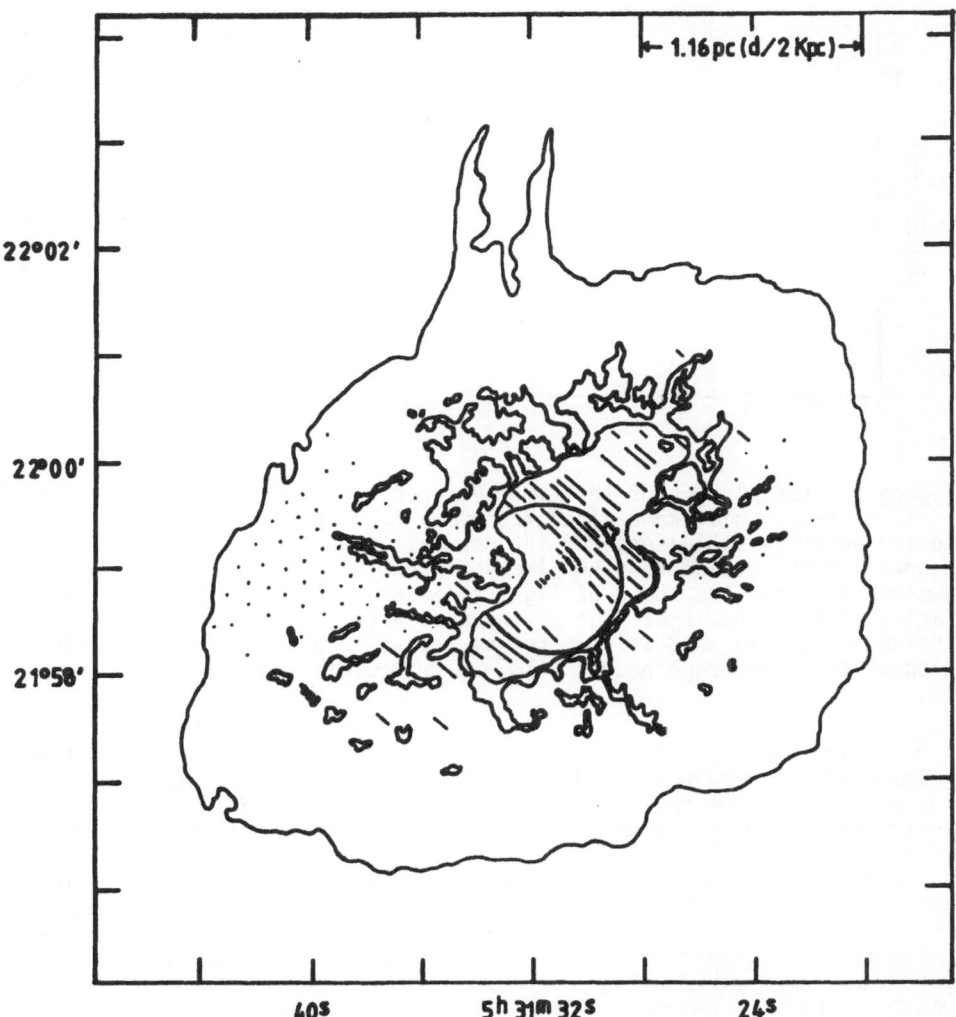

<u>Figure 8</u>. Sketch of the morphology of the Crab nebula as seen in opti-
cal continuum and line radiation ([O III]), taken from refs. 72 and 77.


In any case, it takes surprise that there is a 'central
hole', between the (central) pulsar and the wisps, both at
optical and soft X-ray frequencies. It shows that the pulsar
wind is almost loss-free before being stalled: the charges
perform an $\vec{E} \times \vec{B}$-drift. A similar behaviour is found for the
twin-jets of the extragalactic radio sources.

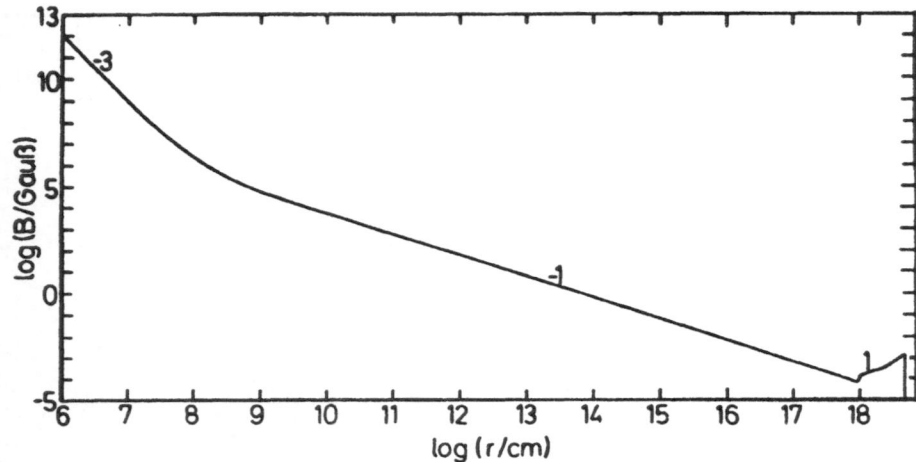

<u>Figure 9</u>. Radial drop of the time-averaged (over one period) magnetic field of the Crab at some fixed latitude, from the surface of the PSR to the outer (visible) edge of the nebula, taken from ref. 94. Some re-connection is assumed to take place between the northern and southern hemisphere, near the speed-of-light cylinder. The field increases by a factor of 2 at the inner edge of the nebula ($r=10^{18}$ cm), is compressed inside the nebula, and is screened by the outer boundary layer. A location of the inner edge near $r = 10^{17.5}$ cm would be inconsistent.

Between the inner and outer edge, the gyrating relativistic component loses energy by radiation and gets compressed. This behaviour would not obtain if a significant fraction of the energy were stored in protons, or heavy ions.

At the outer edge, pair plasma pushes the (much heavier) CSM. Theory predicts the boundary to be Rayleigh-Taylor stable for a (temporally) decreasing pressure of the 'piston', a condition which should prevail most of the time but can be marginally violated for some epoque (Kundt, 1981b). Well, there is the famous 'chimney', or 'spur', discovered by Fesen & Gull, through which the pair plasma escapes and relaxes its pressure; see Wilson et al (1985). Cas A has a similar chimney (Kundt , 1983b), and roughly half of all well-mapped SN shells have one-sided 'bumps'. If this apparently one-sided excess outflow phenomenon is a property of the relativistic component, its speed is subsonic, and a single outlet stabilizes the rest of the boundary.

What happens beyond the 'outer edge'? The outer edge, as we see it, is certainly the outer edge of the relativistic component, and no non-thermal radiation is expected from beyond even if there are (circumstellar) magnetic fields. The thermal component, on the other hand, should be present out to much larger radii, as otherwise the Crab would be atypically slow. Why do we not see it? Here is a simple, speculative answer: Beyond the outer edge is the low-density windzone of the progenitor star, a blue giant, of wind density

$$n_{wind} = 10^{-3 \pm 0.5} \text{ cm}^{-3} \; r_{19}^{-2}, \tag{19}$$

expanding at $\leq 10^{8.3}$ cm s$^{-1}$ out to distances beyond $r_{CSM}$ $\gtrsim 10^{19.2}$ cm. (The corresponding mass loss is $\dot{M} = 10^{-6.5 \pm 0.5}$ $M_\odot$/yr). Filaments beyond the visible outer edge find themselves inside this low-density CSM where they explode sideways because of inhomogeneous pressures $\leq \rho \, v_{rel}^2/2$ and adjust to a radially decreasing wind density, at invisibly low densities. Ram-pressure squeezing implies $n_{wind} \, k \, T_{wind}$ $\leq n_{fil} \, k \, T_{fil} \leq \rho_{wind} \, (v-v_{wind})^2/4$, so that for a wind temperature of $10^{5.5 \pm 1}$ K and filament temperature of $\leq 10^4$ K, one finds $n_{fil}/n_{wind} = 10^{2 \pm 0.5}$, or $n_{fil} = 10^{-1 \pm 1}$ cm$^3$. Such low-density filaments, even if strongly illuminated, are difficult to detect. In other words: all the outer Crab ejecta are presently invisible, both in emission and in absorption.

## 3. THE BLACK-HOLE CANDIDATES

ABSTRACT. Black holes are long-predicted objects, though not correctly described before Einstein's Theory of General Relativity. They form a three-parametric set, determined by their mass, spin, and electric charge. They are magnetically neutral and evaporate slowly if sufficiently isolated.

It is not clear, however, whether any black hole in the Universe has been reliably identified. Mass determinations of massive X-ray binaries can be unreliable if the system has a strong wind, such as Cyg X-1. Even the X-ray nova A0620-00 does not behave like encorporating a black hole.

### 3.1 BLACK HOLES

Within Newton's theory of gravitation and the corpuscular interpretation of light, a black hole can be defined as a (celestial) body which is so massive (compact) that its escape velocity exceeds the speed of light: the body can no longer inform the outside world. In this sense, black holes were first considered by Michell (1784) and Laplace (1795), see also Shapiro & Teukolsky (1986).

But under conditions of strong gravity, the (more) correct theory is Einstein's General Relativity. In it, pressure has a weight, and it can be shown that beyond a certain compactness, no equation of state can support a body against its self-attraction; the body collapses without halt, it forms a black hole. It collapses so fast that even light signals escape more slowly than space contracts: no information can leave a black hole. Its radius of no return (to the outside world) is Schwarzschild's $R_s = 2\,GM/c^2 = 3$ Km $M_{(0)}$, defined through the surface area $A = 4\pi R_s^2$. During collapse, the progenitor liberates an equivalent of its rest energy $Mc^2$.

Among the surprising properties of black holes is the 'no-hair' theorem, proven stepwise by a number of people, most notably Werner Israel (1967), Brandon Carter (1970), and Stephen Hawking (1972). It says that black holes have much less structure than their progenitor bodies: all their higher multipole moments are determined by those of lowest order, mass M, charge Q, and spin J, in the form

$$M_n = M\,a^n \ , \ Q_n = Q\,a^n , \tag{20}$$

with $a := J/Mc$ = specific angular momentum $\leq R_s/2$. I.e. all inertial and heavy, electric and magnetic properties are determined by those of lowest order; deviations from them of the progenitors are radiated away during collapse. When a black hole is spun down, its mass shrinks to the 'irreducible' amount

$$M_{irr} = M [1 - (2 \alpha / R_s)^2 - Q^2 / GM^2]^{1/2} . \tag{21}$$

A contracting celestial body reduces its (non-relativistic, non-degenerate) entropy

$$S = k \sum_j N_j [5/2 - \ln (n_j \lambda_j^3)] , \tag{22}$$

in which $\lambda := h/(2\pi mkT)^{1/2}$ is the thermal de Broglie wavelength; $\lambda$ measures the extent of a particle. In the extremely relativistic limit with vanishing chemical potential, the dimension-less entropy per particle $s := S/Nk$ approaches $2 \pi^4/45 \zeta(3) = 3.60$ for bosons and 7/6 that much for fermions (Kundt, 1971). All in all, the body's entropy S decreases under contraction whereby the excess entropy is radiated towards infinity. By continuity, a forming black hole should have a non-zero entropy which is lower than that of its constituents (Kundt, 1976a).

Hawking proved in 1974 that a black hole, when treated quantum mechanically, should have a (non-zero) temperature

$$T = \hbar c^3 / 8\pi GMk = 10^{-7.2} K/M_{(o)} . \tag{23}$$

The corresponding black-body radiation, of preferred wavelength $\hbar c/kT = 4\pi R_s$, would reduce its mass to zero in a little over $10^{67}$ yr if it were sufficiently isolated; ('at present', a solar-mass black hole would still grow by swallowing part of the 2.7 K background radiation). It would thereby produce an enormous number of photons, of order $N_* \approx \pi GM^2/\hbar c = 10^{76.4} M_{(o)}^2$, and a correspondingly large entropy, $S = 4\pi G M^2 k/\hbar c$ - which is Hawking's black-hole entropy.

Isolated black holes are stationary, apart from the feeble black-body radiation just mentioned, hence they do not radiate. Black holes can only make themselves felt by disturbing their environment, e.g. by forcing another body into a Keplerian orbit, or by accreting radiative matter. They would interact strongly with surrounding plasma if their net charge Q could be distinctly different from zero; but this is unexpected, because neutralization by a compensating charge would be immediate. For vanishing Q, their magnetic dipole moment $Q_1 = Q\alpha$ vanishes likewise. Black holes cannot even distort an ambient magnetic field because their magnetic suspectibility vanishes (King et al, 1975).

Do we expect to find black holes in the Universe? Among the proposed formation processes are the total collapse of the core of a massive star, or the accretion-induced collapse of a neutron star after more than $10^6$ years' feeding at the (maximum) Eddington rate of $10^{-8} M_\odot/yr$. Matter of $\lesssim$ nuclear density can stabilize a star of at least two, more likely three solar masses by its degeneracy pressure (Shapiro & Teukolsky, 1983). Stellar-mass black holes should therefore weigh more than three solar masses.

## 3.2 THE PROPOSED CANDIDATES

How to find a stellar-mass black hole? A promising strategy has been to search for X-ray binaries whose donor star betrays a massive companion by its fast (radially projected) orbital motion. With this criterion, Cyg X-1 was identified as a black-hole candidate in 1971 and has remained one of the best candidates ever since, cf. Ilovaisky (1987), Liang & Dermer (1988), Miyamoto et al (1988). It shows rapid X-ray flickering and an ultrasoft X-ray spectrum, two properties which were thought to be independently characteristic of accreting black holes, i.e. of accreting objects whose hard radiation is formed by the inner accretion disk (White & Marshall, 1984). Similar massive systems are LMC X-1 and V 861 Sco.

More than 15 years of careful observations have revealed a number of remarkable properties of the X-ray binaries, among them spin periods, orbital periods and precession periods, quasi periods, power-law flickering, dipping, bursting, bimodality, polarized (optical) light curves, and erratic radio and optical light curves. Clearly, black holes should not show any good periods, perhaps not even quasi periods. But apart from this very insensitive criterion, none of the (other) properties have allowed to separate the sources uniquely into black-hole and non-black-hole candidates. The only distinguishing criterion has remained the high estimated mass of the X-ray accretor (Kundt & Fischer, 1989).

How reliable is the mass estimate? Given the periodic orbital Doppler shift of at least one spectral line from the donor star and an approximate distance, one can estimate the mass of the X-ray source via Kepler's laws. This method is one of the most powerful tools in astrophysics. A difficulty arises, however, when the wind of the donor star is so strong that the whole windzone is opaque in all the available spectral lines ($\dot{M}/4\pi r_* \, v_\infty \gtrsim 10^{-4}$ g cm$^{-2}$). One then does not measure the (projected) orbital velocity of the star but rather some average motion of the windzone at large distances from it. This windzone can be inhomogeneous, and corotate with the system out to some significant distance, see figure 10. It makes itself felt by a large noise in the measured radial velocities, by orbitally changing spectral line profiles, by different systemic velocities and phase shifts of different lines, and by phase offsets from the (periodic) light curve. In such a situtation, it is not inconceivable that Doppler shift amplitudes differ from stellar Doppler shift amplitudes by a factor of a few, and mass determinations get uncertain by the square of this

factor. In particular, the unseen mass in the Cyg X-1 system
need not exceed that of a neutron star (Indulekha et al).

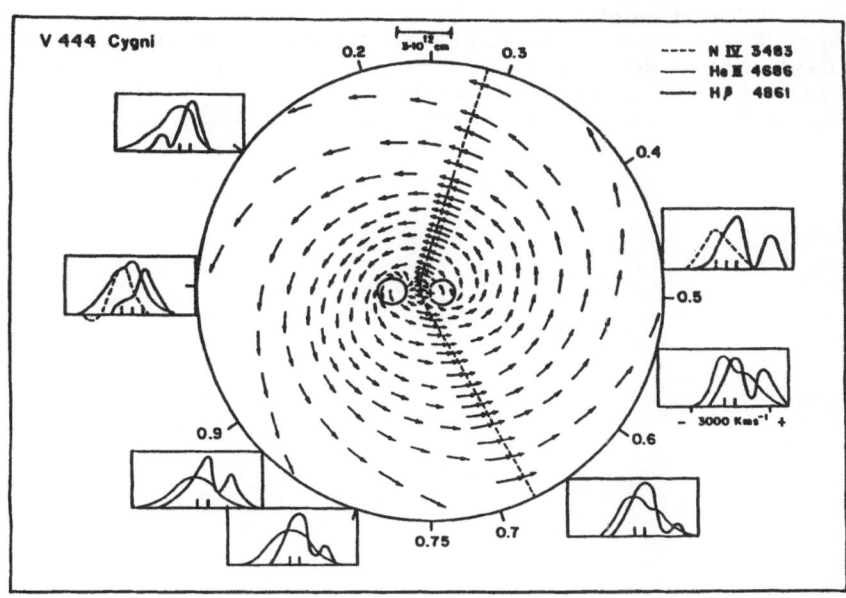

**Figure 10**. Sketch of a hypothetical cut through the orbit plane of the
(massive) Wolf-Rayet binary V 444 Cygni, taken from ref. 55. The winds
of the (more massive) O-star and the (more strongly blowing) W.-R. star
meet in a quasi-hyperboloidal boundary layer. Magnetic viscosity is
assumed to enforce corotation of the winds out to several stellar radii.
Note the orbitally varying shapes of the (three) drawn emission lines
for which the two stars are 'in the fog'.

There is, however, a more recent black-hole candidate, the
1975 X-ray nova A 0620-00, whose radial velocities look
'clean', suggesting a mass of $\gtrsim$ 5 $M_\odot$ of the compact
companion (Mc Clintock & Remillard, 1986). Are we facing the
unique unchallenged stellar-mass black hole?

I am not convinced. Reasons against a massive black hole
in A 0620-00 are (i) the peak X-ray brightness during
outburst, which did not exceed the Eddington limit of a 1.4
$M_\odot$ accretor; (ii) the extremely low present X-ray
brightness, $L_x < 10^{32}$ erg s$^{-1}$, despite the presence of an
accretion disk in the spectrum: why does the hole not
swallow?, (iii) the shape of the optical light curve, which
is far from reflection-symmetric w.r.t. its two minima, a
property required if the system were mirror-symmetric w.r.t.
the plane through both mass centers and perpendicular on the
orbit plane; and (iv) the strong and broad Balmer emission

(Johnston et al, 1989), with $\Delta v \gtrsim 10^3$ Km s$^{-1}$, which signals the presence of an extended, high-velocity windzone reminiscent of a magnetic centrifuge.

If A 0620-00 contained a neutron star - as several other low-mass X-ray novae are believed to do - how can we explain the high inferred mass? To me, a viable possibility is a self-gravitating accretion disk, rigidly rotating around a 1.4 M$_\odot$ neutron star (Kundt, 1979; Krolik, 1984).

# 4. NEUTRON-STAR ACCRETION

ABSTRACT. Matter spiralling through an accretion disk towards a magnetized neutron star has three ways to go: it can (i) be centrifugally re-expelled, (ii) join a 'blade'to spiral all the way in, or (iii) reach a 'downward'leading magnetic field line by evaporation. Predominance of channel (ii) over channel (iii) may give rise to non-pulsing sources. Channel (i) may be responsible for the QPO phenomenon.

Whenever the neutron star's spin is not exactly aligned with that of the binary orbit, magnetic pressures will tilt the inner disk. The mode of decomposition of the disk by the corotating magnetosphere, at its inner edge, should depend not only on the accretion rate and illumination but also on the spin inclination and obliquity of the magnetic dipole moment. An oscillating tilt angle may cause the superorbital period.

## 4.1 THE INNER ACCRETION DISK

Most compact, strong galactic X-ray sources are thought to be accreting neutron stars, or white dwarfs. (Known exceptions are a few young pulsars). The X-rays are emitted as a consequence of matter falling into a deep potential well and hitting a hard surface. Best understood are the pulsing X-ray sources, cf. table 2. Here the literature distinguishes between 'disk-fed' and 'wind-fed' sources, depending on whether or not the accreting matter is intermittently stored in an accretion disk.

Accretion disks are not easily detected in the light curves, or spectra of neutron-star binaries because of their (relatively) low radiated power. Weak emission lines, sometimes held characteristic of a disk, may rather be characteristic of a wind. Better indicators of a disk may be (i) the presence of a superorbital period in the X-ray light curve, and (ii) the detailed spinup history of a pulsing source, i.e. the large noise in the spin history $\omega(t)$ which often dominates a linear spinup. In other words: we see excursions of the angular velocity $\omega$ around a linear increase, on all observable timescales between a week and ten years, with a white-noise power spectrum in angular acceleration (torque) for Vela X-1 (Epstein et al, 1986). Short-time spinup rates are 3 to $10^2$ times higher than average, suggesting that the accreting neutron star exchanges angular momentum with a fly wheel, via its magnetosphere. It could be difficult for a directly accreting wind to mimic the same power spectrum, in particular the high-frequency noise.

In what follows, I restrict considerations to a 1.4 $M_\odot$ star accreting steadily from an accretion disk. Matter in the disk spirals inward because of a viscous loss of angular momentum, according to

$$\dot{M} = 3\pi\sigma\nu \ , \tag{24}$$

where $\dot{M}$ = mass flow rate, $\sigma$ = mass surface density, and $\nu$ = (effective) kinematic viscosity, probably due to toroidally strained magnetic fields. How does the matter reach the stellar surface?

There are three ways for disk matter to go, cf. figure 11. Disk matter can (i) enter the corotating magnetosphere diffusively outside of the corotation radius

$$r_{cor} = (GM/\omega^2)^{1/3} = 10^{8.2} \ \text{cm} \ P_0^{2/3}, \tag{25}$$

($P:=2\pi/\omega$ = spin period); in this case, it will be centrifugally re-expelled towards larger distances (Camenzind, 1986). Once disk matter has crossed $r_{cor}$, it can (ii) continue spiralling in all the way, in the form of clumps of increasing density (due to increasing magnetic confinement), or (iii) evaporate and slide down to the polar caps along magnetic field lines.

The last possibility, a sliding down to the polar caps, has been preferentially considered in the literature because it provides an intuitive explanation of the pulsing sources with their hard spectra, emitted by a small fraction of a neutron star's surface. It is not clear, though, that evaporation would be fast enough in all cases to convert the disk matter into an ionized spray. There is the alternative possibility that most of the accreting matter reaches the neutron star's surface along quasi-Keplerian orbits, in the form of heavy, metallic blades (Kundt et al, 1987). The precise branching ratio, blades versus evaporation, may sensitively depend on the inner disk's thermal structure.

The re-expelled wind, on the other hand, will be intermittently stored in the corotating magnetosphere. It screens the central X-ray source. Torsional oscillations of the plasma-loaded magnetosphere are likely to cause quasi-periodic flux modulations (Kundt & Fischer, 1989).

In any case, there will be an innermost radius of the disk at which it is decomposed, either partially, into clumps, or completely, into spray. This radius is usually called Alfvén radius and evaluated as the distance from the stellar surface at which magnetic pressures equal material ram pressures. Even though this criterion is rather insensitive to details of the geometry, it may be misleading because to first order, centrifugal (ram) forces are balanced by gravity and besides, the stronger the magnetic pressure the denser the confined matter, independent of distance. With

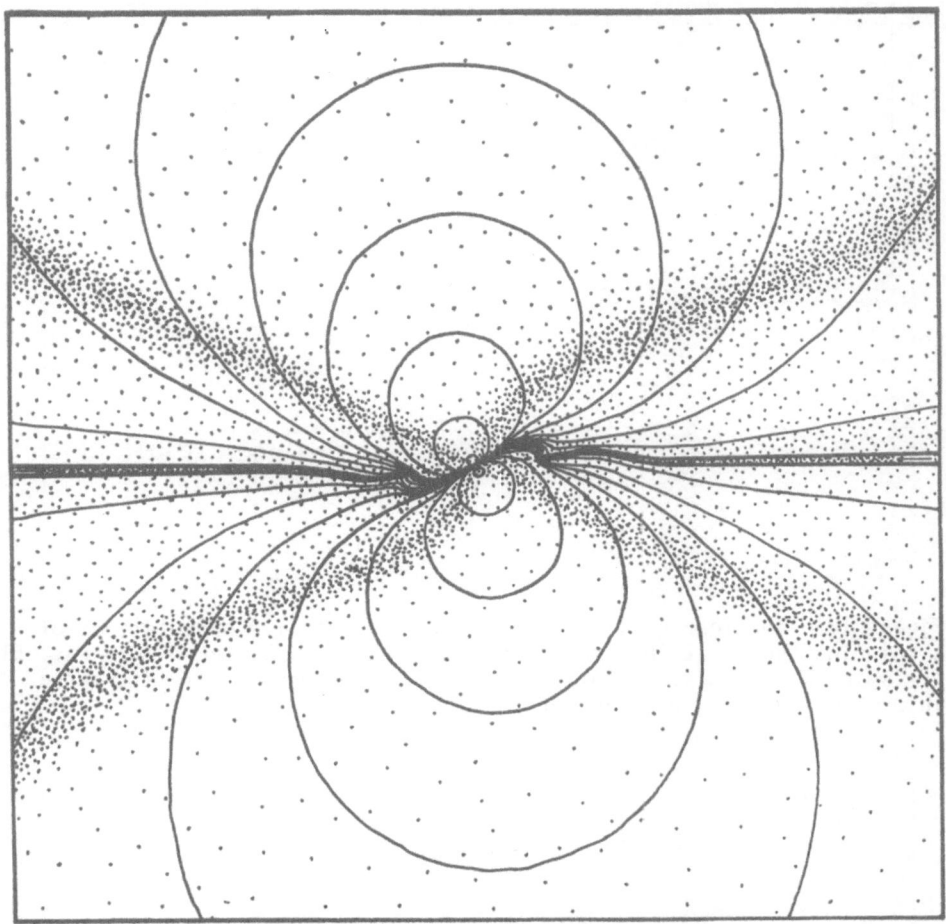

Figure 11. Calculated cut through the magnetosphere of an inclined (2-dim) dipole inside a warped accretion disk, taken from ref. 51. There is one neutral point on either side of the center; the torque corresponds to a weaker force on the side of the neutral point. Matter spiralling in through the disk can be (i) centrifugally re-ejected by the corotating magnetosphere, (ii) reach the central star along quasi-Keplerian orbits, in the form of heavy metallic blades, or (iii) reach the star's polar caps if evaporated in time, by sliding 'down' along magnetic field lines.

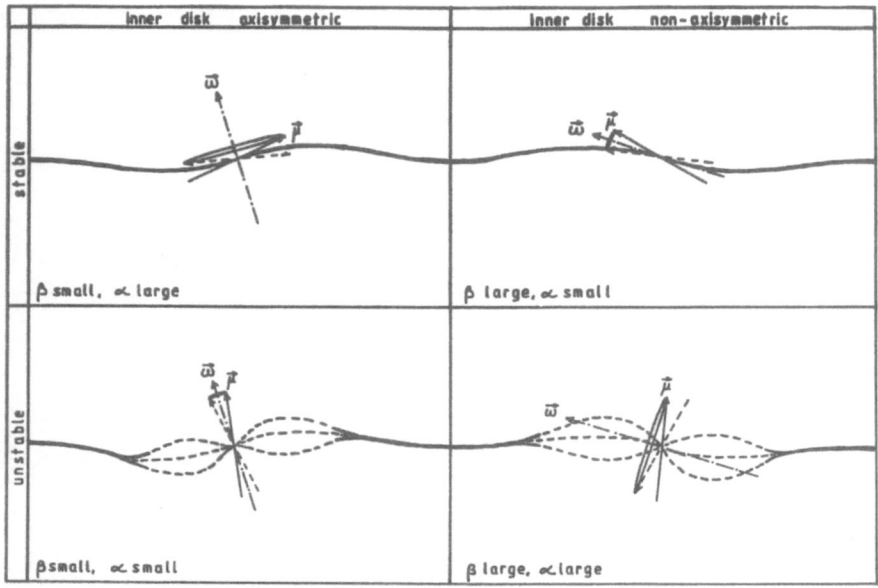

**Figure 12.** The four qualitatively different relative orientations of a spinning magnet at the center of an accretion disk, taken from ref. 51. $\vec{\omega}$ = spin, $\vec{\mu}$ = magnetic dipole. The disk will be more easily disrupted into clumps in the two cases of the bottom row.

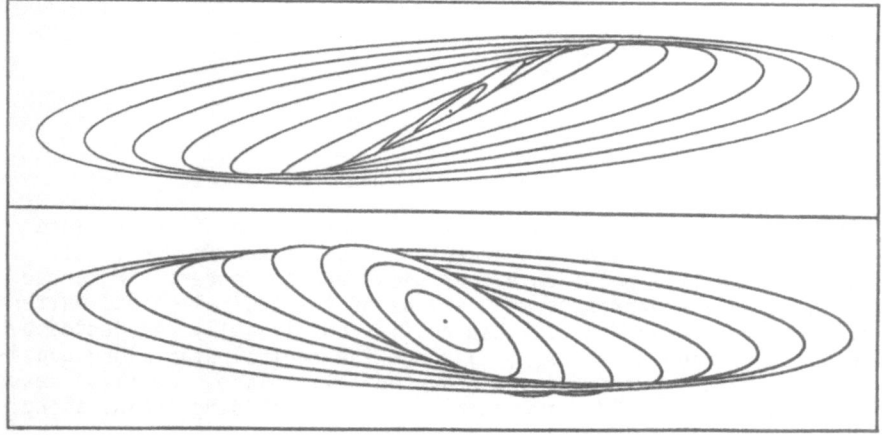

**Figure 13.** Two aspects of a (magnetically) tilted accretion disk, taken from ref. 51. Both the screwing and tilting are exaggerated by weakening the magnetic pressure on convex portions of the disk and by choosing the radial scale logarithmic.

increasing magnetic pressure, matter gets increasingly degenerate, and pressure balance can hold all the way down to the surface.

The inner radius of the disk can depend on a number of conditions, among others on the accretion rate and outside illumination, but also on the strength and relative orientations of the neutron star's spin and magnetic moments w.r.t. each other and w.r.t. the outer disk. Whenever the star's spin is inclined, the inner disk will be tilted (Lipunov & Shakura, 1980). For a typical magnetic moment, this magnetic tilting is much stronger than the general-relativistic Lense-Thirring effect and reaches out beyond $10^3$ corotation radii (Horn & Kundt, 1989). Four qualitatively different relative orientations must be distinguished; they are sketched in figure 12. In two of them, the disk is expected to decompose much more easily than in the two complementary ones, because of rapidly changing magnetic accelerations.

Figure 13 is an exaggerated plot of a magnetically tilted disk. It is exaggerated in two ways: the radial scale has been drawn logarithmically, and magnetic pressure fluctuations have been reduced by more than a factor of two (in order to make the screwing visible). On the other hand, clumping and evaporating are not shown. As soon as the disk gets decomposed, the accretion flow will occupy a much larger range of orbital inclinations.

There remains the question of which of the four scenarios of figure 12 applies to which class of sources. My mind has not yet been made up. I feel certain, though, that an absence of pulsations need not imply an absence of a strong magnetic dipole moment: the reasons were given in section 1.2 above. In particular, strong magnetic fields are likely required for jet formation, VHE pulses, QPOs, flickering, centrifugal ejection of wind matter, and the forcing of a superorbital X-ray period. Strong magnetic dipole moments should not be dismissed too easily.

# 5. SUPERNOVA EXPLOSIONS AND THEIR EJECTED SHELLS

ABSTRACT. There is an increasing awareness that SN explosions may not be driven by neutrinos and/or photons but by the spin motion of the collapsing core, transferred via magnetic viscosity to the shell. On reconnection, the magnetic spring will convert its remaining energy into pair plasma. It is not clear at this time whether SNe of type Ia should be excepted from this scheme: there simply are too many neutron stars.

Supernovae have mechanical energies of $10^{51 \pm 0.5}$ erg, some 2% of the collapsing core's maximal spin energy. Their time-integrated bolometric radiation amounts to $10^{49.5 \pm 0.5}$ erg; similar energies are radiated by old SN shells. Most of the matter is ejected at speeds near $10^{8.8 \pm 0.3}$ cm s$^{-1}$.

If supernovae are eventually propelled by pair plasma, pressure balance forces their shells into small-filling-factor fragments, or filaments. Such filamentary shells have a high photon-storage capacity. Once they are transparent in the continuum, they are still highly opaque near all the resonances. Photon leakage via diffusion can give rise to an exponentially declining light curve. The 'photospheric'and 'nebular' stage can thus be understood as the stages in which the shell is opaque and transparent in the continuum, respectively.

The universal powering of the exponential tails by radioactive decay is not without problems. The different SN types can perhaps be explained by hydrogen or helium stars exploding as red or blue supergiants. A radio flare can only be observed when the progenitor star's late windzone is sufficiently transparent.

## 5.1 SN EXPLOSIONS

An examination of the literature shows that our understanding of SN explosions is still evolving. In his accompanying lecture, Branch presents the majority view since the mid 1980s which allows for two different explosion mechanisms: (i) liberation of the neutron star's binding energy (for high-mass stars), and (ii) nuclear chemical explosion (for low-mass stars). In the first case, it is not clear whether it is the liberated heat energy (neutrinos) or the transfer of the neutron star's spin energy which gives rise to the mass ejection (heat: Woosley & Weaver, (1986); spin: Hoyle (1946), Kardashev (1970), Bisnovatyi – Kogan et al (1976), Kundt (1976), Srinivasan (1987)). My own preference is for the latter, for all types of SNe; in this lecture I shall explain why.

Blaauw (1985) has shown that in a 0.5 Kpc neighbourhood of the Sun, a replenishment of the pulsar population requires (at least) all stars of mass $\gtrsim 6$ $M_\odot$ as progenitors, with about 90% in the mass range between 6 and 10 $M_\odot$. He has not considered binary neutron stars, which would lower the required mass range to (at least) $\gtrsim$ 5 $M_\odot$. A factor of two in mass, in this range, means a factor of $\gtrsim$8 in luminosity and a factor of $\lesssim$ $10^{-2}$ in the number of progenitors (figure 5). This estimate conflicts with mass estimates for young SN shells (like that of SN 1987 A) which tend to exceed 15 $M_\odot$ (Dopita, 1988b; McCray, 1989). In other words: SN shell models tend to overestimate the ejected mass by a factor of $\gtrsim$ 5 (because a star born with 6 $M_\odot$, after wind losses and core contraction, would hardly eject more than 3 $M_\odot$).

SN 1987 A is considered typical: we should have failed to detect many similar events due to their low luminosity. But if it is typical, so should be the mass distribution in its shell. The often-quoted model 10 HM by Pinto & Woosley has more than one quarter of the mass at (radial) velocities $v_r$ $<10^{1.6}$ Km $s^{-1}$ whereas the historical SN shells tend to have most of their mass in a spherical zone around $10^{3.8\pm0.3}$ Km $s^{-1}$, the Crab probably included (see section 2.3). Again, I see a problem for the SN model.

Further concern has already been expressed in section 1.3. Among others, (i) neutron star masses don't appear to scatter considerably, hence the SN piston must sweep efficiently; (ii) all SN shells look filamentary, reminiscent of magnetic Rayleigh-Taylor instabilities during their acceleration; (iii) neutron-star spin rates at birth tend to be slow, much slower than expected by conservation of angular momentum during core collapse; and (iv) the equal evolution time scales of SN lightcurves of types Ia and Ib imply equal masses of their shells, unless their chemistries were vastly different; (for spectral analysis of type Ia SNe see Kirshner et al (1973) and Branch et al (1982), for type Ib see Wheeler et al (1987)).

A certain homogeneity of the class of all SN explosions is also expressed by their similar energetics: Estimates of their mechanical energy yield $E_{kin}$ = $10^{51\pm0.5}$ erg - some 1% of a neutron star's binding energy, $E_{bind}$ = $10^{53\pm0.5}$ erg - corresponding to a mass of $\Delta M$ = 3 $M_\odot$ expanding at a speed of $10^{8.8}$ cm $s^{-1}$:

$$E_{kin} = \Delta M \ v^2 /2 = 10^{51} \ erg \ (\Delta M/3M_\odot) \ v_{8.8}^2 \ . \qquad (26)$$

This estimate is in accord with slightly higher velocities ($10^9$ cm $s^{-1}$) of the (less massive) type I events. It may mean that only some 2% of the core's spin energy, $E_{rot}$ $\lesssim$ $10^{52.7}$ erg, are converted into radial expansion.

Time-integrated bolometric luminosities of SNe are

$$\int L \ dt = 10^{49.5\pm0.5} \ erg \ , \qquad (27)$$

some 3% of the mechanical energy. SN shells emit $10^{37.5\pm0.5}$ erg $s^{-1}$ throughout $\lesssim$ $10^4$ yr, or even $\lesssim$ $10^5$ yr (in the IR:

Arendt, 1989), integrating again to $\int L dt = 10^{49.5 \pm 0.5}$ erg. Still, a SN explosion is an inefficient lamp; most of its energy is converted into re-expansion of the galactic disk.

## 5.2 THE SN PISTON

The 'piston' of a SN is the medium that transfers the energy of the explosion to the ejected shell. By definition, the piston is complementary to the shell; it falls back onto the central remnant if it is heavy. If the central remnant is to be a neutron star, the mass of the piston must not exceed a neutron star's mass; the piston must be light-weight. Possible SN pistons are hot gas, neutrinos, photons, magnetic fields, and/or relativistic $e^{\pm}$-pair plasma.

In the literature, one can often read that SNe are radiation driven (Chevalier, 1976; Mc Cray, 1989). What is meant is that throughout the outer parts of the exploding star, photons dominate over other particles, and the equation of state is almost that of a photon gas. Because of a (Planckian) photon density of

$$n_\gamma = 8\pi \cdot 2.404 \ (kT/hc)^3 = 10^{28.3} \ cm^{-3} \ T_9{}^3 \qquad (28)$$

for $T = 10^9$ K, photon dominance means $T_9{}^3 \gg n_{28.3}$.

But the outer parts of the star form the projectiles, not the piston. At Bad Honnef (1988b) I concluded that the SN piston has a relativistic sound speed. The conclusion was based on the argument that a non-relativistic piston would cool adiabatically, in proportion to $r^{-2}$, throughout the 'runway' which extends from the core radius, $r_c = 10^{6.5 \pm 0.5}$ cm, to the effective shell radius, $r_s = 10^{12 \pm 1}$ cm. Its sound speed would thereby fall below the required ejection speed, $v_{SN} = 10^{8.8 \pm 0.3}$ cm s$^{-1}$. The piston would lose pressure contact with the shell; the shell could not achieve its observed speed.

A critic has argued that pressure contact throughout the runway is not required for a compressible (fluid) projectile because the latter can store energy elasticly and escape faster than being pushed. What can be stored elasticly, however, is energy, not (radial) momentum: a compressed projectile cannot accelerate without pressure contact to a ramp. Besides, the elastic storage capacity of a shocked gas is bounded by $[3(\varkappa-1)]^{1/2}/(\varkappa+1)$ times the shock speed, or 53% for $\varkappa := c_p/c_v = 5/3$. It cannot enhance the escape speed (of most of the matter) by more than half the piston speed, which is negligble for the present purposes.

The temperature of the collapsing core of a pre-SN-star cannot exceed $10^{11}$ K for long because neutrino cooling from the core's surface would be immediate:

$$t_{cool} = 10^{0.5} \ s \ r_6{}^{-2} \ T_{11}{}^{-4} \ . \qquad (29)$$

At temperatures below $10^{11}$ K , ordinary gases are non-relativistic. Photons are plentiful, according to eq. (28), but so are nuclei, electrons, and (thermally created) $e^{\pm}$ pairs. Such a piston has a non-relativistic sound speed and therefore cannot achieve the explosion. This estimate agrees with numerical simulations for which the entropy per baryon takes values near 1 k in the core region, cf. Woosley & Weaver (1986) and the text below eq. (22).

As far as I can see, there is only one way out: the piston is non-thermal. A transfer of the collapsing core's spin energy - via magnetic shear forces - satisfies all the constraints: Only a few percent of the core's maximal spin energy have to be converted into radial expansion. There are no speed problems. Soon after the transfer, the enhanced magnetic fields recombine and create a highly relativistic pair plasma - via vacuum discharges - which continues the boosting. The pair plasma is essentially weightless: an energy of order $10^{52}$ erg, which the piston has to transfer, can be stored in a rest mass of only $10^{23}$ g/$\gamma_8$ if the average Lorentz factor $\gamma$ is of order $10^8$; (it may well be higher initially.) And the magnetized pair plasma is expected to be an efficient sweep, because of its high pressure and (magnetic) viscosity.

## 5.3 YOUNG SN SHELLS

If the SN piston consists of (extremely relativistic) pair plasma, during its relaxation stage, there are a few immediate consequences. Pressure balance with the thermal (projectile) component squeezes the latter into small-filling-factor fragments, or filaments, of volume-filling-factor f, because

$$f = p_{therm}/p_{rel} = \xi \, T_{9.5} \, \Delta M_{(0.5)} \ll 1 \qquad (30)$$

holds for the pressure ratio of the two components if both occupied the same volume. Here it has been assumed that $p_{therm} = 2 n_{therm} kT$, that $\Delta M_{(0.5)} := \Delta M/10^{0.5} M_{\odot}$ is the shell mass in units of 3 $M_{\odot}$, and that $10^{52\pm0.5}$ erg have been converted into pair plasma. The correction factor $\xi$ takes care of the fact that the pair plasma relaxes during expansion in proportion to $r^{-4}$, as long as it is confined, whereas the thermal gas relaxes faster, in proportion to $r^{-5}$; $\xi$ should therefore be smaller than one. The temperature of the boosted gas shell must temporarily rise to the temperature of a strong shock,

$$T_{shock} = 2m \, v^2 \, (\varkappa-1)/(\varkappa+1)^2 k = 10^{9.4} K \, (m/m_p) \, v_9^3 \qquad (31)$$

(for $\varkappa = 5/3$), i.e. T will reach $10^9$ K for a hydrogen shell boosted to $10^9$ cm s$^{-1}$, and cool quickly thereafter. With $\xi < 1$

and $T_9 < 1$, f in eq. (30) is $<< 1$ (unless $\Delta M/3M_\odot$ were much larger than one, in which case the inequality would be reached at a somewhat larger radius).

It has thus been shown that a SN shell is torn and squeezed into bits and pieces right during the explosion event. Once the shell is torn, it can be overtaken by the pressurized pair plasma, only resisted by the late windzone of the progenitor star. If the latter is thin - as suggested above for the Crab nebula - we need not be surprised to encounter $\gtrsim 10^4$ filaments inside a pair plasma bubble still $10^3$ years later.

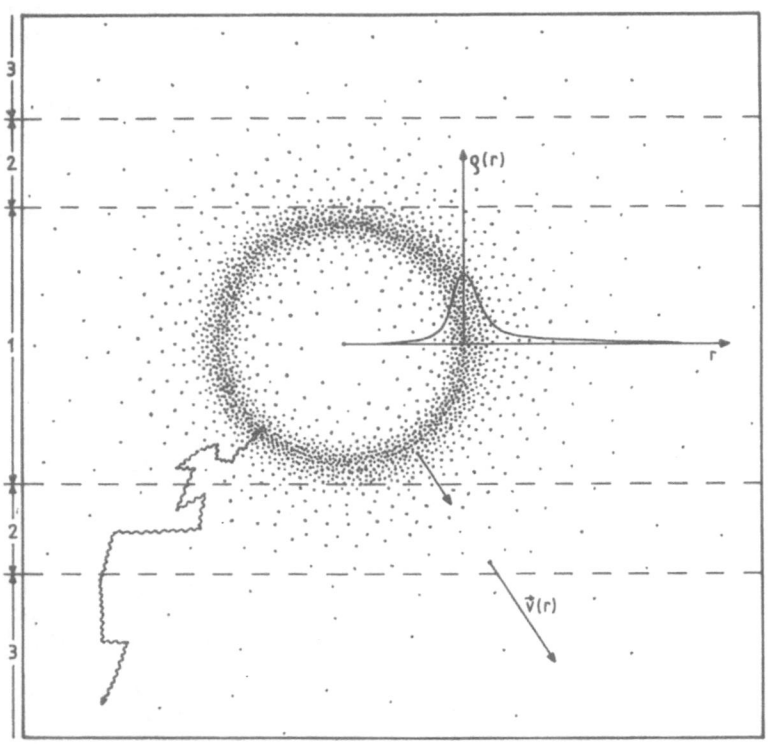

Figure 14. Expected cross section through a young SN shell, after ref. 90. The dots symbolize filaments of small volume-filling factor; their density $\varrho(r)$ decreases to both sides from the peak, as steep powers of radial distance (not drawn quantitatively). Once collisions have become ignorable, the velocities $\vec{v}(\vec{r})$ of individual splinters are proportional to $\vec{r}$. Line-photons are trapped between the filaments for many months before leaking out. the three ring zones 1,2,3 are optically thick, thin and quasi homogeneous, and almost transparent respectively.

Figure 14 is a sketch of what a young SN shell may look like. Dots are representative for filaments. The space between the dots is filled with (almost weightless) pair plasma, bounded at some larger distance by the former windzone. There is a spread in radial velocities around some average because the ejected shell has a high pressure during the acceleration event, exerted mainly by photons, which will push off its outer layers even if they had all received the same velocity initially by the piston. It can thus be understood that the fastest filaments are radiatively accelerated to ≳ three times average velocity and that there is an analogous class of trailing filaments. The resulting fat, expanding shell is a photon bag with a diffusive leakage lifetime (near resonances) of years.

## 5.4 SUPERNOVA LIGHTCURVES

Figure 15 shows a representative set of SN lightcurves, taken from Barbon et al (1984), Doggett & Branch (1985), and Schaefer (1987). Most of the curves are from the blue visual frequency window; bolometric lightcurves would be preferable. Uncertainties enter through an uncertain distance and reddening as well as through the alignment of their peaks. But the impression is real that the curves form a heterogeneous set. This impression would be strengthened if the exotics had not been left out, such as SN 1961 V, or $\eta$ Carinae (Kundt, 1987 d). A similar heterogeneity - even among events of type Ia - shows up in their spectra (Branch, 1987). Remarkably, though, all SN lightcurves have similar time-integrated luminosities, $\int L dt = 10^{49.5 \pm 0.5}$ erg, as though photon bags of standard initial energy were discharged.

Supernovae of type II have hydrogen shells whereas hydrogen is missing in those of type I. Apparently, their progenitors are hydrogen and helium stars, respectively. Moreover, type II events tend to be more massive and (hence) moving somewhat slower. They take longer to reach their peak luminosity mainly because they insulate better.

Type Ib lightcurves look like type Ia, scaled down (in intensity) by a factor of four, but with the same evolutionary time scales. (The spectra near maximum of type Ib look premature: like those of 20d older ones of type Ia.) If the timescale is a measure of shell mass, as will be argued below, the only other obvious parameter is the progenitor's radius. Clearly, if the same energy in photons is provided at a smaller initial stellar radius $r_*$, adiabatic expansion losses of the photon bath are larger in proportion to $r_*^{-1}$. Lightcurves of different total energy would thus find an explanation in terms of exploding red and blue supergiants, respectively.

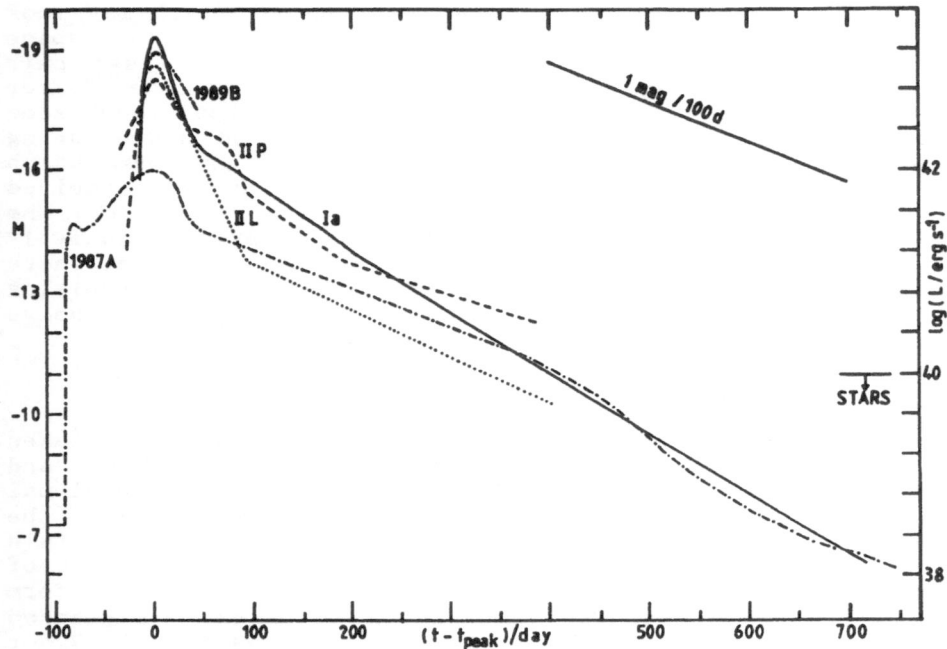

<u>Figure 15</u>. A collection of SN lightcurves — with peak fluxes aligned — taken from refs. 5,31,136 plus those of SN 1987A and 1989B, the latter provided by D. Fischer. Uncertainties enter through ill-known distances and bolometric corrections. Only in few cases do we know the explosion date; but our knowledge is consistent with steeper slopes correlating with younger ages. Note in particular the large scatter in the (slopes of the) exponential tails.

The different types of SNe, Ia, Ib, II. L-P, can thus be tentatively attributed to a different (shell) mass ($\Delta M = 10^{0.5 \pm 0.3}$ $M_\odot$) and hydrogen richness of the progenitor, whereby the brighter SNe come from the more extended (redder) stars ($r_* = 10^{13 \pm 0.5}$ cm). What are the common characteristics of their lightcurves?

A simplified lightcurve is drawn in figure 16, together with the effective colour temperature $T_{eff}$. It shows an initial rise to maximum plus subsequent decline during which the spectrum is continuum-dominated ('photospheric'), and thereafter a more or less exponential tail during which the spectrum is line-dominated ('nebular'). Whereas $T_{eff}$ drops monotonically from its high initial value, probably above $10^{5.5 \pm 0.5}$ K, until the onset of the exponential tail, it stays essentially constant afterwards, at $T_{eff} = 10^{3.8 \pm 0.1}$ K, for all types (Kirshner & Kwan, 1975; Kirshner, 1977; Della Valle et al, 1988). During the exponential-tail stage,

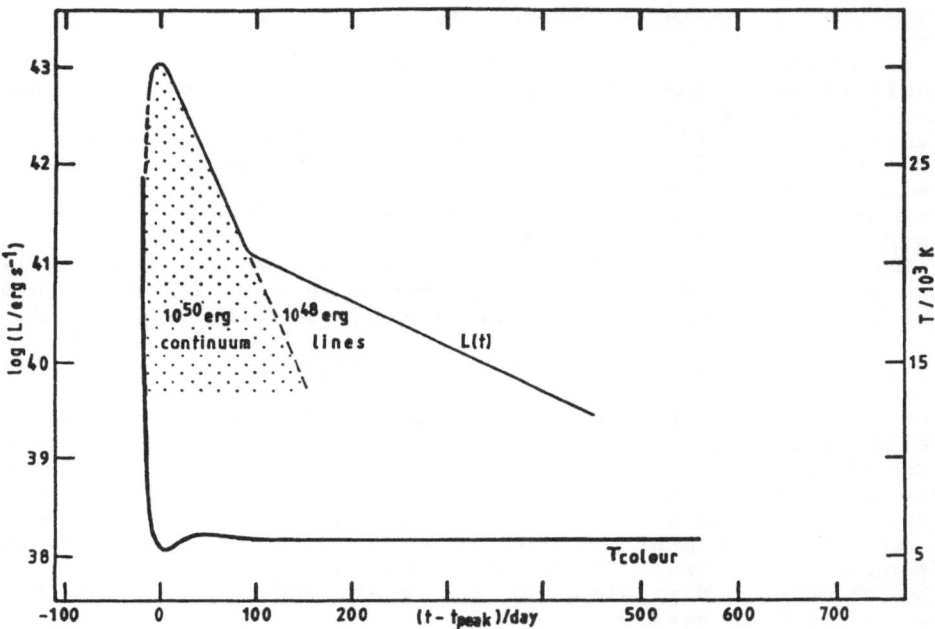

**Figure 16.** A 'typical' SN lightcurve and simultaneous colour-tempera-
ture evolution ($T_{eff}(t)$). There is a continuum-dominated ('photospheric')
rise to maximum and beyond during which $T_{eff}$ drops monotonically, and a
line-dominated ('nebular') quasi-exponential tail of constant $T_{eff}$. The
total radiated energy is $10^{49.5 \pm 0.5}$erg, that of the tail some $10^{-1.3 \pm 0.7}$
thereof.

the size of the continuum photosphere is (likewise) constant
in SN 1987 A (Karovska et al, 1988). In other words: the
luminous evolution consists of two qualitatively different
epochs, (i) a continuum epoch of falling temperature, and
(ii) a line epoch of (more or less) constant temperature and
continuum size.

In Bad Honnef (1988b) I proposed a leaky-box model for SN
cooling according to which these properties can be
(essentially) explained by photon diffusion. Initially, a
growing surface ($\sim r^2$) of the expanding shell gives rise to a
rising luminosity but is counteracted by a falling
temperature. SN 1987A has shown that the product $r^2 T^4$ can
have two maxima (within the first $10^2$ days) even though both
factors are essentially monotonic during this interval. The
shell gets transparent in the continuum, $\tau \lesssim 1$, when the column
number density of particles $N = \Delta M/4\pi r^2 m$ drops below $\sigma_{cont}^{-1}$,
where $\sigma_{cont}$ ( $\gtrsim \sigma_T$) is the continuum scattering cross
section. $r = vt$ and $\sigma_{cont} \lesssim 10^{-24}$ cm$^2$ imply

$$t_{tr} \approx 10^7 \text{ s } (\Delta M_{(0.5)} \, m_p/m)^{1/2}/v_9 \, , \tag{32}$$

i.e. a 'transition' time of order $10^2$ days for a $3M_\odot$ hydrogen shell at bulk speed $10^9$ cm s$^{-1}$. $t_{tr}$ is of the order of the transition time from continuum-dominated ('photospheric') to line-dominated ('nebular'). For uniform kinetic energy, $t_{tr}$ scales as $\Delta M/\sqrt{m}$ with the shell mass $\Delta M$ and mean particle mass m; it can shrink to two weeks - for a one-solar-mass helium shell - and grow towards one year, for an unusually large shell mass ($\Delta M \gg 3 \, M_\odot$).

Once the shell gets transparent in the continuum, it still has an optical depth $\tau$ of $\lesssim 10^{12}$ near resonances because the frequency-averaged scattering cross section $\sigma_{line}$ reads

$$\sigma_{line} = gf \pi e^2 \lambda/\beta m_e c^2 = 10^{-12} \text{ cm}^2 (gf\lambda_{-4.5}/\beta_{-4.5}) \tag{33}$$

for visible wavelengths $\lambda$ of order $10^{-4.5}$ cm and (random) velocity dispersions $\Delta v =: \beta c = 10^6$ cm s$^{-1}$ (corresponding to $T \approx 10^4$ K). Of course, much smaller cross sections apply to resonance lines whose absorbing levels are depopulated; but there should be a sufficient number of resonances to make even small filaments opaque in a practically dense set of frequency intervals. Moreover, coherent scattering preserves the frequency distribution: the scattered photons are not red-shifted on average.

Consequently, a sufficiently dense set of filaments (to avoid Doppler offsets) can act as a medium of perfect scatterers, with a high photon-storage capacity (which has not been available to the "Schildbürger"). Some photon thermalization may happen, but only for photons penetrating deeply into filaments; its effect on the radiation colour temperature need not be large. And escape times from the shell can be delayed by $\lesssim \tau_{fil}^2$ compared with transit times r/c:

$$t_{esc} \approx \tau_{fil}^2 \, r/c \gg \text{yr } r_{16} \, , \tag{34}$$

the latter for $\tau_{fil} \gg 10$, where $\tau_{fil}$ is an effective opacity for hitting filaments. Most of the stored photons are expected to escape from the expanding shell when it has thinned sufficiently. As shown in (1988b), diffusive photon escape yields an exponentially falling lightcurve for a non-varying escape radius; which has been observed, in the meantime, by Karovska et al (1989) for SN 1987A. The time constant $\Delta t$ of the exponential tail is of order

$$\Delta t \approx \Delta r/v = 10^7 \text{ s } (\Delta r)_{16}/v_9 \, . \tag{35}$$

Analogous line-photon delays have been observed in variable AGN sources, cf. Kundt (1988c).

In the literature, this time constant $\Delta t$ has instead been identified with that of radioactive decay of $^{56}$Co to $^{56}$Fe even though figure 15 shows a wide and quasi-systematic scatter of the slopes. Certainly, some radioactive decay of $^{56}$Ni via $^{56}$Co is expected, and even indicated by the IR- and X-ray spectrum of SN 1987A (Meikle et al, 1989). But the $10.52\mu$m line from [Co II] blends with $10.68\mu$m from [Ni II] , and there clearly is a large amount of stable Ni inside SN 1987A. The radioactive decay hypothesis has not really stood its quantitative test.

To my mind, radioactive heating is not an absolute requirement. The overall energy in an exponential tail is small, some $10^{-1.3\pm0.7}$ of the total, and photon diffusion causes a delay in the fading. Besides, there are at least three further energy reservoirs which must not be ignored in the treatment of a SN lightcurve tail. They are:

(1) The heat energy of the newly formed neutron star, some $10^{49.2}$ erg $T_{10}^2$, which does not escape fast enough by conduction (Romani, 1987) but may well escape fast enough by convection, via volcanos; the star's surface temperature would be of order $10^8$ K;

(2) Magnetobremsstrahlung by the pair plasma which has served as the piston of the SN explosion, of initial energy $10^{52\pm0.5}$ erg, almost all of which has been converted into mechanical expansion but a small fraction of which may survive throughout hundreds of years; and:

(3) The huge kinetic energy of the shell, of order $10^{51\pm0.5}$ erg, part of which can be tapped by overtaking crashes between filaments of different velocities.

As long as the shell is opaque in most frequency intervals, it is difficult to tell the relative contributions of the various energy inputs.

A final remark concerns the radio lightcurves of SNe. They tend to be explained by crashes of the ejecta into circumstellar matter followed by shock acceleration of electrons to relativistic energies. I doubt the required high efficiency of the shock booster (Kundt, 1984b). If the SN piston consists of pair plasma, radio emission is expected right from the beginning unless the surrounding CSM is opaque to it. Now the thermal free-free opacity

$$\tau_{ff} = (\int n_e^2 ds)_{26.5}/T_5^{3/2} \nu_9^2 = \begin{cases} 10^{-2.5} \ r_{15}^{-3} \ (\dot{M}_{(-6)}/v_8)^2 \nu_9^{-2} \\ 10^{0.5} \ r_{16}^{-3} \ (\dot{M}_{(-5)}/v_6)^2 \nu_9^{-2} \end{cases} \quad (36)$$

is $\ll 1$ for the fast, low-density wind of a blue giant (with $\dot{M} = 10^{-6}$ $M_\odot$/yr and v $= 10^8$ cm s$^{-1}$, at $\nu = 1$ GHz) already at a distance of $10^{15}$ cm, whereas $\tau_{ff} \gg 1$ still holds for the slow, high-density wind of a red giant at a distance of $10^{16}$

cm. In other words: we expect to see an early radio burst from SNe produced by blue giants but hardly by red giants.

This expectation is consistent with the fact that the low-luminosity - and 'hence' compact - SNe (of type Ib, and SN 1987A) are radio emitters long before peak luminosity whereas the high-luminosity SNe are seen after several months (type II L) or not at all during the first years (types Ia, II P), (Weiler & Sramek, 1988). In all cases, we have to wait for the non-thermal radiation until the sky clears, which takes a great deal longer for SNe resulting from progenitors with high-density winds than with low-density winds.

Acknowledgements:

During the year of this ASI, I have enjoyed special advice and information from Norbert Bartel, Klaas de Boer, Daniel Fischer, Wolfgang Hillebrandt, Pramesh Rao, and Tom Richtler. All of the non-conservative ideas have been tested on, and much clarified by the members of my seminar, in particular Reinhold Schaaf and Axel Jessner. To all of them go my thanks.

## 6. ALTERNATIVE SCENARIOS RELATING TO NEUTRON STARS

My interest in neutron stars was stimulated by Hans Heintzmann in 1971. Ever since then, I have actively pursued the progress made both observationally and theoretically, in understanding these compact astrophysical objects whose proper description cannot avoid Einstein's General Theory of Relativity. In the course of those studies, I could not always be convinced of the reality of assumptions made by some of my colleagues. The 'alternatives' grew in number. They often brought me into conflict with referees. And they survived through the four discussion sessions at Erice. Here they are:

| | References |
|---|---|
| n* form from (evolved) stars with M≥ $M_\odot$(e.g. white dwarf) | 161, 43 |
| n* form from (evolved) stars with M≥ 3 $M_\odot$ | 122, 82 |
| n* derive from often single stars. | 151 |
| n* derive from at least binary stars. | 82, 145, 9 |
| PSR beams are pencil beams. | 104 |
| PSR beams are fan beams. | 82 |
| PSR radio pulses come from polar caps | 148 |
| PSR radio pulses come from speed-of-light cylinder | 82, 160 |
| Pulsars die by flux decay | 105, 148 |
| Pulsars die by alignment | 74, 17, 18, 89 |
| n* magnetic dipole moments do decay (within $10^{10}$ yr). | 105, 148 |
| n* magnetic dipole moments do not decay (within $10^{10}$ yr). | 74, 95, 89, 113, 22 |
| PSR winds consist of pair plasma, post-accelerated by certain fields | 4 |
| PSR winds consist of pair plasma, post-accelerated by strong outgoing wave | 84, 87, 94 |
| Pair-plasma winds (jets) are generated by exceptional n*. | |
| Pair-plasma winds (jets) are generated by all n*. | 78, 92 |
| The mapped jets of SS 433 consist of local galactic matter | 107 |
| The mapped jets of SS 433 consist of pair plasma | 81, 86 |
| Cosmic rays are accelerated by I.S. shocks | 35, 53 |
| Cosmic rays are accelerated by young binary n*. | 50,76,79,80,135,38 |
| The ms-PSRs are spun up by accretion | 152, 153, 147 |
| The ms-PSRs are born fast | 82 |
| The Crab PSR wind is strong enough to post-accelerate the filaments. | 131, 59 |
| The Crab PSR wind is not strong enough to post-accelerate the filaments. | 94, 72 |
| A common envelope can form around a n*. | 151 |
| A common envelope cannot form around a n*. | 82 |
| The black-hole candidates involve black holes. | 156 |
| The black-hole candidates involve n*. | 71, 54, 92 |

| Statement | Refs |
|---|---|
| Radial-velocity oscillations from binaries { do / need not } yield the mass function. | 41 / 55 |
| Accretion onto n** takes place { along magnetic field lines, as a plasma flow / sometimes quasi-Keplerian, as heavy blades }. | 143 / 95 |
| Accreting X-ray sources are { sometimes wind-fed / always disk-fed }. | 143 / 82 |
| QPOs stem from { accretion flow from the inner edge / corotating magnetosphere, near the speed-of-light surface }. | 98 / 95 |
| UHE bursts (pulses) stem from relativistic { protons / electrons }. | 49 / 95 |
| SNe are driven by { neutrinos / a magnetic spring, converting to pair plasma }. | 161 / 7, 70, 142, 90 |
| SN explosions give birth to { B.H. or ∅ (occasionally) / n* }. | 161 / 82 |
| SNe behave like { pressure / splinter } bombs, or { shock waves / shrapnel }. | 161 / 85, 90 |
| SN light curves are powered by { radioactive decay / $\&ns$ cooling & ram pressure & $e^{\pm}$-radiation }. | 161 / 85, 88, 90 |
| SN shells are { multiple shock waves / ram-pressure confined filaments } when propagating through non-relativistic CSM. | 37 / 83, 45, 77, 93, 90 |
| SN shells receive their relativistic electrons (and positrons) { in situ / at birth }. | 132 / 73, 79, 85, 88, 90 |
| SN shells lose their kinetic energy to { radiation / galactic disk expansion }. | 37 / 83, 88 |

REFERENCES

1. Adams, T.F. (1975): Astrophys.J. **201**, 350
2. Alcock, Ch. (1989): Nature **337**, 405
3. Arendt, R.G. (1989): Astrophys.J. **339**, 1161
4. Asseo, E., Kennel, C.F. & Pellat, R. (1978): Astron.
   Astrophys. **65**, 401
5. Barbon, R., Capellaro, E. & Turatto, M. (1984):Astron.
   Astrophys. **135**, 27
6. Bartel, N. (1988) in: SNSs and their Birth Events,
   Lecture Notes in Physics **316**, ed. W. Kundt,
   Springer, p. 206
7. Bisnovatyi-Kogan, G.S., Popov, Yu.P.& Samochin, A.A.
   (1976): Astroph.Space Science **41**, 287
8. Blaauw, A. (1985), in: Birth and Evolution of Massive
   Stars and Stellar Groups, eds. Boland & van
   Woerden, Reidel, p. 211
9. Blair, D.C. & Candy, B.N. (1985): Mon.Not.R.astr.Soc.
   **212**, 219
10. Blair, D.G., Candy, B.N., Fabian, A.C. & Kembhavi, A.K.
    (1989): Mon.Not.R.astr.Soc., submitted
11. Blome, H.J. & Kundt, W. (1988): Astrophys.Space Science
    **148**, 343
12. Boer, M., Hameury, J.M. & Lasota, J.P. (1989): Nature
    **337**, 716
13. Branch, D. (1987): Astrophys.J. **316**, L81
14. Branch, D., Buta, R., Falk, S.W., Mc Call, M.L.,
    Sutherland, P.G., Uomoto, A., Wheeler, J.C. &
    Wills, B.J. (1982): Astrophys.J. **252**, L 61
15. Brinkmann, W. & Ögelman, H. (1987): Astron.Astrophys.
    **182**, 71
16. Camenzind, M. (1986): Astron.Astrophys. **162**, 32
17. Candy, B.N. & Blair, D.C. (1983): Mon.Not.R.astr.Soc.
    **205**, 281
18. Candy, B.N. & Blair, D.C. (1986): Astrophys.J. **307**, 535
19. Caraveo, P.A., Bignami, G.F. & Goldwurm, A. (1989):
    Astrophys.J. **338**, 338
20. Caswell, J.L. (1988), in: SNRs and the ISM, IAU 101, eds
    Roger & Landecker, Cambridge Univ. Press, p. 269
21. Caswell, J.L., Kesteven, M.J., Komesaroff, M.M.,
    Haynes,R.F., Milne, D.K., Stewart, R.T. & Wilson,
    S.G. (1987): Mon.Not.R.astron.Soc. **225**, 329
22. Chanmugam, G.& Brecher, K. (1987): Nature **329**, 696
23. Coroniti, F.V. & Kennel, C.F. (1985), in: The Crab
    Nebula and related SN remnants, eds. M.C. Kafatos
    & R.B.C. Henry, Cambridge Univ. Press, p. 25
24. Cheng, K.S., Ho, C. & Ruderman, M. (1986): Astrophys.J.
    **300**, 500 and 522
25. Chevalley, R.A. (1976): Astrophys.J. **207**, 872
26. Clifton, T.R. & Lyne, A.G. (1986): Nature **320**, 43

27. Cowsik, R., Ghosh, P. & Melvin, M.A. (1983): Nature 303
    308
28. Davidson, K. & Fesen, R.A. (1985): Ann.Rev.Astron.
    Astroph. 24, 119
29. Della Valle, M., Capellaro, E., Ortolani, S. & Turatto,
    M. (1988): The Messenger 52, 16
30. Dewey, R.J., Maguire, C.M., Rawley, L.A., Stokes, G.H.
    & Taylor, J.H. (1986): Nature 322, 712
31. Doggett, J.B. & Branch, D. (1985): Astron.J. 90, 2303
32. Dopita, M.A. (1988a): Nature 331, 506
33. Dopita, M.A. (1988b): Space.Sci.Rev. 46, 225
34. Dowthwaite, J.C. et al (7 authors)(1984): Nature 309,
    691
35. Drury, L.O'C. (1983) Rep.Progr.Phys. 46, 973
36. Epstein, R.I., Lamb, F.K., Priedhorsky, W.C. & Eckhardt,
    R.C. (1986): Astrophysics of Time Variability in X-
    ray and $\gamma$-ray Sources, Taos workshop report, Los
    Alamos Science
37. Falle, S.A.E.G. (1981): Mon.Not.R.astr.Soc. 195, 1011
38. Falle, S.A.E.G. & Giddings, J.R. (1987):Mon.Not.R.astr.
    Soc. 225, 399
39. Fesen, R.A., Becker, R.H., Blair, W.P. & Long, K.S.
    (1989): Astrophys.J. 338, L 13
40. Fürst, E., Reich W. & Sofue, Y. (1987): Astron.
    Astrophys.Suppl. 71, 63
41. Gies, D.R. & Bolton, C.T. (1986): Astrophys.J. 304, 371
    and 389
42. Goldman, I. (1979): Astron. & Astrophys .78, L15
43. Grindlay, J.E. & Bailyn, C.D. (1988): Nature 336, 48
44. Gwinn, C.R., Cordes, J.M., Bartel, N., Wolszczan, A. &
    Mutel, R.L. (1988): Astrophys.J. 334, L 13
45. Hamilton, A.J.S. (1985): Astrophys.J. 291, 523
46. Harnden, F.R., Grant, P.D., & Seward, F.D. (1985):
    Astrophys. J. 299, 828
47. Heintzmann, H., Kundt, W. & Lasota, J.P. (1975): Phys.
    Rev. A 12, 204
48. Henrichs, H.F. (1983), in: Accretion-driven stellar
    X-ray sources, eds. W.H.G. Lewin & E.P.J. van den
    Heuvel, Cambridge Univ. Press, p. 393.
49. Hillas, A.M. (1984): Ann.Rev.Astron.Astrophys. 22, 425
50. Holloway, N., Kundt, W. & Wang, Y.-M. (1978): Astron.
    Astrophys. 70, L 23
51. Horn, S. & Kundt, W. (1989): Astrophys.Space Science
    158, 205
52. Hoyle, F. (1946): Mon.Not.R.astr.Soc. 106, 343
53. Illarionov, A.F. & Sunyaev, R.A. (1975): Astron.
    Astrophys. 39, 185

54.  Ilovaisky, S.A. (1987), in: Variability of Galactic and
     Extragalactic X-ray Sources, ed. A. Treves,
     Associazione per L'Avanscamento dell' Astronomia,
     Milano, p. 175
55.  Indulekha, K., Kundt, W. & Shylaja, B.S. (1988): J.A.A.,
     subm. in January; to appear in Astrophys.Space Sci.
56.  Johnston, H.M., Kulkarni, S.R. & Oke, J.B. (1989):
     Astrophs.J., submitted
57.  Kardashev, N. (1970): Sovj. Astron. 14, 375
58.  Karovska, M., Koechlin, L., Nisenson, P., Papaliolios,
     C. & Standley, C. (1989): Astrophys.J. 340, 435
59.  Kennel, C.F. & Coroniti, F.V. (1984): Astrophys.J. 283,
     694 and 710
60.  King, A.R., Lasota, J.P. & Kundt, W. (1975): Phys.Rev.D.
     12, 3037
61.  Kirshner, R.P. (1977): Ann. N.Y. Acad. Sci. 302, 81
62.  Kirshner, R.P. & Kwan, J. (1975): Astrophys.J. 197, 415
63.  Kirshner, R.P., Oke, J.B., Penston, M.V. & Searle, L.
     (1973): Astrophys.J.185, 303
64.  Kirshner, R.P., Winkler, P.F. & Chevalier, R.A. (1987):
     Astrophys.J. 315, L 135
65.  Krolik, J.H. (1984): Astrophys.J. 282, 452
66.  Krotscheck, E. & Kundt, W. (1978): Commun.Math.Phys. 60,
     171
67.  Kulkarni, S.R. & Hester, J.J. (1988): Nature 335, 801
68.  Kundt, W. (1971): Springer Tracts in Modern Physics 58,1
69.  Kundt, W. (1976a): Nature 259, 30
70.  Kundt, W. (1976b): Nature 261, 673
71.  Kundt, W. (1979): Astron.Astrophys. 80, L7
72.  Kundt, W. (1980a): Ann. New York Acad.Sci. 236,429
73.  Kundt, W. (1980b): Nature 284, 246
74.  Kundt, W. (1981a): Astron.Astrophys. 98, 207
75.  Kundt, W. (1981b): IAU Symp. 95, eds. W. Sieber & R.
     Wielebinski, Reidel, p. 57
76.  Kundt, W. (1983a): Astrophys.Space Science 90, 59
77.  Kundt, W. (1983b): Astron.Astrophys. 121, L15
78.  Kundt, W. (1984a): Astrophys.Space Science 98, 275
79.  Kundt, W. (1984b): J.Astroph.Astron. 5, 277
80.  Kundt, W. (1984c): Advances in Space Research (COSPAR)
     4, 381
81.  Kundt, W. (1985a): Astron.Astrophys. 150, 276
82.  Kundt, W. (1985b): Bull.Astron.Soc.India 13, 12
83.  Kundt, W. (1985c), in: 'The Crab Nebula and related SN
     remnants', eds. M.C. Kafatos, R.B.C. Henry,
     Cambridge Univ.Press, p. 151
84.  Kundt, W. (1986), in: 'Cosmic Radiation in Contemporary
     Astrophysics', NATO ASI C 162, ed. M.M. Shapiro,
     Reidel, pp. 57,67
85.  Kundt, W. (1987a), in: 'Interstellar Magnetic Fields',
     eds. R. Beck & R. Gräve, Springer, p. 185
86.  Kundt, W. (1987b): Astrophys. & Space Science 134, 407

87.  Kundt, W. (1987c), in: 'Astrophysical Jets and their
        Engines', ASI C 208, ed. W. Kundt, Reidel, p. 1
88.  Kundt, W. (1987d): ESO workshop on SN 1987A, ed. J.
        Danziger ESO, p. 633
89.  Kundt, W. (1988a): Comments on Astrophysics 12, 113
90.  Kundt, W. (1988b): SN shells and their birth events,
        Lecture Notes in Physics 316, Springer
91.  Kundt, W. (1988c): Astrophys.Space Science 149, 175
92.  Kundt, W. & Fischer, D. (1989): J.A.A. 10, 119
93.  Kundt, W. & Gopal-Krishna (1984): Astron.Astrophys. 136,
        167
94.  Kundt, W. & Krotscheck, E. (1980): Astron.Astrophys .83,
        1
95.  Kundt, W. Özel, M. & Ercan, E.N. (1987): Astron.
        Astrophys. 177, 163
96.  Kundt, W. & Robnik, M. (1980): Astron.Astrophys. 91, 305
97.  Kuzmin, A.D., Dagkesamanskaya, I.M. & Pugachev, V.D.
        (1984): Pis'ma v Astronomicheskii Zhurnal 10, 854
98.  Lewin, W.H.G., van Paradijs, J. & van der Klis, M.
        (1988): Space Sci.Rev. 46, 273
99.  Liang, E.P & Dermer, C.D. (1988): Astrophys.J. 325, L39
100. Lipunov, V.M., Moskalenko, E.I. & Shakura, N.I. (1982):
        Astrophys.Space Sci. 85, 459
101. Lipunov, V.M. & Shakura, N.I. (1980): Sov.Astron.Lett.
        6 (1), 14
102. Long, K.S., Blair, W.P. & van den Bergh, S. (1988):
        Astrophys.J. 333, 749
103. Lyne, A.G. (1987): Nature 326, 569
104. Lyne, A.G. & Manchester, R.N. (1988): Mon.Not.R.astr.
        Soc. 234, 477
105. Lyne, A.G., Manchester, R.N. & Taylor, J.H. (1985): Mon.
        Not.R.astr.Soc. 213, 613
106. Manchester, R.N. (1987): Astron.Astrophys. 171, 205
107. Margon, B. (1984): Ann.Rev.Astron.Astrophys. 22, 507
108. Mathewson, D.S., Ford, V.L., Dopita, M.A., Tuohy, I.R.,
        Long, K.S. & Helfand, D.J. (1983): Astrophys.J. 51,
        345
109. Mc Clintock, J.E. & Remillard, R.A. (1986): Astrophys.J.
        308, 110
110. Mc Cray, R. (1989), in: Molecular Astrophysics, ed.T.
        Hartquist, Cambridge Univ.Press, p. 1
111. Meikle, W.P.S., Allen, D.A., Spyromiko, J. & Varani,
        G.-F. (1989): Mon.Not.R.astr.Soc. 238, 193
112. Melrose, D.B. (1980): Plasma Astrophysics, Gordon &
        Breach
113. Michel, F.C. (1987): Nature 329, 310
114. Michel, F.C. (1989): Nature 337, 236
115. Michell, J. (1784): Phil.Trans.Royal Soc. 74, 35
116. Mills, B.Y., Turtle, A.J., Little, A.G. & Durdin, J.M.
        (1984): Austr.J.Phys. 37, 321
117. Milne, D.K. (1987): Austr.J.Phys. 40, 771

118. Miyamoto, S., Kitamoto, S., Mitsuda, K. & Dotani, T. (1988): Nature 336, 450
119. Murakami, T. et al (13 authors) (1988): Nature 335, 234
120. Narayan, R. (1987): Astrophys.J. 319, 162
121. Narayan, R. & Vivekanand, M. (1983): Astron.Astrophys. 122, 45
122. Oemler, A. & Tinsley, B.M. (1979): Astron.J. 84, 985
123. Paczyński, B. (1989): Nature 337, 689
124. Parker, R.A.R., Gull, T.R. & Kirshner, R.P. (1979): An emission survey of the Milky Way, NASA
125. Pelling, R.M., Paciesas, W.S., Peterson, L.E., Makishima, K., Oda, M., Ogawara, Y. & Miyamoto, S. (1987): Astrophys.J. 319, 416
126. Priedhorsky, W.C. & Holt, S.S. (1987): Space Science Reviews 45, 291
127. Protheroe, R.J. & Clay, R.W. (1985): Nature 315, 205
128. Ramana Murthy, P.V. & Wolfendale, A.W. (1986): Gamma-ray Astronomy, Cambridge Univ. Press
129. Rankin, J.M. & Gil, J.A. (1989): Comments on Astrophysics 14, 1
130. Rankin, J.M., Campbell, D.B., Isaacman, R.B. & Payne, R. R. (1988): Astron.Astrophys. 202, 166
131. Rees, M.J. & Gunn, J.E. (1974): Mon.Not.R.astron.Soc. 167, 1
132. Reynolds, S.P. & Chevalier, R.A. (1984): Astrophys.J. 281, L33
133. Ritter, H. (1986), in: The evolution of Galactic X-ray binaries, NATO ASI C 167, eds. J. Trümper, W.H.G. Lewin & W. Brinkmann, Reidel, p. 271
134. Romani, R.W. (1987): Astrophys.J. 313, 718
135. Sarris, E.T. & Krimigis, S.M. (1985): Astrophys.J. 298, 676
136. Schaefer, B.E. (1987): Astrophys.J. 323, L51
137. Schwarzschild, B. (1988): Physics Today, Nov., p. 17
138. Seward, F.D. (1985): Comments Astrophys. 11, 15
139. Shapiro, S.L. & Teukolsky, S.A. (1983): Black Holes, White Dwarfs, and Neutron Stars, John Wiley
140. Shapiro, S.L. & Teukolsky, S.A. (1986): Highlights in Modern Astrophysics, Concepts and Controversies, John Wiley
141. Shitov, Yu.P. & Malofeev, V.M. (1985): Sov.Astron.Lett. 11, 40
142. Srinivasan, G. (1987): Current Science 56, 245
143. Stella, L., White, N.E. & Rosner, R. (1986): Astrophys. J. 308, 669
144. Stokes, G.H., Segelstein, D.J., Taylor, J.H. & Dewey, R. J. (1986): Astrophys.J. 311, 694
145. Stone, R. (1979): Astrophys.J. 232, 520
146. Strom, R.G., van Paradijs, J. & van der Klis, M. (1989): Nature 337, 234

58

147. Taam, R.E. & van den Heuvel, E.P.J. (1986): Astrophys.J.
     305, 235
148. Taylor, J.H. & Stinebring, D.R. (1986): Ann.Rev.Astron.
     Astrophys. 24, 285
149. Taylor, J.H. & Weisberg, J.M. (1989): preprint on the
     binary PSR 1913+16
150. Trümper, J., Kahabka, P., Ögelman, H., Pietsch, W. &
     Voges, W. (1986): Astrophys.J. 300, L63
151. van den Heuvel, E.P.J. (1983), in: 'Accretion-driven
     stellar X-ray sources', eds. W.H.G. Lewin & E.P.J.
     van den Heuvel, Cambridge Univ.Press, p. 303
152. van den Heuvel, E.P.J. & Taam, R.E. (1984): Nature 309,
     235
153. van den Heuvel, E.P.J., van Paradijs, J.A. & Taam, R.E.
     (1986): Nature 322, 153
154. Weiler, K.W. & Sramek, R.A. (1988): Ann.Rev.Astron.
     Astrophys. 26, 295
155. Wheeler, J.C., Harkness, R.P., Barker, E.S., Cochran,
     A.L. & Wills, D. (1987): Astrophys.J. 313, L69
156. White, N.E. & Marshall, F.E. (1984): Astrophys.J. 281,
     354
157. Wilson, A.S., Samarasinha, N.H. & Hogg, D.E. (1985):
     Astrophys.J. 294, L 121
158. Wilson, D. & Rees, M. (1978): Mon.Not.R.astr.Soc. 185,
     297
159. Winkler, P.F., Kirshner, R.P., Hughes, J.P. & Heathcote,
     S.R. (1989): Nature 337, 48
160. Wolszczan, A. & Cordes, J.M. (1987): Astrophys.J. 320,
     L35
161. Woosley, S.E. & Weaver, T.A. (1986): Ann.Rev.Astron.
     Astrophys. 24, 205

# RADIO PULSARS: NEW and OLD

Shrinivas R. Kulkarni
Division of Physics, Mathematics and Astronomy
California Institute of Technology
Pasadena, CA 91125, USA

## ABSTRACT

The field of radio pulsars is now a very exciting field. New pulsar surveys have uncovered millisecond pulsars in the Galactic disk and the globular cluster system. Sensitive high frequency surveys have penetrated the inner Galaxy and discovered a large number of predominantly young pulsars. Combination of radio and optical observations have resulted in a reasonable understanding of the evolution of magnetic field in pulsars from birth to very late stages. The discovery of three new pulsar-SNR associations have vastly improved our understanding of the formation of neutron stars by Type II supernovae. The discovery of new millisecond and binary pulsars, the so-called 'recycled' pulsars, has offered new insights in the evolution of their progenitors, the low mass X-ray binaries. However, the kinematics of the millisecond pulsars is still a major puzzle. New observations especially of kinematics of X-ray binaries and binary pulsars are needed to make further progress. It was only two years ago that the first pulsar was discovered in the globular cluster system. The current score stands at eight and it appears that pulsars will soon eclipse the X-ray binaries as a probe of the formation and evolution of neutron stars in this most ancient part of the Galaxy. Thus, on the whole, observations of radio pulsars are shedding considerable light on the formation and evolution of neutron stars.

## 1. Introduction

The main aim of this article is to review the important strides that have been made in the field of radio pulsars. The past six years has been a period of phenomenal progress. New discoveries have dramatically showed the close evolutionary link between Galactic X-ray binaries, supernova remnants and radio pulsars. Given this close evolutionary connection, I will also discuss the related areas of supernova remnants and X-ray binaries as well.

W. Kundt (ed.), Neutron Stars and Their Birth Events, 59–90.
© 1990 Kluwer Academic Publishers.

The beginning of the renaissance in the field of radio pulsars started towards the end of 1982 with the discovery of the first millisecond pulsar, 1937+214 (Backer et al. 1982). This discovery confirmed our theoretical ideas of spin-up by accretion and opened up a parameter space (in period) larger than had been searched before. Before the discovery of PSR 1937+214, pulsar surveys had sharply reduced sensitivities for pulsars with periods less than a few tenths of a second and *no* sensitivity to pulsars spinning faster than the Crab pulsar. Searches for millisecond pulsars are now in full swing. So far, less than 10% of the sky has been searched for millisecond pulsars. Clearly, we can look forward to more exciting discoveries.

Preceeding the discovery of PSR 1937+214 important developments were taking place in the field of X-ray astronomy. The launching of the X-ray satellite *Uhuru* in 1970 resulted in the discovery of X-ray pulsars, bright Low Mass X-ray Binaries (LMXBs) and X-ray sources in globular clusters; the latter two categories appeared similar in their X-ray properties including a lack of pulsed emission. The X-ray pulsars were immediately identified with strongly magnetized neutron stars that were accreting matter from a companion and the LMXBs and the cluster sources were interpreted to be weakly magnetized accreting neutron stars. Observations of the changes in rotation rates as a result of variable accretion rate pointed to the importance of accretion in spinning up (or down) neutron stars in X-ray binaries. Theoretical models were developed and satisfactorily accounted for the phenomena of period changes. A direct consequence of these models was that the LMXBs, being weakly magnetized neutron stars, were supposed to be spun up to high rotation rates. Consequently, there were efforts to look for rapid X-ray pulsations in the LMXBs, with no success.

The discovery of PSR 1937+214 took the entire astrophysical community by surprise. Given the historical background it is difficult to understand this. I think that this was due to the isolation of the radio pulsar and the X-ray binary community. Perhaps the biggest effect of the discovery of PSR 1937+214 was that it lead to breakdown of the artificial barrier between these two communities.

The next milestone was the discovery of a pulsar in the globular cluster M28 (Lyne et al. 1987). This discovery, anticipated on the basis of stellar evolution scenarios, reaffirmed the strong evolutionary link between low mass X-ray binaries and millisecond pulsars. The searches for pulsars in clusters have just begun and already eight pulsars have been reported so far. Indeed, in the cluster M15 alone three pulsars have been found. In contrast, the entire cluster system contains only 10 LMXBs. Thus, it is quite clear that very soon radio pulsars will become our prime probe of neutron stars in globular clusters.

The most important recent discovery has been that of the eclipsing binary millisecond pulsar, 1957+20 (Fruchter, Stinebring and Taylor 1988). In this system, there is clear evidence that some of the pulsar spindown energy is resulting in the ablation of the companion. Over due course, the companion may be completely ablated leaving a solitary millisecond pulsar like PSR 1937+214.

It is believed by most workers but not all that the binary pulsars and the millisecond pulsars are the end products of the X-ray binaries. These pulsars are referred to as 'recycled' pulsars. The nomenclature comes from a model of initial spindown by electromagnetic torque and subsequent spinup by angular momentum accretion. The recycled pulsars are now believed to be the oldest neutron stars in the Galaxy. There are now a total of 18 recycled pulsars, the salient parameters of which can be found in Table 1.

Table 1. Binary, Millisecond and Globular Cluster Pulsars

| NAME | ROTATION PERIOD (ms) | PERIOD DERIVATIVE ($10^{-18}$ s$^{-1}$) | DISPERSION MEASURE (pc cm$^{-3}$) | ORBITAL PERIOD (d) | MASS FUNCTION ($M_\odot$) | ECCEN-TRICITY | ASSOCIATION (Classification) |
|---|---|---|---|---|---|---|---|
| 0021-72A | 4.479 | : | 65 | 0.022 | 1.6E-8 | 0.33 | 47 TUC (NGC 104) |
| 0021-72B | 6.127 | : | 65 | 51 | : | : | 47 TUC |
| 0655+64 | 195.671 | 0.68 | 8.7 | 1.03 | 0.0712 | <.00005 | field (HMBP) |
| 0820+02 | 864.873 | 103.9 | 23.7 | 1232.47 | 0.0030 | 0.01187 | field (LMBP) |
| 1620-26 | 11.076 | 0.82 | 62.9 | 191.44 | 0.0080 | 0.02532 | M4 (NGC 6121) |
| 1639+36 | 10.378 | : | 31. | -- | -- | -- | M13 (NGC 6205) |
| 1820-11 | 279.828 | 1378. | 428.4 | 357.76 | 0.068 | 0.79462 | field (LMBP) |
| 1821-24 | 3.054 | 1.62 | 120 | -- | -- | -- | M28 (NGC 6626) |
| 1831-00 | 520.947 | 14.3 | 88.3 | 1.81 | 1.2E-4 | 0.0001 | field (LMBP) |
| 1855+09 | 5.362 | 0.017 | 13.3 | 12.33 | 0.0056 | 0.00002 | field (LMBP) |
| 1913+16 | 59.030 | 8.64 | 171.6 | 0.32 | 0.1323 | 0.61713 | field (HMPB) |
| 1937+21 | 1.558 | 0.11 | 71.0 | -- | -- | -- | field (LMBP) |
| 1953+29 | 6.133 | 0.030 | 104.6 | 117.35 | 0.0024 | 0.00033 | field (LMBP) |
| 1957+20 | 1.607 | 0.016 | 29.1 | 0.38 | 5.2E-6 | 0.000 | field (LMBP) |
| 2127+11A | 110.664 | -20.0 | 67.2 | -- | -- | -- | M15 (NGC 7078) |
| 2127+11B | 56.133 | : | 67. | -- | -- | -- | M15 |
| 2127+11C | 30.529 | : | 67. | 0.335 | 0.15 | 0.68 | M15 |
| 2303+46 | 1066.371 | 569.3 | 61 | 12.340 | 0.2455 | 0.65838 | field (HMBP) |

There has been some progress in our understanding of the relationship between neutron stars and supernova remnants (SNR). Three new pulsar-SNR associations have been found: CTB 80, W44 and the Gem-Mon Ring. While these discoveries are not as dramatic as the recyled pulsars they show the importance of diligent searches for pulsars in SNRs. These associations show that there is a wide range of the parameters (magnetic field strength and rotation rate) of pulsars at birth.

The discovery of a nebula around PSR 1957+20 (Kulkarni and Hester 1988) and that of the 40-ms pulsar in CTB 80 have forcefully drawn our attention to the interaction of pulsars with the ambient interstellar medium (ISM). Careful observations designed to detect the interaction of pulsar spindown energy with the ambient medium especially around energetic pulsars, young and old, are now being carried out. These observations offer us new diagnostics of the ambient medium as well as excellent probes of the nature of the relativistic wind emanating from pulsars.

The reader is directed to the proceedings of four conferences for recent review articles on related topics: (1) *Physics of Neutron Stars and Black Holes*, Ed. Y. Tanaka, (Universal Academy Press:Tokyo); (2) *Timing Neutron Stars*, Eds. H. Ögelman and E. P. J. van den Heuvel, NATO ASI Series, (Kluwer:Boston); (3) *The Physics of Compact Objects*, Eds. N. E. White and L. G. Filipov, Adv. in Space Research 8, (Pergamon:Oxford); and (4) *The Origin and Evolution of Neutron Stars*, Eds. D. J. Helfand and J.-H. Huang, IAU Symposium No. 125, (Reidel:Dordrecht). In addition, review articles on most of these issues can also be found in the *Annual Reviews of Astronomy and Astrophysics*.

I have purposely left out the discussion of the radio emission phenomenology for two reasons: it is not my area of specialty and the availability of recent excellent reviews. I refer the interested reader to a particularly clear review article by Smith (1989) on this topic as well as a seminal paper by Lyne and Manchester (1988). See also Taylor and Stinebring (1986).

This review is an update of my earlier review (Kulkarni 1988).

## 2. Pulsar Surveys

After a lull, many new pulsar surveys have been undertaken. While the earlier surveys were designed to find pulsars, the newer surveys have been specifically designed to overcome the selection effects that invariably plague pulsar surveys:

1. *Minimum Period.* By the sampling theorem, the minimum detectable period is twice the sampling interval. However, since pulsar signals have small duty cycles, the sensitivity is increased considerably if the sampling interval is a fraction (1/8 or 1/16 will suffice) of the pulsar period, with concomitant increase in computing load.

2. *Dispersion.* Dispersion of the radio signal by the interstellar medium results in smearing of the pulsed signal which is proportional to the width of the channels in the filter bank. Thus, for a given channel width, there is a maximum dispersion measure (which varies directly with the pulsar period) to which the pulsar survey is sensitive. For this reason and (1) above, the computing and memory requirements of a pulsar survey increase quadratically with the minimum pulse period of a pulsar survey.

63

3. *Observing Frequency.* Radio telescopes are single pixel instruments with the pixel size being equal to $\lambda/D$, where $D$ is the diameter of the radio telescope and $\lambda$ is the observing wavelength. Thus the lower the frequency the faster can the sky be covered. However, the Galactic synchrotron background increases with decreasing frequency. These considerations have led to compromise frequency between 400 MHz and 800 MHz.

4. *Scattering.* The rays from pulsars undergo multipath propagation in the ionized interstellar medium. Owing to this, the pulsar signal "twinkles", much like stars at night. If the path lengths along the different paths are sufficiently large, the pulsed signal will get quenched and the effective sensitivity of the pulsar survey is reduced. Since the refractive index of the ionized interstellar medium is proportional to $\lambda^2$, this problem can be overcome by operating at higher radio frequencies.

5. *Pulsars in Binary Systems.* Pulsar surveys are also insensitive to pulsars in tight eccentric orbit binary systems because the pulsar velocity keeps changing during the interval over which data is collected. None of the surveys, past and present, allow for variable pulsar velocity (i.e. acceleration) and thus there is a small but very interesting region of parameters (small rotation period and short orbital period; Figure 1) which have been excluded from all the large pulsar surveys carried out so far.

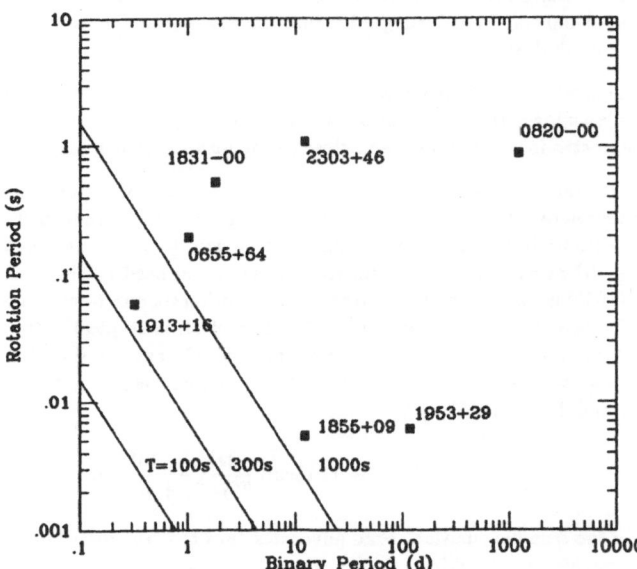

Figure 1. Binary pulsars in a rotation period - orbital period diagram. Slanting lines in lower left corner represent upper limits to parameters of pulsars detectable with the integration times marked; the assumption is that the fourth harmonic is smeared by the velocity change over the integration time. The binary system is assumed to consist of two neutron stars in circular orbit. The fourth harmonic is assumed (from Backer 1987).

The Green Bank survey by Dewey et al. (1985) was designed explicitly to determine the pulsar luminosity function. With a coverage of 1.8 sr and a minimum detectable flux density of about 2 mJy at 400 MHz, this survey searched a larger volume of space for low luminosity pulsars than any previous survey. This search discovered 34 new pulsars and 40 previously known pulsars. In addition, the survey discovered the 1.8 d binary pulsar 1831-00. In accordance with previous surveys, this survey did not find any pulsars with intrinsic luminosity $L_{400}$ less than 0.3 mJy kpc$^{-2}$; here $L_{400} \equiv S_{400}d^2$ where $S_{400}$ is the mean or the time-averaged flux density of the pulsar at 400 MHz and $d$ is the distance (estimated from dispersion measure and a model for the galactic electron density) in kpc. Statistical modelling of the survey detection process by means of Monte Carlo simulations suggests that the luminosity function of pulsars drops below $L_{400} \lesssim 3.5$ mJy kpc$^{-2}$ (Dewey 1984). The existence of this turnover makes it possible to reliably estimate the local surface density of pulsars to be $30^{+55}_{-10}$ kpc$^{-2}$. Extrapolating to the rest of the Galaxy yields a total of $1.2 \times 10^4$ active pulsars.

The Jodrell Bank survey by Clifton and Lyne (1986; hereafter CL) was specifically designed to overcome the deleterious effects of scattering in the interstellar medium. The survey, done at an observing frequency of 1400 MHz and a sampling rate of 2 ms, covered the inner Galaxy: $-4° \leq l \leq 105°$, $|b| < 1°$. The survey resulted in the detection of 62 pulsars of which 40 were new. The longitude distribution of the new pulsars (almost all were found in the inner Galaxy, $l < 50°$) and the dispersion measure distribution (dispersion measure extending over to 1000 cm$^{-3}$ pc with a broad peak around 400 cm$^{-3}$ pc) is suggestive that the new pulsars are mainly located in the star forming "5-kpc molecular" ring.

Timing observations of the CL sample (McKenna 1989) reveal that the CL pulsars are an order of magnitude younger than the typical sample. This is quite expected since the survey was biased to pulsars in star forming regions and close to the disk i.e. young pulsars.

Statistical analysis of the CL survey requires a knowledge of the distances to the new pulsars. The distances can be estimated from the observed dispersion measure if an electron distribution can be assumed. Alternatively, HI absorption data can be used to obtain an independent kinematic distance estimate. HI measurements are sensitivity limited and need bright pulsars. Clifton et al. (1988) and Frail (1989) have    managed to obtain absorption spectra towards six and two pulsars, respectively. Unfortunately, the Arecibo sample of Clifton et al. is atypical of the CL pulsars in the sense that these six pulsars are not located deep in the inner Galaxy. Frail (1989) has analyzed the new and old HI absorption data and concludes that the standard electron density distribution (cf. Lyne Manchester and Taylor 1985)

$$n_e(R, z) = [0.025 + 0.015 exp \frac{-|z|}{70pc}][\frac{2.0}{1 + R/R_0}] \qquad (1)$$

is a lower limit to the true distribution. Frail advocates that the true distribution is a factor of 1.5 larger than that specified by Equation (1); here $R_0$ has been assumed to be 10 kpc.

The next best handle we have on Equation (1) comes form a statistical study of the CL sample of pulsars done by Clifton, Lyne and Jones (1989) who assume a variety of forms for $n_e(R, z)$ and then solve for $\rho_P(R)$, the density distribution function of pulsars consistent with the observed distribution in the sky. For a variety of realistic electron distribution functions, Clifton et al. (1989) find that (a) the electron density of the low scale height component (the second term in Equation 1) is about a factor of three higher in the inner Galaxy, and (b) $\rho_P(R)$ peaks in the range $R = 4-6$ kpc and falls for $R \lesssim 3$ kpc i.e. *pulsars have the same radial distribution as the other Population I tracers such as molecular gas.* While this is expected, it is reassuring to find observation agree with our prejudice!

The pulsars in the disk can provide limited information about the vertical electron density distribution because of their limited $z$-heights. Thus the diffuse component, i.e. the first term in Equation 1, is really not well constrained. However, the discovery of pulsars in globular clusters has enabled us to constrain the vertical extent of the diffuse component: a one-sided vertical scale height of $\sim 1500$ pc (Reynolds 1989) and a vertical surface density of 2 $M_\odot$ pc$^{-2}$ which is 25% of the HI layer (Kulkarni and Heiles 1988).

Following the discovery of PSR 1937+214, several surveys for fast pulsars have been carried out. The Princeton group carried out one search at Green Bank (Stokes et al. 1986) and two at Arecibo (Stokes et al. *ibid*; Fruchter 1988). The Green Bank survey covered nearly 3700 square degrees of sky (between Galactic longitudes 15° and 210°) and maintained resonable sensitivity to periods as short as 30 ms. Twenty new pulsars were discovered but not a single pulsar with a period below 100 ms was detected. The first Arecibo survey covered 290 square degrees and had sensitivity to pulsars with periods as short as 5 ms. Five new pulsars were detected of which one pulsar (1855+09; Segelstein et al. 1986) had a period of 5 ms. The second Arecibo survey covered a similar area as the first one but with increased sampling rate (2 kHz) and finer frequency resolution. The outstanding discovery of this survey was the 1.6 ms eclipsing binary pulsar, 1957+20 (Fruchter, Stinebring and Taylor 1988). The Berkeley/Caltech group used a custom pipeline processor in a large survey from Green Bank (using the now defunct 300-ft telescope) at frequencies of 400 MHz and 1400 MHz. No bright millisecond pulsars akin to PSR 1937+214 were discovered.

The straightforward conclusions from these surveys are (1) PSR 1937+214 is a particularly luminous millisecond pulsar, (2) pulsars with periods between 10 ms and 100 ms are intrinsically rare objects and (3) the millisecond pulsars form a distinct group of pulsars, constituting about 10% of the sample in any flux limited sample. The implication of conclusion (3) is further discussed in section 4. Another important conclusion is that it now appears that the recycled pulsars, especially the millisecond pulsars, have a very steep spectrum, $\alpha \sim 2.5$ (Fruchter 1988) rather than $\alpha \sim 1.5$, typical of the standard pulsars; here, flux density, $S_\nu \propto \nu^{-\alpha}$. Thus it is advantageous to search for the recycled pulsars at low frequencies – a conclusion being borne out by the high discovery rate of pulsars in globular clusters by searches at low frequencies as compared to the searches at higher frequencies. This is somewhat unfortunate in that the distant millisecond pulsars may be hidden from us due to dispersion and scattering by the intervening medium.

Currently several large surveys are in progress: at Arecibo by the Berkeley/Caltech group (anti-center region), by the Caltech/Berkeley group (intermediate latitudes), Jodrell Bank (U. K.) and Parkes (Australia). All these surveys record data on magnetic tapes and process them on supercomputers. The two Arecibo surveys also use a 'fast pulsar search machine'(a custom pipeline processor) to detect millisecond pulsars in quasi-real time.

## 3. Magnetic Field of Single Pulsars

It is commonly accepted that radio pulsars are rotating, highly magnetized neutron stars. Both rapid rotation and large magnetic field strengths are needed to generate large potential differences which in turn result in the generation of high energy particles and pulsed electromagnetic radiation. The loss in the rotational energy *i. e.* the power radiated away, $I\omega\dot{\omega} \propto B^2\omega^4$. Thus the secular evolution of the magnetic field strength, $B$, crucially decides the fate of pulsars. Here we review observations which shed some light on this issue.

It is currently thought pulsars are born with large surface magnetic fields. According to the simple dipole model, the surface field strength $B_{12} = B/10^{12}$ G can be estimated from

$$B_{12}^2 sin^2\alpha = P\dot{P}_{-15} \qquad (2)$$

where $\alpha$ is the angle between the rotation axis and the magnetic dipole axis and the period derivative, $\dot{P}$, is expressed in convenient units of $10^{-15}$ s s$^{-1}$. Thus in the simple dipole model, the product $P\dot{P}$ is a measure of the equatorial component i.e. $Bsin\alpha$. In the more sophisticated particle-acceleration models the product $B_{12}^2 = P\dot{P}$ i.e. independent of $\alpha$. Regardless, it is traditional to ignore $sin\alpha$ in estimating field strengths from $P$ and $\dot{P}$.

Various statistical analyses of pulsars and pulsar surveys agree that the spread in the initial field strength is small; however, the estimate of the median field strength at birth ranges from $B_{12} = 3$ (Stollman 1987; Narayan and Schaudt 1988) to a value four times smaller (Lyne et al. 1985); here $B_{12}$ is the surface magnetic field of the pulsar in units of $10^{12}$ G. The X-ray properties of the plerionic remnants favor the Lyne et al. distribution of initial field strengths (see section 4 for further discussion).

The primary source of energy of a radio pulsar is its store of rotational energy. Thus pulsars slow down as they age. The observed rate of period lengthening, $\dot{P}$, allows us to obtain a simple estimate of the dynamical age of a pulsar, the so-called characteristic age, $\tau_c = P/2\dot{P}$. Measured characteristic ages range from $10^3$ y to $5 \times 10^9$ y; for many pulsars, the characteristic age greatly exceeds $10^7$ y. In contrast, another estimator of the age, the so-called kinematic age, $\tau_z \equiv |z|/|v_z|$ is less than between $10^5 - 10^7$ y for a sample of two dozen pulsars for which proper motion data exists; here $|v_z|$ is the velocity of the pulsar perpendicular to the Galactic plane. The kinematic age is derived on the assumption that (i) pulsars are Population I objects and born close to the Galactic plane, an assumption which is certainly true of all pulsars except the recycled pulsars (see sections 6,7); and (ii) pulsars are not a dynamically relaxed system like the disk stars. Proper motion studies carried out by Lyne and colleagues (see Lyne 1987) show that (ii) is a good assumption since no pulsar is seen coming back towards the Galactic plane and furthermore, they have a mean $z$ speed, $\overline{|v_z|} \sim 100$ km s$^{-1}$.

Lyne and coworkers found that for $\tau_c \sim 5 \times 10^6$ y, $\tau_z \sim \tau_c$, whereas for large $\tau_c$, $\tau_c \gg \tau_z$. This behavior can be accounted for by a model in which the pulsar magnetic field strength decays exponentially with time. In such a model, the current value of $\dot{P}$ reflects the current rather than the intial field strength and hence $\tau_c$ is no longer a measure of the true age of the object. Instead, the " magnetic age", $\tau_B$, age can be estimated by

$$\tau_B = 2.3\tau_d(12.5 - logB) \, y \qquad (3)$$

where $\tau_d$ is the timescale of exponential field decay. Various statistical analyses have yielded $\tau_d$ ranging from $2 \times 10^6$ y (Radhakrishnan 1982) to $9 \times 10^6$ y (Lyne et al. 1985). We adopt $\tau_d \sim 5 \times 10^6$ y.

The notion of field decay is not universally accepted. Theoretically the field is expected to arise in the crust and decay owing to the finite conductivity of the crust. Sang and Chanmugam (1987) point out that the field not only diffuses outward from the crust but also inwards, where the conductivity is much larger than the crust. In detail, Sang and Chanmugam find that the magnetic field strength cannot decay by more than a factor of 100 over Hubble time scale. However, Jones (1988) suggests that Hall drift could account for $\tau_d$ as small as $10^7$ y. Clearly, the theoretical situation is still murky.

Another objection against Equation (3) is that what is observed is *torque decay* and not the decay of field strength. The relation between the torque and $B$ depends upon the particular pulsar emission model. In particle-acceleration models such as the Julian-Goldreich model the orientation of the magnetic field axis with the rotation axis is irrelevant and the observed torque decay can be unambiguously interpreted as field decay. However, as stated above, in other models the angle $\alpha$ does matter and the observed torque decay can be interpreted to be a gradual alignment between the magnetic and rotation axes.

In a recent and important paper Lyne and Manchester (1988; hereafter LM) have attacked this issue of alignment by doing an in depth study of the properties (spectral index, and polarization) of pulse profiles from which they can infer $\alpha$. LM show that alignment indeed does take place on a time scale similar to that of the observed torque decay time scale. LM also point out that some young pulsars have seemingly large magnetic field strengths and nearly aligned dipole field. Thus such pulsars have either truly large field strengths or the braking torque is insensitive to $\alpha$. Thus LM conclude that pulsar fields align and decay at the same rate. Clearly, we need more data on young pulsars to confirm LM's hypothesis of simultaneous alignment and field decay.

The existence of the recycled pulsars is an independent and strong argument for field decay since it is clear that recycled pulsars have truly low field strengths (sections 6, 7). Hence either recycled pulsars are either born with low field strengths or some kind of field decay must be taking place. The birthrate of recycled pulsars is already embarassingly high and appealing to a different birth channel makes matters only worse (section 7). Thus we are forced to conclude that some kind of field decay must be taking place. We will continue this discussion in section 7.

## 4. Pulsar Wind Nebulae

As pulsars spin down, energy is liberated at a rate of $\dot{E}_R = I\omega\dot{\omega}$ where $\omega = 2\pi/P$ is the angular rotational frequency and $I \sim 10^{45}$ gm cm$^2$ is the moment of inertia. Only a small fraction of this energy, $10^{-4}$ or so, appears in the radio window. For a long time astronomers have wondered about the precise channel through which this rotational energy is lost: low frequency Poynting flux or $e^+$-$e^-$ pairs or high energy photons. In one of the earliest papers published on this issue, Blandford et al. (1973) predicted that old pulsars should be surrounded by a radio emitting synchrotron nebula ("radio halos") powered by the 'wind' from the pulsar. Numerous radio searches were conducted with no success.

With the launch of the *Einstein* X-ray satellite, Helfand and co-workers surveyed 40 radio pulsars in an effort to detect the putative synchrotron nebulae (see Helfand 1983; also Seward and Wang 1988). Soft X-ray emission was detected around five old pulsars (0355+54, 1929+10, 1055-52, 0950+08, 1642-03). Because of the faintness of the sources, the shapes and the spectra were not very well determined. Consequently, alternative interpretations such as thermal emission from the neutron star surface, especially for the unresolved sources, were advanced (Alpar et al. 1987; Brinkmann and Ogelman 1987). The synchrotron nebula interpretation is probably secure for the extended emission associated with PSR 1055-52, which also happens to be the brightest source in the above sample.

68

The ease with which the optical line emission from the shocks driven by the nebula around PSR 1951+32 in CTB 80 could be detected convinced us that narrow band imaging at optical wavelengths is an ideal method to detect such nebula, second only to studying the X-ray emission. Consequently, we started a survey of optical observations of all pulsars with large $\dot{E}_R$.

Our first detection was a nebula around the eclipsing, binary millisecond pulsar 1957+20 (Kulkarni and Hester 1988; Figure 2). PSR 1957+20, like other recycled pulsars (sections 6, 7) is thought to be an old pulsar ($10^9$ y or older). Thus this is the first detection of a nebula around an old pulsar. The nebula is visible in the H$\alpha$ line which is thought to arise in the shock where the relativistic pulsar wind reaches momentum balance with the ISM. The distinctive cometary shape of the nebula indicates that the ram pressure dominates over the static ISM pressure. Indeed, interstellar scintillation observations of the pulsar indicate a transverse velocity of about 100 km s$^{-1}$. It is clear from Figure 2 that the transverse motion must be along the South West direction. Analysis of the optical data allow us to show that the mean nebular pressure is $1.5 \times 10^{-10}$ erg cm$^{-3}$ – nearly three orders of magnitude higher than the static ISM pressure. The high pressure is attributed to a relativistic thermal plasma filling the nebula.

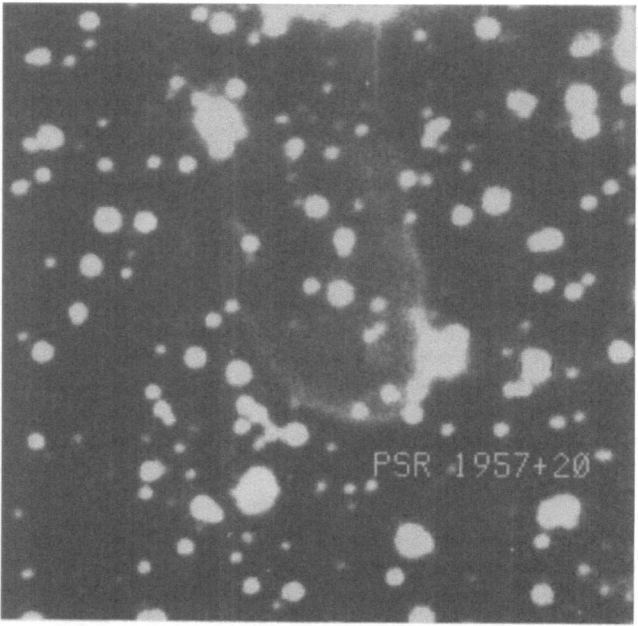

Figure 2. H$\alpha$ nebula surrounding the eclipsing, millisecond binary pulsar 1957+20. The image was obtained using narrow band filters centered on and off the H$\alpha$ nebula at Palomar Observatory.

Optical spectroscopy of the nebula reveals only the H$\alpha$ line and no detectable forbidden lines like [S II]. The Balmer decrement (the ratio H$\beta$ to H$\alpha$ emission) exceeds 11! Such "Balmer dominated" spectra are seen in a number of supernova remnants such as Tycho and Cygnus Loop. The Balmer emission is thought to arise from the ambient pre-shock H atoms drifting into the post-shock region where they are excited by collisions with the post-shock hot plasma and by charge exchange with the hot ions (Chevalier and Raymond 1978). This theoretical model along with the planned high resolution optical spectrum would enable us to directly measure the shock speed.

So far we have surveyed well over 30 pulsars, both in the Northern and Southern sky with no new detections. Our low rate of detection at optical wavelengths means that either (a) the typical ambient ISM density is low i.e. below 1 cm$^{-3}$ so that the ram pressure is weak and cannot confine the nebula, or (b) the ambient ISM is predominantly ionized in which case H$\alpha$ emission is suppressed since there are no ambient H atoms to excite.

The discovery of the nebula around PSR 1957+20, here after PWN 1957+20 (Pulsar Wind Nebula) shows the great value of optical observations in determining the density and the ionization fraction of the ambient ISM, the space motion (including direction) of the pulsar etc.

Observations of PWN also offer new diagnostics of the nature and composition of the relativistic winds from pulsars. Standard models of pulsar magnetospheres suggest that the spindown luminosity is carried primarily by electromagnetic wave and Poynting flux. However, observations of the Crab nebula seem to indicate a flow of relativistic particles ($e^+$-$e^-$ pairs; Kennel and Coroniti 1984). Coroniti (1989) reconciles these two views in a new model in which the electromagnetic wind gives its energy to embedded particles as it travels outwards, increasing the mean particle Lorentz factor $\gamma$. However, considerable theoretical uncertainty exists in pulsar wind models and multi-wavelength observations of PWN are our *only* tools in determining the properties of high energy emission from pulsars. For example, in PSR 1957+20 system, detection of X-ray emission from the shocked wind near the companion would require Lorentz factors, $\gamma \sim 10^5$; detection of nebular X-rays, $\gamma \sim 10^{7.5}$; and nebular radio emission, $\gamma \sim 10^4$. Clearly the search for more PWNs is worth the effort.

Perhaps the best way to search for PWNs is in the X-ray window since, rather than in the optical window, X-ray emission does not critically depend upon the specifics (viz. ionization fraction) of the ambient gas. The launch of the X-ray satellite *Rosat* is thus eagerly awaited.

## 5. Pulsars and Supernova Remnants

Of the 150 Galactic supernova remnants, about half or 75 are expected to result from Population I progenitors, the majority of which are expected to leave a neutron star stellar remnant. Despite this expectation, for a long time we knew only two *firm* neutron-star SNR associations: Crab and Vela. There was no real progress in this area till the launch of the *Einstein* X-ray satellite. The *Einstein* observations revealed point sources in about 20 remnants (see Seward 1989 for a recent review). However, in only two cases the central point sources were proved to be neutron stars beyond doubt: PSR 1509-58 in the remnant MSH 15-*52* and the 7-s X-ray pulsar 1E2259+586 in the remnant CTB 109. (SS 433 in the remnant W50 is almost certainly a binary neutron star and the central X-ray source in RCW103 is probably a bare neutron star; however, these contentions have yet to be firmly established). In the last two years, three new associations have been found, primarily as a result of diligent radio observations: PSR 1951+32 in CTB80, PSR 1853+01 in W44 and PSR 0656+14 in G201.2+8.2. I think the high rate of these recent successes bodes well for future studies of young pulsars. In this section I discuss the implications of the detection of the new associations.

Supernova remnants can be divided into three categories (see Becker 1987 for a review): "shells" (surface brightness in the form of a shell); "Crab-like" or "Plerions" (surface brightness peaking in the center of the remnant); and "Crab-shells" or "composites" or "combination" (a Crab-like remnant embedded in a shell). The composites and plerions are supposed to house an active source which is either a radio pulsar or an accreting source. The presence of an active source is indicated by observations of a polarized, flat spectrum radio nebula and/or a compact X-ray source. In contrast, the shell remnants are not supposed to contain any stellar remnant. For the purpose of the discussion here, we will lump the composites and the plerions into one category: "active" remnants. Such a lumping also makes sense since, as Becker points out, there is no obvious distinction other than morphology between the composites and the plerions. The number of sources which can be classified as active remnants vary between 15 (Becker 1987) to 23 (Narayan and Schaudt 1988), or about 10% of the remnants. However, remnants with active sources are difficult to find. Thus the true number of active sources could well be much higher, a point repeatedly emphasized by Helfand and Becker. They hope to uncover many more SNRs, especially plerions by systematically surveying the Galactic plane using the VLA. The X-ray satellite *Rosat* and the planned *AXAF* satellite are also expected to discover more active remnants.

I now summarize the observational data of three new pulsar-SNR associations:

**PSR 1951+32 and CTB80.** Till recently, the origin of the radio nebula CTB 80 has been a great puzzle. At centimeter wavelengths, CTB 80 consists of three steep-spectrum ridges of length $\sim 30$ arcmin, at the intersection of which lies a 10 arcmin $\times$ 6 arcmin plateau with a flatter spectral index (Angerhofer et al. 1980). The latter feature was used to classify the remnant as an active remnant. The large size of the ridges and the inferred low shock velocities there suggest an age comparable to $10^5$ y. A number of authors (Strom 1987 and refs. therein) citing the compact core and the associated X-ray emission have suggested an age of less than $10^3$ y.

Interest in this nebula heightened following the discovery of a steep spectrum, highly polarized point source at the brightest region of the plateau by Strom (1987). Since such characteristics are peculiar only to pulsars, searches were immediately conducted. These searches finally led to the discovery of a 40 ms pulsar (1951+32) by Kulkarni et al. (1988). Observations at Arecibo soon revealed $\dot{P} = 5.9 \times 10^{-15}$ s s$^{-1}$ and a transverse motion of about 300 km s$^{-1}$ (Fruchter *et al.* 1988). The derived magnetic field strength is $B_{12} = 0.5$, nearly an order of magnitude smaller than that of the Crab pulsar. Not surprisingly, the nebula is not as luminous as the Crab nebula.

Kulkarni et al. (1988) proposed that the pulsar was born in a supernova explosion at least $10^4$ y or some time ago comparable to the characteristic age of the pulsar, $P/2\dot{P} \sim 10^5$ y. The pulsar acquired a large transverse velocity either due to a kick resulting from the formation process or the disruption of the progenitor binary system and began a westward motion (Figure 3). Hester and Kulkarni (1988, 1989) have further elaborated on this model and show that their detection of the narrow filamentary emission in the region containing the ridges supports this model. In particular, the filamentary emission is due to momentum driven shocks generated following the explosive birth of the pulsar.

Fesen, Shull and Saken (1988) reported the discovery of an infrared shell of a size comparable to the ridges which they interpret as yet more support for this model. Furthermore they argue that the pulsar has now finally caught up with the shell that was swept up following the explosion. The plateau is thus a pulsar wind nebula similar to that around PSR 1957+20 (section 4). The only difference is that the ambient ISM around PSR 1951+32 is probably a factor of 100 denser than that around PSR 1957+20. The larger ambient density and the rapid motion of the pulsar result in a larger confining ram pressure and hence a nebula considerably brighter than PWN 1957+20.

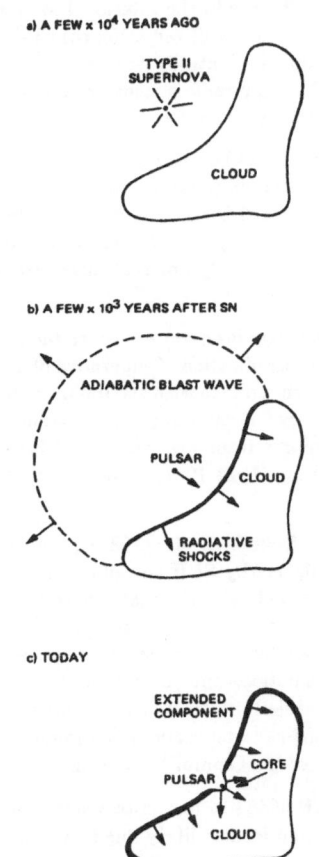

Figure 3. A scenario for the formation of CTB 80 and the relationship between the pulsar, the core, the plateau and the extended ridges.

Clearly several coincidences are necessary for the latter to happen *viz.* high pulsar space velocity and a narrow window of time of shell-pulsar interaction due to the small thickness of the shell. Thus, Fesen *et al.* predict only a few such remnants at any given time and propose that G 5.4–1.2 is another CTB 80 type remnant.

**PSR 1853+01 and W44.** W44 is a shell type remnant in the radio but a filled remnant at X-ray wavelength (Smith et al. 1985). *Einstein* observations found no resolved sources in the most central region. The X-ray spectrum of W44 shows emission lines indicating that much of the X-ray emission is thermal (see Seward 1989). Smith et al. (1985) suggested that the remnant is in its late adiabatic stage expanding into a cloudy ISM, with an age of $10^4$ y. The shell has now cooled to a point where X-rays are no longer radiated. The gas in the interior, however, being less dense than that in the shell, takes longer to cool and hence is still radiating copiously at X-ray wavelengths. This eminently reasonable hypothesis explains the centrally peaked X-ray emission without appealing to a synchrotron origin. Indeed, the lack of any radio synchrotron nebula and the detection of X-ray emission lines definitely rule out a synchrotron source.

Given this backdrop, W44 would not be a prime candidate for pulsar searches. Fortunately, Wolszczan and colleagues included W44 in their pulse search survey of SNRs visible from the Arecibo radio telescope. Soon, the discovery of a 0.267 s pulsar within the radio confines of W44 was announced (Wolszczan et al. 1988). With a period derivative of $208 \times 10^{-15}$ s s$^{-1}$ the inferred magnetic field strength and the characteristic age of the pulsar are $7.4 \times 10^{12}$ G and $2 \times 10^4$ y, respectively.

The characteristic age of the pulsar is an *upper* limit to the true age of the pulsar. Thus the remnant cannot be older than $2 \times 10^4$ y. Application of supernova blast wave model to the size of the nebula yield a nebular age of $10^4$ y, in concurrence with the limits set by the characteristic age of the pulsar. The pulsar appears to be offset by 5 arcmin from the centroid of the approximately spherical shell. This can be explained by assuming a transverse velocity of 215 km s$^{-1}$, corresponding to the median transverse velocity of field pulsars (Lyne 1987). Thus a self-consistent picture emerges from these observations.

**PSR 0656+14 and G201.2+8.2.** Nousek et al. (1981) proposed that a region of enhanced X-ray emission in Gemini was an old, nearby SNR. In addition, they pointed out that a 0.38 s period pulsar, PSR 0656+14, was located close to the center of this remnant, hereafter G201.2+8.2. However, the then known value for the period derivative of PSR 0656+14 led to a characteristic age of $4 \times 10^6$ y – much too high to be associated with the remnant whose age is estimated to be $6 \times 10^4$ y. Five years later, the error in $\dot{P}$ was noticed and the actual value of $\dot{P}$ was found to be 30 times the previous value (Dominique et al. 1986). Independently, Cordova et al. (1987) found a bright X-ray source coincident with PSR 0656+14 and realized the importance of the coincidence of the revised characteristic age and the age of the Gemini X-ray ring.

With a $\dot{P}$ of $54 \times 10^{-15}$, PSR 0656+14 has quite a strong magnetic field strength: $B_{12} \sim 4.5$. The X-ray emission is unresolved and is most likely due to a compact synchrotron nebula much like PSR 1055-52 (section 4).

The pulsar's characteristic age is $1.1 \times 10^5$ y and its distance, as deduced from dispersion measure data is 400 pc – both of which are in agreement with those of the X-ray ring. These coincidences argue for the association of the pulsar with the SNR. The scintillation velocity of the pulsar is quite low ($\sim 40$ km s$^{-1}$). Thus the maximum distance that the pulsar is expected to move since birth is 4 pc, much smaller than the 50 pc radius of the ring. This naturally explains the location of the pulsar close to the center of the remnant.

The key issue in the study of young pulsars is the following: what is the distribution of the rotational periods and the magnetic field strengths at birth? These two parameters are important because the rotational luminosity of pulsars is $\propto \omega^4 B^2$.

The simplest hypothesis is that all pulsars are born like the Crab pulsar i.e. $B_{12} \sim 3$ and $P_i \sim 10$-$20$ ms. Limited support for this simple hypothesis is offered by a multi-wavelength study of nearly a dozen combination sources by Helfand and Becker (1987). The total energy contained in the plerionic nebula of these combination remnants can be inferred from the multi-wavelength observations of the synchrotron nebula. This energy must have been supplied by the central pulsar and hence must be equal to $1/2I\omega_i^2 - 1/2I\omega^2$ (the radiative losses are small over the lifetime of these young remnants); here $\omega$ is the rotational angular frequency of the pulsar at the current time and $\omega_i = 2\pi/P_i$ is that at birth. For most cases, $\omega \ll \omega_i$. Helfand and Becker conclude "it is clear from the total amount of energy stored in these sources that the pulsars which created them must have all been born with relatively short periods ($< 50$ ms)". Thus, observationally, there is proof that most combination remnants house rapid pulsars. A subsidiary conclusion is that the large variation that is seen in the combination remnants (surface brightness of the plerionic component, size of the shell, etc.) must be due to other parameters such as the magnetic field strength of the pulsar and the interstellar medium surrounding the progenitor star.

The simple hypothesis is in contradiction to the conclusions arrived from (some) statistical analysis of pulsar surveys. In particular, Narayan and coworkers (see Narayan 1987) find that the birthrate of rapidly rotating pulsars is considerably smaller than that of the slower pulsars ($P > 0.5$ s). They propose that most pulsars are born with $P_i \gtrsim 0.4$ s, instead. This hypothesis is often referred to as the "injection" hypothesis. Here we will expand the hypothesis to include pulsars being born rotating slowly or/and possessing weak magnetic fields. Stokes *et al.* (1986) cite the non-detection of any new pulsar in the period interval from 10 ms to 100 ms in their two recent surveys (section 2) explicitly designed to have high sensitivity for fast pulsars as observational evidence in favor of the injection hypothesis. However, the injection hypothesis remains controversial since other statistical analyses (Lyne et al. 1985, Stollman 1987) do not come to the same conclusion as Narayan and coworkers.

Narayan and Schaudt (1988) present new evidence for the injection hypothesis based on the rate of detection of pulsars in the known active remnants. They assume that all the active remnants contain a young pulsar with initial period, $P_i \sim 20$ ms and an initial magnetic field strength drawn from a Gaussian distribution with $\overline{\log B} \sim 12.5$ and $\sigma = 0.3$ (obtained from statistical analysis of field pulsars; section 3). The assumed values of $P_i$, $B_i$, the distance to and the age of the remnant enable them to estimate the probability to detect a pulsar in one of the existing surveys. Their principal conclusions are:

1. The current rate of successful detections would require a beaming factor close to unity for young pulsars. This lends weight to another hypothesis advocated by Narayan and coworkers that the beaming fraction, $f$, is a function of $P$ with $f \sim 0.2$ for long period pulsars and approaching unity for short period pulsars.

2. The lack of *any* detection of a pulsar in a shell remnant clearly implies that shells do not harbor rapidly rotating, highly magnetized pulsars. Instead, shells may contain may contain either a slowly rotating or weakly magnetized pulsars. This conclusion is in agreement with the lack of any intense radio or X-ray nebula in the shell sources (Srinivasan, Bhattacharya and Dwarakanath 1984; Helfand and Becker 1987).

Evidence in favor of conclusion (2) appears to be slowly accumulating with the discovery of a slow pulsar (0.267 s) in the young remnant W44 and the discovery of a fast but relatively weakly magnetized ($B_{12} = 0.7$) pulsar in the middle-aged remnant CTB80.

The situation with the active remnants is unclear. Active remnants have been targets of numerous pulsar searches. Despite that the success rate has been rather low; in fact, so far only one detection viz. PSR 1951+32 in CTB 80. Thus either our searches are still not sufficiently sensitive or the pulsars in most active remnants are intrinsically underluminous either due to long period or weak magnetic field strength. However, Helfand and Becker's study rule out the possibility of long period pulsars and hence we are forced to conclude that most active remnants contain rapidly rotating but weakly magnetized pulsars. PSR 1951+32 in CTB80 fits this description and hence it may be the prototypical young pulsar in active remnants. Clearly, only future observations especially detection of additional such pulsars can substantiate this statement.

To summarize, shell remnants may harbor slowly rotating ($P \gtrsim 0.4$ s) but highly magnetized pulsars ($B_{12} \sim 3$) whereas active remnants are expected to contain rapidly rotating ($P \sim 10$ to 20 ms) but weakly magnetized ($B_{12} \sim 0.7$) pulsars. Since the bolometric luminosity of a pulsar is $\propto \omega^4 B^2$, both these pulsars are underluminous and hard to detect.

So far we have not discussed what determines the initial field strengths of young radio pulsars. The tight clustering of the initial field values around $B_{12} \sim 1$ suggests that simple flux conservation of the progenitor star is not a good model. Blandford, Applegate and Hernquist (1983) have elaborated on earlier suggestions of field generation by thermo-electric processes in order to explain the observed field strengths in young pulsars. In their model, the pulsar becomes visible in the radio sky only after the field has grown ($10^4$ y). Thus the observed absence of active radio pulsars in the centers of all Type II supernova remnants is naturally explained since by the time the pulsar turns "on" the nebula has expanded and essentially disappeared. However, the high magnetic field strengths of the Crab pulsar (age $\sim 10^3$ y) and the 50 ms pulsar (age $\sim 2 \times 10^3$ y) in the LMC is certainly a problem for this model. Incidentally, if PSR 1951+32 is a young pulsar whose field is still growing, then timing observations over the next two years will conclusively confirm the Blandford et al. hypothesis.

## 6. Millisecond and Binary Pulsars

We now discuss the group of single millisecond pulsars and binary pulsars (Table 1). In this section we confine the discussion to the pulsars in the disk of the Galaxy. Pulsars in the globular cluster system are discussed in section 8. At the current time, of the nearly 500 known pulsars, nine are in binary systems and one is a lone millisecond pulsar. The binary pulsars and the single millisecond pulsar are now believed to be old pulsars which have been spun-up by accretion from a companion. We will collectively refer to this group as "recycled" pulsars (RPSR). The RPSRs appear to be quite different from the bulk of the radio pulsars in several ways:

1. All the rapid pulsars (say with period below Crab pulsar) are certainly recycled pulsars.

2. The recycled pulsars have very low field strengths, typically around $10^9$ G unlike the field pulsars for which $B \gtrsim 10^{11}$ G.

3. The ages of the recycled pulsars are inferred to be $\gtrsim 10^8$ y (section 7) whereas the field pulsars are believed to be young objects ($\lesssim 10^7$ y).

4. The mean $|z|$ of the recycled pulsars is 200 pc, nearly half that of the field pulsars.

The parameters of the nine recycled pulsars are displayed in Table 1. Based on the measured mass functions, the binary pulsars can be conveniently divided into two groups: (i) the high mass binary radio pulsars (HMBPs) and (ii) the low mass binary radio pulsars (LMBPs). The lone millisecond pulsar is assumed to be an LMBP with a zero mass secondary. The classification is shown in the last column of Table 1. Note the big jump in the value of the mass function. The HMBPs are characterized by small orbital periods (0.3 to 10 d) and (sometimes) eccentric orbits and the LMBPs by invariably highly circular orbits and longer orbital periods (days to years). According to stellar evolutionary scenarios, the secondary companion in an HMBP is expected to be either another neutron star or a standard mass ($\gtrsim 0.6 M_\odot$) white dwarf. In contrast, the secondary companion in an LMBP system is a low mass ($\lesssim 0.4 M_\odot$) white dwarf.

van den Heuvel (1987, 1988) reviews the stellar evolutionary models for binary pulsars. Here we present a brief summary. HMBPs are believed to be the descendants of high mass main sequence binaries, the OB binaries. In the standard scenario, the primary star with mass $M > 8M_\odot$ evolves first, explodes and becomes a neutron star. If the system began as a close binary, the primary would have transferred a significant part of its envelope to the secondary and the binary will not be disrupted. Eventually the secondary evolves, either to another neutron star via a supernova explosion or to a standard mass white dwarf, depending on whether its mass is greater than or less than $8M_\odot$. If a neutron star is formed, the binary is most often disrupted and we have two neutron stars in a highly eccentric orbit. Clearly, we do not expect to detect any optical counterpart of such systems. If instead, the secondary forms a white dwarf, then the system will invariably remain bound and the orbit will circularize owing to tidal forces. In this case, we do expect to detect the white dwarf secondary at optical wavelengths.

LMBPs are somewhat of an enigma. The naive model for the LMBPs starts off with a binary system consisting of a neutron star and a solar mass main sequence star. In due course, the secondary evolves and, under suitable conditions, starts dumping matter onto the neutron star primary. The accretion of matter onto the neutron star results in the generation of copious X-rays and the system becomes visible as a bright X-ray source. It is now commonly accepted that the progenitors of LMBPs are the low mass X-ray binaries (LMXBs). In addition to the X-rays, the orbital angular momentum of the accreting matter spins up the neutron star and the binary separation increases. The accretion phase can last a long time, up to a billion years, since the nuclear evolutionary time scale for the low mass secondary is $\gtrsim 10^9$ y. Again, owing to tidal circularization, the orbit is expected to be highly circular. Once the accretion ends and the debris settles down, the highly spun up primary becomes visible as a radio millisecond pulsar. The secondary should be visible as a white dwarf which slowly cools with time.

The problem with the naive model is that it is difficult to understand how the (weakly bound) binary system still remains bound during the explosive formation of the primary neutron star when at least a few solar masses of the envelope is suddenly ejected form the system. Consequently, it has been suggested that the primary star was a massive white dwarf. If sufficient mass is transferred then the white dwarf can be converted to a neutron star when its mass exceeds the Chandrashekar limit. This is the so-called Accretion Induced Collapse (AIC) model. The subsequent evolution of this system is the same as outlined above. Even this sophisticated model has its own share of problems, all of which are related to the uncertainty of the formation of a neutron star from a white dwarf.

A much simpler model invokes a spiral-in phase of the binary (eg. Verbunt, this volume). The primary star, before collapsing to a neutron star, has to go through a super-giant phase during which phase the secondary could suffer from frictional drag from orbiting in the extended atmosphere of the primary. The secondary then "spirals in" and the final result is a system similar to the one invoked by the naive model. This model has not received much consideration because the physics of the spiral-in phenomenon is not well understood. However, it is clear that a large number of massive binaries will undergo one or more spiral-ins. Given this certainty and the many uncertainties of the AIC model, my own preference lies with the more conventional spiral-in model rather than the AIC model.

The biggest observational difference between the AIC and the standard model lies in the kinematics of the binary pulsars. In the standard model, the formation of a neutron star is necessarily accompanied by a sudden mass loss and hence the system acquires a "kick" velocity, whereas in the AIC model the mass loss is small ($0.1\ M_{\odot}$, equal to the gravitational binding energy of a neutron star) and hence the resulting binary system is expected to have a low systemic velocity. Indeed, this was one of the original motivations for invoking the AIC model (Helfand, Ruderman and Shaham 1983).

The discovery of PSR 1957+20 by Fruchter, Stinebring and Taylor (1988) has dramatically highlighted the role played by a spun-up pulsar in the evolution of LMXBs. The system consists of a 1.6-ms pulsar in a 9.17-h orbit around a low mass ($0.02\ M_{\odot}$) companion. The pulsar is eclipsed once each orbit for about 50 min. It is now widely believed that a small fraction of the pulsar's bolometric luminosity is evaporating the companion and the resulting wind from the companion is responsible for the radio eclipse (eg. Kluzniak, Ruderman and Tavani 1988; Phinney et al. 1988). It is unclear whether the pulsar will completely ablate the companion or not. However, the existence of this system shows that single millisecond pulsars like PSR 1937+214 can possibly evolve from binary systems like the 1957+20 system.

The correlation of low magnetic field strength and the pulsar rotation rate seen in Table 1 can be neatly explained by the "spin up by mass transfer" model, alluded above *viz.* the transfer of mass and angular momentum from an evolving secondary to the primary neutron star can spin up the neutron star to rapid rotation rates. It can be shown that the limiting period to which the neutron star can be spun up is directly proportional to the strength of the dipole magnetic field, $B_{12}$,

$$P_{eq} \sim B_{12}^{6/7}\text{s}, \tag{4}$$

where the accretion rate has been assumed to be the Eddington rate. This line is referred to as the rebirth line or spin-up line. It is a great triumph of theory that all RPSRs lie between the rebirth line and the death line. Recently White and Stella (1988) have pointed out that the simple spin-up theory (*i. e.* Equation 4) may be incorrect since radiation pressure (expected to be intense in the inner regions of the accretion disk) has been ignored in previous theoretical work. Thus Equation (4) should be considered more like a derivation based on dimensional arguments.

What the above model does not explain is why most of the binary pulsars have such low magnetic field strengths. In the next section we discuss optical data which sheds some light on this issue and also provide evidence that most recycled pulsars are old objects with ages comparable to their characteristic ages $10^8$ to $10^9$ y, significantly older than standard field pulsars. Given this it is hard to understand why the recycled pulsars are found so close to the Galactic plane. In Table 2, we present the Galactic latitude ($b$), the distance in kpc ($d$), the vertical height in pc ($|z|$), the magnetic field strength (Eq. 2), the characteristic age ($\tau_c = P/2\dot{P}$), the transverse motion $v_t$ (from ISS measurements or otherwise) and the kinematic age $\tau_z \equiv |z|/v_t$ of the recycled pulsars. For distances we have either used estimates from dispersion measures or other estimates (PSR 0820+02, Kulkarni and Narayan 1988). The updated $\dot{P}$ for PSR 1831-00 is from Dewey et al. (1988b).

The transverse velocity estimates were obtained from the following sources: 1855+09 (Dewey et al. 1988a), 0655+64 (Lyne 1984); 0820+02 (Cordes 1986); 1913+16 (Dewey et al. 1988a) and PSR 1957+20 (Kulkarni and Hester 1988). For PSR 1937+214, Rawley, Taylor and Davis (1988) measure a scintillation velocity of about 40 km s$^{-1}$ and a transverse motion, in the reference frame of the solar system ephemerides, of less than 15 km s$^{-1}$ from the timing observations. The observed transverse motion is the sum of the motion of the pulsar in *its* local standard of rest and the motion due to Galactic rotation expected from an object at the distance and the direction of the pulsar. After accounting for Galactic rotation, the intrinsic motion of the pulsar is inferred to be 100 km s$^{-1}$. (In contrast to the timing observations, the scintillation observations, to first order, are immune to Galactic rotation and hence are a measure of the intrinsic transverse motion of the pulsar). However, this result rests heavily upon assuming that the frame of the solar system ephemerides is truly inertial, and Rawley et al. (1988) suggest that the reference frame may, in fact, rotate approximately with the Galaxy. In that case, the velocity of the pulsar with respect to its local standard of rest cannot exceed 25 km s$^{-1}$. To summarize, scintillation observations favor a transverse motion of 40 km s$^{-1}$ while the timing observations indicate either 25 km $^{-1}$ or 100 km s$^{-1}$. Unfortunately, the scintillation observations lack precision and the interpretation of the timing data is problematic. Thus, the transverse velocity of this important pulsar remains uncertain.

Consider the HMBPs first. The HMBP system 1913+16 certainly has a large transverse velocity. The z height of the HMBP 2303+46 is consistent with its age of $\sim 10^7$ y and a z velocity of 50 km s$^{-1}$. A similar velocity is indicated for PSR 1820-11. Thus apart from 0655+64, we can argue that HMBPs are short lived objects born of massive stars.

Consider next the LMBPs. All the rapid pulsars in this group are within 50 pc of the Galactic plane. This would immediately suggest that the z-velocity of the fast LMBPs is small. However, this could be entirely due to selection effects since all these objects were found in systematic surveys from Arecibo and were limited to the Galactic plane. (The considerations leading to the discovery of both PSR 1937+214 and PSR 1953+29 were their close location to the Galactic equator). Thus the z-distribution does not yield any worthwhile constraint on the progenitor population. Indeed, so convinced I am that this is a selection effect, that our group has initiated a millisecond pulsar survey at Arecibo towards intermediate latitudes.

The velocity data are more revealing. The large observed transverse motion of PSR 1957+20 definitely rules out the AIC model for at least this pulsar. For PSR 1937+214, the situation is unfortunately ambiguous. The larger velocity (100 km s$^{-1}$) decisively rules out any kind of AIC/merger origin, whereas the smaller velocities would pose some problems for the AIC model but not rule it out. On the other hand, the observed low velocities of PSR 1855+09 and PSR 0820+02 support the AIC model. Clearly we need more kinematical data before we can arrive at a firm conclusion. However, one conclusion is clear: the AIC model is ruled out for PSR 1957+20 and perhaps PSR 1937+214.

Table 2. Kinematics of Recycled Pulsars

| PSR | $b$ (°) | $d$ (kpc) | $|z|$ (pc) | $B$ (G) | $\tau_c$ ($10^6$ y) | $v_t$ (km $s^{-1}$) | $\tau_z$ ($10^6$ y) |
|---|---|---|---|---|---|---|---|
| 1855+09 | 3.1 | 0.35 | 20 | $3.3 \times 10^8$ | $4.5 \times 10^3$ | $\lesssim 25$ | $\gtrsim 1.3$ |
| 1953+29 | 0.4 | 3.50 | 25 | $4.4 \times 10^8$ | $3.2 \times 10^3$ | ? | – |
| 1937+21 | −0.3 | 4.80 | 30 | $3.9 \times 10^8$ | $2.7 \times 10^2$ | 40 or 100 | 0.3 to 0.8 |
| 1957+20 | −4.2 | 0.73 | 55 | $1.6 \times 10^8$ | $1.7 \times 10^3$ | 100 | 0.5 |
| 1831−00 | 3.7 | 3.00 | 195 | $8.8 \times 10^{10}$ | $5.8 \times 10^2$ | ? | – |
| 0820+02 | 21.2 | 1.80 | 650 | $2.9 \times 10^{11}$ | $1.4 \times 10^2$ | $\sim 13$ | $\sim 50$ |
| 0655+64 | 25.0 | 0.28 | 120 | $1.0 \times 10^{10}$ | $5.2 \times 10^3$ | 10 or 50 | 5 or 28 |
| 1820−11 | 0.9 | 11. | 170 | $6.3 \times 10^{11}$ | 3.2 | ? | – |
| 1913+16 | 2.1 | 5.20 | 190 | $2.2 \times 10^{10}$ | $1.2 \times 10^2$ | 50 to 300 | 0.6 to 4 |
| 2303+46 | −12.0 | 2.30 | 480 | $6.5 \times 10^{11}$ | $4.5 \times 10^1$ | ? | – |

## 7. Recycled Pulsars: Confrontation with Models

In this section, I review the observations of recycled pulsars and confront them with the "spin-up" hypothesis. As mentioned in section 4, all the recycled pulsars lie between the death line and the rebirth line, which is a resounding success of the spin-up hypothesis. Optical observations allow us to confront the stellar evolutionary scenarios. Specifically:

1. In eccentric HMBP systems, no optical counterpart is expected since the secondary star is another neutron star.

2. The secondary star in HMBP systems with circular orbit is expected to be a standard mass white dwarf which should be visible.

3. All LMBP systems are expected to have low mass and hence low luminosity secondary white dwarfs which may be detectable at optical wavelengths.

In order to assess quantitatively the luminosity of the white dwarf counterparts it is important to estimate the age of the white dwarfs. It is well known that white dwarfs shine by radiating away their internal heat. Thus the luminosity of white dwarfs is an age indicator. In all the stellar evolutionary scenarios discussed, for both LMBPs and HMBPs, the white dwarf is formed *after* the neutron star primary. In the context of an exponentially decaying field, the age of the neutron star can be estimated from Equation (2) and this provides an *upper limit* to the age of the white dwarf. Incidentally, "age" here refers to the time elapsed since the formation of the compact object in question.

The above reasoning in conjunction with the cooling curves for white dwarfs (eg. Iben and Tutukov 1984) enable us to estimate the luminosity of the secondary counterparts. For $B_{12} = 3 \times 10^{-4}$, typical of the LMBPs, an upper limit to the age of the white dwarf companion is $5 \times 10^7$ y. However, a standard mass white dwarf that young is expected to be *bright and hot*, eg. the white dwarf LB 1497 in the Pleiades has a luminosity of $\sim 10^{-1} L_\odot$ and an estimated $\tau_{cool} \sim 3 \times 10^7$ y. For a given $\tau_{cool}$, the luminosity decreases with mass of the white dwarf (eg. Sweeney 1976) and hence the LMBP white dwarfs are expected to be less luminous than the HMBP white dwarfs.

In Table 3, we summarize the optical observations for five recycled pulsars: $m_R$ refers to the apparent magnitude of the companion or the limit to which no counterpart was found; $T_{BB}$, the equivalent blackbody temperature of the counterpart assuming it to be a white dwarf; $\tau_B$ is specified by Equation (3); $\tau_{cool}$ the cooling time scale estimated from white dwarf cooling theories; and finally $\tau_c$, the characteristic age, $P/2\dot{P}$. The optical observations for the remaining four pulsars (HMBP 1913+16, HMBP 1953+29, HMBP 1820-11 and LMBP 1831-00) do not provide much information owing to the obscuration caused by the Galaxy towards these sources and are therefore not listed in the table.

The optical data were drawn from the following sources: 0655+64 (Kulkarni 1986), 0820+02 (Kulkarni 1986), and 1957+20 (Djorgovski and Evans 1989). The optical candidate for PSR 1855+09 is somewhat controversial. Wright and Loh (1986) identified a candidate. The finding chart of Wright and Loh (1986) is exceedingly poor and given the poor seeing and crowded field it has been difficult for others to firmly identify the candidate star. Kulkarni, Djorgovski and Klemola (1989c) claim to have identified a candidate which lies close to the radio source, but Callanan et al. (1989) do not detect any candidate at the nominal radio position. Kulkarni et al. show that the extinction towards the pulsar is small ($A_v \sim 0.4$) from HI 21-cm absorption and CO 2.6-mm emission spectroscopic observations and from their optical spectroscopic observations derive a black body temperature of 4000 K – consistent with a cooling white dwarf.

The optical companion of PSR 1957+20 is under much investigation currently. As remarked in section 6, the companion, a degenerate brown dwarf is heated up by the particles/photons from the pulsar. The orbital modulation of the companion has been seen (eg. Djorgovski and Evans 1989). Undoubtedly this object will be intensely studied over the years to come.

Table 3. Optical Observations

| PSR | $m_R$ | $T_{BB}$ | $\tau_B$ $(10^6\ y)$ | $\tau_{cool}$ $(10^6\ y)$ | $\tau_c$ $(10^6\ y)$ |
|---|---|---|---|---|---|
| LMBP 1855+09 | $\sim$23.0 | $\sim$4,000 | 40 | $\gtrsim 1,000$ | 5,300 |
| LMBP 0820+02 | 22.8 | > 20,000 | 6 | $\lesssim 20$ | 140 |
| LMBP 1957+20 | 19.5 | $\sim$5,800 | 65 | – | 1,700 |
| HMBP 2303+46 | > 26. | No id. | 2 | – | 160 |
| HMBP 0655+64 | 22.1 | $\sim$8,000 | 21 | $\gtrsim 1,000$ | 4,800 |

To first order, the optical observations vindicate the conventional stellar evolutionary scenarios:

1. No optical counterparts have been found for the eccentric HMBP 2303+46, strongly suggesting that the secondary star is another neutron star.

2. An optical counterpart has been found for a circular orbit HMBP 0655+09 system and is a white dwarf, as expected.

3. Optical counterparts have been found for three out*five LMBPs. In LMBP 0820+02, a white dwarf is clearly seen. In LMBP 1855+09, there is some disagreement about the secondary star. If Kulkarni et al. (1989c) are correct, then this system also contains a cool white dwarf.

To second order, there are some agreements and disagreements. In 0820+02, the white dwarf is young and hot and $\tau_B$ is comparable to $\tau_{cool}$, both as expected. However, the white dwarfs in 0655+64 and 1855+09 appear to be cold and hence old, an unexpected conclusion. More quantitatively, $\tau_{cool}$ vastly exceeds $\tau_B$. Indeed $\tau_{cool}$ is comparable to $\tau_c$ suggesting that the magnetic field in the recycled pulsars does not decay, contrary to the situation with field pulsars (section 3).

In an effort to resolve this discrepancy, Kulkarni (1986) suggested that the magnetic fields of pulsars consist of two components: a high field ($B_{12} \sim 3$) component which decays exponentially with time ($\tau_d \sim 5 \times 10^6$ y) and a low field component ($B_{12} \lesssim 10^{-2}$) which does not decay with time. In this picture, most field pulsars with their high $B$-field strength and PSR 0820+02 form one group of pulsars and the low $B$-field pulsars like the single millisecond pulsar, PSR 1937+21 and weak $B$-field binary pulsars like 0655+64, 1855+09 and 1953+29 form another group. In the former group, the surface field is dominated by the decaying component and in the latter group the decaying component has decayed to a point where the dominant field is the second component $i.\ e.$ the lower field strength steady component.

Triggered by these observations van den Heuvel, van Paradijs and Taam (1986) and independently Bhattacharya and Srinivasan (1986) arrived at a similar conclusion. The authors of both these papers assumed that LMXBs are the progenitors of the LMBPs and that, in steady state, the two birthrates must equal each other. Given the rather small number of LMXBs and their estimated lifetimes ($\sim 10^8$ y), they were forced to increase the ages of the known millisecond pulsars in order to match the two birthrates.

The implications of this "two-component" or "asymptotic decay" hypothesis are far reaching. Under this hypothesis, the age of pulsars like PSR 0655+64 and PSR 1855+09 is given by the characteristic age $\tau_c = P/2\dot{P}$; this is typically $\sim 10^9$ y for the spun-up pulsars. Thus a large fraction of binary pulsars and all the millisecond pulsars belong to the old population, unlike the field pulsars.

In order to get a handle on the relation between LMXBs and LMBPs Kulkarni and Narayan (1988; hereafter KN) have reinvestigated the issue of birthrates of LMXBs and LMBPs. In steady state, these two birthrates should be equal if the spin-up by accretion model is correct. The birthrates have been estimated taking into account the selection effects in radio surveys, the incompleteness of the X-ray surveys and the vastly different space sampled by the radio surveys (a few kpc around the sun) and the X- ray surveys (essentially the entire Galaxy). In addition, KN take into account of the steep orbital period dependence of the X-ray lifetime of LMXBs.

To first order, KN find that the birthrate of LMBPs exceed that of LMXBs by a factor of about 10. Equality of the birthrates can be obtained by appealing to small number statistics. However, the observed distribution of the orbital periods of the LMBPs cannot be satisfactorily explained. Specifically, KN find that the birthrate of the small orbital period systems like 1855+09, i.e. close binaries, exceeds the birthrate of the progenitor LMXB systems by a factor of 100. Even with pessimistic estimates of the uncertainties, this discrepancy cannot be reduced to below a factor of 10. Coté and Pylyser (1989) come to a similar conclusion.

The origin of this disagreement could be due to small number statistics or poorly known X-ray lifetimes. The agreement for the long period systems suggests that small number statistics is probably not the culprit. The discovery of PSR 1957+20 shows that the mass transfer in close binaries is more complicated than assumed by stellar evolutionary models. Also, the X-ray emission from the accretion disk can affect the companion, leading to an increase in the mass transfer rate. Thus, there are good physical reasons to believe that the mass transfer rate in the close binaries is considerably higher than predicted by simple models – alleviating the birthrate discrepancy problem. In section 8, we show that a similar discrepancy exists in globular clusters. Since most cluster LMXBs are short orbital period systems we invoke a similar explanation to explain the discrepancy in the cluster population.

## 8. Pulsars in Globular Clusters

Following the discovery of the first millisecond pulsar, 1937+214, a number of authors proposed the now accepted hypothesis that millisecond pulsars are old pulsars spun up by accretion from a companion star. In particular, Alpar *et al.* (1982) suggested that the LMXBs are the progenitors of the millisecond pulsars. Motivated by this suggestion and noting the much higher incidence of LMXBs in globular clusters, Hamilton, Helfand and Becker (1985) surveyed a dozen nearby, dense globular clusters with the VLA. Pulsars differ from extragalactic sources in several ways: pulsed emission, extremely compact structure, high linear or circular polarization and steep spectral index. Hamilton et al. detected 21 sources within ten core radii of the cluster centers. All the 21 sources could be explained satisfactorily as background extragalactic sources. However, one source, a point source was found in the core of the globular cluster M28. Further observations at 6 cm showed that the M28 source had a spectral index $\alpha > 1.1$ (where $S_\nu \propto \nu^{-\alpha}$), which already made it a somewhat unusual extragalactic source. Subsequent low frequency (30-60 MHz) observations at the Clark Lake Observatory established that the source was certainly unusual with $\alpha \sim 2.5$ (Mahoney and Erickson 1985). Further VLA observations at 327 MHz showed that this steep spectrum source had a high degree of linear polarization – highly suggestive of a pulsar. Several pulse search efforts were undertaken between 1985 and 1987 but with no success. Finally, in 1987 an international collaboration between Jodrell Bank, Los Alamos, Caltech and U. C. Berkeley discovered a 3 ms single pulsar (1821−24) towards this cluster (Lyne et al. 1987).

Following the discovery of PSR 1821-24 a number of searches were initiated at several observatories:

(1). *Jodrell Bank Survey.* This extensive survey was done at Jodrell Bank by Lyne and collaborators (Biggs, Lyne and Brinklow 1989; Lyne and Biggs 1989) using the 76-m Lowell telescope. Well over two dozen nearby and rich clusters were observed in one or more observing configurations: central frequency, $\nu_{RF} = 610$ MHz and a $32 \times 1$ MHz filter bank ("low Dispersion Measure" search); $\nu_{RF} = 610$ MHz, $32 \times 250$ KHz ("high Dispersion Measure" search); $\nu_{RF} \sim 1400$ MHz, the so-called "L-band" and a $32 \times 1$ MHz filter bank. Further details of the L-band search can be found in Lyne et al. (1987) and that of the 610 MHz search in Biggs et al. (1989). The L-band search uncovered a 3-ms pulsar, 1821-249 in M28 (Lyne et al. 1987) and the 610 MHz "high Dispersion Measure" search discovered a 11-ms binary pulsar, 1620-264 with $P_{orb} \sim 191$ d in the cluster M4 (Lyne et al. 1988).

(2). *Arecibo Survey.* All the rich clusters visible from the Arecibo telescope ($-1° \leq \delta \leq 38°$) were searched at L-band. An isolated pulsar, 2127+11A with a period of 110 ms was discovered (Wolszczan et al. 1989); see also this reference for the parameters of the survey.

(3). *Green Bank Survey.* Survey conducted using the NRAO 100-m telescope at Green Bank at $\nu_{RF} = 1400$ MHz (Backer and Dey 1988). No pulsars were seen to the limit of the survey (2 mJy).

(4). *Parkes Survey.* The CSIRO group led by Ables have searched a number of Southern globular clusters using the Parkes telescope in Australia. The details of this survey have not yet been published. So far, the group has announced the discovery of two pulsars in the cluster 47Tuc (Ables et al. 1988; preprint, 1989). The pulsar 0021-72A has a rotational period of 4.5 ms and is also in a binary system with an orbital period of 32 min! 0021-72B has a period of 6.1 ms and is in a binary system (parameters are unknown at the time of this writing). The relatively small size of the Parkes dish and the faintness of the pulsars combine to make it difficult to consistently detect these pulsars. Consequently, independent confirmation of the pulsars are urgently needed.

(5). *VLA Survey.* Hamilton, Helfand and Becker (1985) used the VLA to carry out an *imaging* search, at L-band, for point sources within the cores of a dozen, nearby clusters. In comparison to the pulsed searches, imaging searches have the advantage that they are insensitive to many of the selection effects that plague pulsar surveys: scattering by the interstellar plasma, rotation period of the pulsar, magnitude of dispersion measure (see section 2). An important advantage of imaging surveys is their uniform sensitivity to pulsars in binary systems, unlike pulsed signal searches (section 2; Figure 1). Thus non-detection of point sources below a threshold level place firm upper bounds on the flux density of any undiscovered pulsar. This survey drew attention to an interesting point source in M28 which eventually led to the discovery of the first pulsar in a globular cluster.

(6). *Deep VLA Survey.* Kulkarni et al. (1989b) present results from deep VLA observations of four clusters at L-band. The 1-$\sigma$ point source flux density values in the images of these clusters are: GC 1339+286 (M3; $\sigma_S \sim 45\mu$ Jy); GC 1620-264 (M4; $\sigma_S \sim 60$ $\mu$Jy); GC 1715+432 (M92; $\sigma \sim 24$ $\mu$ Jy); and GC 1821-249 (M28; $\sigma_S \sim 50$ $\mu$Jy). Kulkarni et al. contend that a point source at the level of $3.5\sigma_S$ would have been readily detected. To this sensitivity limit a total of three sources were found within the core regions of the four clusters that were mapped. The three sources include the pulsars in M4 and M28 and an intriguing point source in M3 with a flux density of 180 $\mu$Jy (Figure 4). Analysis of our Arecibo M3 data (both at 1400 MHz and 430 MHz) rule out an isolated pulsar with a comparable flux density at 1400 MHz and an appropriately scaled flux density at 430 MHz. However, a pulsar in a binary orbit of less than a week cannot be ruled out and the search (in supercomputing centers and not observatories!) continues. The value of an imaging search is best illustrated by the deep VLA image of the cluster M92 which shows that the cluster, at 1400 MHz, cannot contain any pulsar with flux greater than 80 $\mu$Jy – a rather sensitive limit that is challenging even for the Arecibo telescope.

(7). *Second Arecibo Survey.* A new pulsar survey is now in progress at Arecibo by the Caltech/Arecibo team. This survey differs from the previous Arecibo survey in two ways: observations at 430 MHz and/or faster sampling rate. The second survey has been very productive. Two new pulsars have been discovered in the cluster M15: PSR 2127+11B, a 56-ms single pulsar (Anderson et al. 1989a) and PSR 2127+11C, a 30-ms pulsar in an

8-h binary system (Anderson et al. 1989b). A 10-ms single pulsar has been reported in the cluster M13 (Anderson et al. 1989c). As remarked in section 2, pulsar surveys are insensitive to rapidly rotating pulsars in tight orbits (Figure 1). This limitation is now being overcome in the second Arecibo search. The search for binaries in M15 is made feasible because the dispersion measure of the cluster is known from PSR 2127+11A. Assuming the dispersion measure of PSR 2127+11A, a search for accelerated pulsars was conducted. Specifically, the assumption is made that the acceleration is constant but unknown over the duration of the observations. With this restriction the problem reduces to a two-parameter search: rotational period and acceleration. While this scheme is a distinct improvement over the standard searches it is still insensitive to pulsars in tight binaries (see Figure 1) where the above assumption breaks down. This scheme with its limitations resulted in the discovery of PSR 2127+11C. Clearly, the search for pulsars in clusters is an exciting and rewarding task. However, immense amounts of computing are needed.

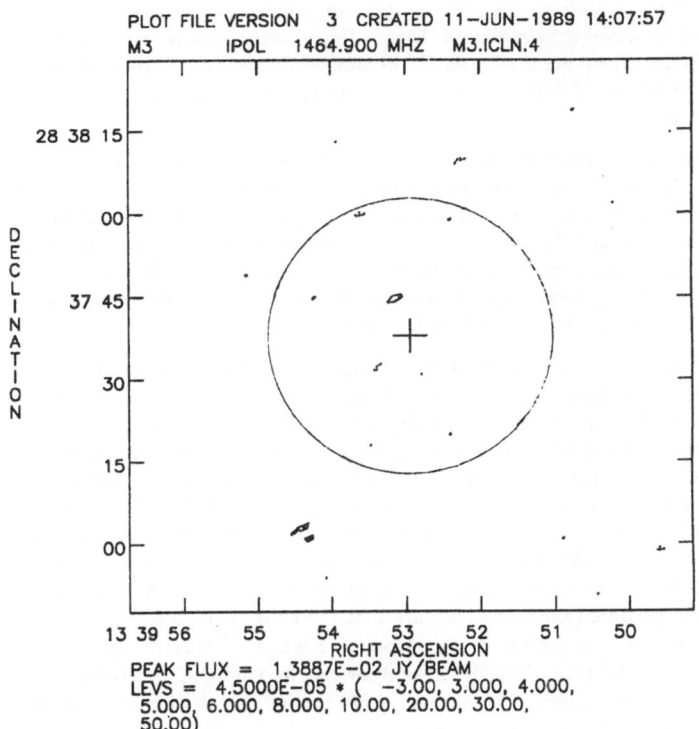

Figure 4. VLA L-band image of the cluster M3. The center of the cluster is marked by a cross and the circle has a radius of one core radius. Note the 180$\mu$Jy point source north of the center of the cluster but well within the core. No pulsations have yet been detected from this source.

The discovery of pulsars in globular clusters raises again the question of the origin of neutron stars in these ancient systems. The conventional view is that these neutron stars were formed as a result of the first wave of star formation. Like the young pulsars observed in the disk of the Galaxy, the primordial neutron stars probably started with large magnetic field strengths, but the fields decayed rapidly to an asymptotic value of $10^8$ G to $10^{10}$ G (section 7) and the pulsars switched off their radio emission. Subsequently, some fraction of the neutron stars, aided by the high stellar density, tidally captured field stars (Fabian, Pringle and Rees 1975). The newly formed binaries have separations uniformly distributed from a small separation (head-on collision) to $6R_*$ where $R_*$ is the radius of the field star. The outcome of the head-on collisions is uncertain. Krolik (1984) has speculated that such a collision may lead to a solitary accretion disk which may last long enough to spin up the neutron star. The remaining two thirds of the collisions lead to bound systems. Thus, the orbital period is a function of the evolutionary state of the field star: capture of a red giant would result in an extended binary, whereas a main sequence star would give rise to a binary with initial $P_b < 12$ h. The subsequent evolution is pretty much the same as for the disk LMXBs i.e. the long orbital period ($P_b \gtrsim 1$ d) systems evolve with mass transfer driven by nuclear evolution of the secondary star and the short orbital period systems evolve by magnetic braking aided by gravitational wave radiation. During this phase the binary systems are visible in the X-ray sky as LMXBs. These eventually should become visible as rapidly rotating binary pulsars. I refer the reader to Verbunt (1988a, 1988b) for a summary of the formation and evolution of neutron star tidal binaries.

In dense clusters, the wide binaries can be affected by perturbations caused by the passing field stars. Specifically, field stars can "ionize" (Romani, Kulkarni and Blandford 1987) the binary system or the binary system may undergo an "exchange" (Verbunt et al. 1987). The net result is an isolated spun-up pulsar. In less dense clusters or more tightly bound binaries the passing encounters are not strong enough to break up the cluster but can induce orbital eccentricity (Rappaport, Putney and Verbunt 1989). Indeed, Rappaport et al. show that the measured eccentricity of the M4 binary system is in rather good agreement with their model.

Following the discovery of the first pulsar 1821-24 in the cluster M28 (Lyne et al. 1987), it was immediately realized that there should be many more pulsars in globular clusters. The 400 MHz flux density of this pulsar, $S_{400} \sim 25$ mJy, corresponds to a radio luminosity, $L_{400} = S_{400}d(\text{kpc})^2$, which is nearly three orders of magnitude greater than the minimum luminosity of pulsars, $\sim 1$ mJy kpc$^2$ (Dewey 1984). This coupled with the $L_{400}^{-1}$ luminosity law of the observed pulsars (Lyne et al. 1985) suggested that PSR 1821-24 represented only the "tip of the iceberg". It already boded ill for the standard model since in steady state the LMXB and the pulsar birthrates should be equal and thus a much too high LMXB birthrate was indicated. This suspicion became more concrete with the discovery of the second pulsar in M15 (Anderson et al. 1989a). Based on this discovery, Grindlay and Bailyn (1989) were the first to remark that the number of pulsars far exceeded the number of LMXBs. This issue was dealt in greater depth by Kulkarni, Narayan and Romani (1989a; hereafter KNR). The essential goal of the KNR analysis was an empirical determination of the number of pulsars in the cluster system. To this end, KNR used all the above pulsar surveys as their primary inputs to their analysis (except the second on-going, Arecibo pulsar survey). A number of elements went into their calculations:

(a). *Luminosity Law.* Empirically it has been noted that the luminosity function of the observed pulsars is related to the quantity, $Q = log(\dot{P}_{-15}/P^3)$. This law describes the luminosity of the observed pulsars and is biased towards the stronger pulsars. Narayan and Ostriker (1989), from an extensive analysis of eight flux-limited pulsar surveys, have obtained a form for the distribution of the luminosities of the true population and show that the average observed pulsar is nearly a factor of 20 more luminous than the typical pulsar of the underlying population.

(b). *Selection Effects in Pulsar Surveys.* The selection effects discussed in section 2 prevent some fraction of pulsars from being discovered. Following Narayan (1987), these selection effects were extensively modelled and incorporated in the analysis.

(c). *Cluster Weighting Function.* There is a wide range in the properties of the $\sim 150$ known globular clusters in the Galaxy, which should be reflected in their pulsar content. The simplest weighting function is one of mass or equivalently light: massive clusters may be expected to have more pulsars than less massive clusters. However, in the "standard" scenario the total number of binary interactions plays an important role in which case the weighting is proportional to $n_*^2 r_c^3 / v_{rms}$ where $n_*$ is the stellar density, $v_{rms}$ is the rms velocity in the core and $r_c$ is the core radius. The next level of sophistication, relevant only to the standard model is a weighting which explicitly includes the fraction of primordial neutron stars that are expected to be retained by the shallow gravitational potential of the cluster core. However, the velocity distribution of the primordial neutron stars is unknown which make this weighting function highly uncertain. The calculations were done with all three weighting functions.

(c). *Model of Pulsar Birth and Evolution.* Since the luminosity of a pulsar depends upon $P$ and $\dot{P}$ (see (a) above), a model specifying the distribution of $P$ and $\dot{P}$ is needed. Since the evolution of $P$ depends upon the magnetic field strength what is needed is a model of the initial periods and magnetic fields strengths of pulsars. KNR assumed that cluster pulsars are born with a uniform distribution of $\log B$ in the range of $8 < \log B < 10$ on the rebirth line (section 7). Furthermore, it was assumed that the field strength does not vary during the life of the pulsar. Finally, a constant birthrate over the last $10^{10}$ y was assumed.

The main results of KNR are as follows:

- There are $\sim 10^4$ pulsars in the globular cluster system.

- The birthrate of the pulsars exceeds that of the cluster LMXBs by nearly a factor of 100.

The second result is similar to that obtained for the disk recycled pulsars by Kulkarni and Narayan (1988). KNR argue that these two discrepancies can be resolved provided the X-ray lifetimes of the LMXBs is $10^7$ y, rather than the commonly assumed value of $10^9$ y. Evidence for this change in the standard scenario comes from the theoretical work on accelerated evolution of LMXBs due to ablation by the spun-up pulsar, as in PSR 1957+20 (eg. Kluzniak et al.1988; Ruderman, Shaham and Tavani 1989). However, it is fair to say that while the theoretical work shows that the X-ray lifetimes are shortened, it still remains to be demonstrated that the lifetimes are shortened by a factor as large as 100.

The first result viz. 100 active pulsars per rich cluster is quite unexpected. In the standard model, in a typical rich cluster, only 10% of the primordial neutron stars undergo tidal capture. Thus, we are led to the conclusion that there are nearly $10^3$ *neutron stars* per rich cluster of $10^6$ $M_\odot$. Hence the neutron star efficiency of clusters is $10^{-3}$ neutron stars per $M_\odot$ of ordinary stars – strikingly similar to that in the disk of the Galaxy.

The real problem in understanding this result comes from the recent theoretical studies of the evolution of globular clusters (Chernoff and Weinberg 1989), which find that only clusters with spectrum steeper than the Salpeter IMF survive, whereas clusters with shallow IMF lose so much mass by stellar evolution that they disrupt within the first $10^9$ y. Bailyn and Grindlay (1989) present this as strong evidence against the standard scenario. KNR concur that this is a strong objection but point out that Chernoff and Weinberg's models assume a single exponent IMF and hence all these arguments apply principally to the evolved stars dominating the total mass loss i.e. those born with $M < 5M_\odot$. With a more complicated IMF, the upper main sequence, and hence the production of neutron stars, is quite poorly constrained. All this discussion notwithstanding, it is possible that the standard model is wrong and, as has been argued in the past, recycled pulsars may be the products of accretion induced collapse (AIC) of white dwarfs. As remarked in section 7, there are many theoretical difficulties with the AIC model. Apart from these problems, the main observational problem with the AIC model is the requirement of a large number of suitable O-Ne-Mg white dwarfs. To explain KNR's results, every rich cluster is expected to contain nearly $10^3$ O-Ne-Mg white dwarfs of which 10 are expected to be in accreting binaries at any given time. In addition to these accreting white dwarf systems (i.e. the cataclysmic variables, CVs) a much larger number of CVs with ordinary white dwarfs (i.e. the C-O white dwarfs) are also expected. However, searches for CVs in clusters have been woefully unsuccessful, prompting Verbunt, Lewin and van Paradijs (1989) to question the AIC hypothesis.

The distribution of the orbital periods of the known cluster pulsars is also at variance with the standard model (Romani 1989). Only a few wide orbit systems are expected (those which evolve from captures of giants). Also, most cluster pulsars are expected to be binaries with the singles formed by disruption of wide binaries (operative in dense clusters) or by ablation of the companion a la PSR 1957+20 (short period pulsars). Of the eight cluster pulsars, four are single – an unexpectedly high ratio. It is unclear whether the AIC models can successfully account for the properties of the observed sample, either. Thus, in summary, while some reasonable estimates for the number of pulsars and the rate of their production can be made, we are still unable to identify uniquely their genesis and evolution.

I would like to thank NATO for providing financial support. My research is supported by a U.S. Presidential Young Investigator award and an award from the Alfred P. Sloan Foundation. I thank José Navarro for a careful reading of this manual.

## References

Ables, J. G., Jacka, C. E., Hamilton, P. A. and McCulloch, P. M., *I. A. U. Circ.* # 4602, (1988).

Alpar, M. A., Cheng, A. F., Ruderman, M.A. and Shaham, J., *Nature* 300, 728-730, (1982).

Alpar, A., Brinkmann, W., Kiziloglu, U., Ogelman, H. and Pines, D., *Astr. Astrophys.* 177, 101-104, (1987).

Anderson, S. B., Gorham, P. M., Kulkarni, S. R., Prince, T. A. and Wolszczan, A., *I. A. U. Circ.* # 4762, (1989a).

Anderson, S. B., Gorham, P. M., Kulkarni, S. R., Prince, T. A. and Wolszczan, A., *I. A. U. Circ.* # 4772, (1989b).

Anderson, S. B., Kulkarni, S. R., Prince, T. A. and Wolszczan, A., *I. A. U. Circ.* # 4819, (1989c).

Angerhofer, P. E., Wilson, A. A., Mould, J. R., *Astrophys. J.* **236**, 143-152, (1980).

Backer, D. C., Kulkarni, S. R., Heiles, C., Davis, M. M. and Goss, W. M., *Nature* **300**, 615-618, (1982).

Backer, D. C., in *The Origin and Evolution of Neutron Stars*, Eds. D. J. Helfand and J. -H. Huang (Dordrecht:Reidel), 13-22, (1987).

Backer, D. C. and Dey, A., *Adv. Space Res.* **8**, 353-354, (1988).

Bailyn, C. D. and Grindlay, J. E., *sub. to Astrophys. J.*, (1989).

Becker, R. H., *The Origin and Evolution of Neutron Stars*, Eds. D. J. Helfand and J.-H. Huang (Dordrecht:Reidel), 91-98, (1987).

Bhattacharya, D. and Srinivasan, G., *Curr. Sci.* **55**, 327-330, (1986).

Biggs, J. D., Lyne, A. G. and Brinklow, A., in *Timing Neutron Stars*, Eds. H. Ogelman and E. J. P. van den Heuvel (Dordrecht:Kluwer), 157-162, (1989).

Blandford, R. D., Ostriker, J. P., Pacini, F. and Rees, M. J., *Astr. Astrophys.* **23**, 145-146, (1973).

Blandford, R. D., Applegate, J. H. and Hernquist, L., *M. N. R. A. S.* **204**, 1025-1048, (1983).

Brinkman, W. and Ogelman, H., *Astr. Astrophys.* **182**, 71-74, (1987).

Callanan, P. J., Charles, P. A., Hassal, B. J. M., Machin, G., Mason, K. O., Taylor, T., Smale, A. P. and van Paradijs, J., *M. N. R. A. S.* **238**, 25P-28P, (1989).

Chernoff, D. and Weinberg, M., *sub. to Astrophys. J.*, (1989).

Chevalier, R. and Raymond, J. C., *Astrophys. J.* **225**, L27-L30, (1978).

Clifton, T. R. and Lyne, A. G., *Nature* **320**, 43-45, (1986).

Clifton, T. R., Lyne, A. G. and Jones, A. W., *in prep.*, (1989).

Clifton, T. R. Frail, D. A., Kulkarni, S. R. and Weisberg, J. M., *Astrophys. J.* **333**, 332-340, (1988).

Cote, J. and Pylyser, E. H. P., *sub. to Astron. Astrophys.*, (1989).

Cordes, J. M., *Astrophys. J.* **311**, 183-196, (1986).

Cordova, F., Hjellming, R. M., Mason, K. O. and Middleditch, J., *sub. to Astrophys. J.*, (1989).

Coroniti, F. V., *U. C. L. A. preprint*, (1989).

Dewey, R. J. *Ph. D. Thesis*, Princeton Univ. (1984).

Dewey, R. J., Taylor, J. H., Weisberg, J. M. and Stokes, G. H., *Astrophys. J.* **294**, L25-L29, (1985).

Dewey, R. J., Cordes, J. M., Wolszcan, A. and Weisberg, J. M., in *Radio Wave Scattering in the Interstellar Medium*, Eds. J. M. Cordes, B.J. Rickett and D. C. Backer (New York:AIP), 217-221, (1988a).

Dewey, R. J., Taylor, J. H., Maguire, C. M. and Stokes, G. H.,, *Astrophys. J.* **332**, 762-769, (1988b).

Djorgovski, S. and Evans, C. R., *Astrophys. J.* **335**, L61- L65, (1988).

88

Dominique, D., Rankin, J. M., Weisberg, J. M. and Backus, P. R., *Astron. Astrophys.* **161**, 303-304, (1986).

Fabian, A. C., Pringle, J. E. and Rees, M. J., *M. N. R. A. S.* **172**, 15P-18P, (1975).

Fesen, R. A., Shull, J. M. and Saken, J. M. , *Nature* **334**, 229-231, (1988).

Frail, D. A., *Ph. D. Thesis*, Univ. of Toronto, (1989).

Fruchter, A. S., Taylor, J. H., Backer, D. C., Clifton, T. R., Foster, R. S. and Wolszczan, A., *Nature* **331**, 53-54, (1988).

Fruchter, A. S., Stinebring, D. R., Taylor, J. H., *Nature* **333**, 237-239, (1988).

Fruchter, A. S., *Ph. D. Thesis*, Princeton Univ., (1988).

Grindlay, J. E. and Bailyn, C. D., *Nature* **336**, 48-50, (1989).

Hamilton, T. T., Helfand, D. J. and Becker, R. H. , *Astron. J.* **90**, 606-608, (1985).

Helfand, D. J., in *Supernova Remnants and their X-ray Emission*, Eds. J. Danziger and P. Gorenstein (Dordrecht:Reidel), 471-486, (1983).

Helfand, D. J., Ruderman, M. A. and Shaham, J., *Nature* **304**, 423-425, (1983).

Helfand, D. J. and Becker, R. H., *Astrophys. J.* **314**, 203-214, (1987).

Hester, J. J. and Kulkarni, S. R., *Astrophys. J.* **331**, L121-L125, (1988).

Hester, J. J. and Kulkarni, S. R., *Astrophys. J.* **340**, 362-379, (1989).

Iben, I. and Tutukov, A. V., *Astrophys. J.* **282**, 615-630, (1984).

Jones, P. D. B., *M. N. R. A. S.* **233**, 875-885, (1988).

Kennel, C. F. and Coroniti, F. V., *Astrophys. J.* **283**, 710-730, 1984).

Kluzniak, W., Ruderman, M., Shaham, J. and Tavani, M., *Nature* **334**, 225-227, (1989).

Krolik, J. H., *Astrophys. J.* **282**, 452-465;(1984).

Kulkarni, S. R., *Astrophys. J.* **306**, L85-L89, (1986).

Kulkarni, S. R., Clifton, T., Backer, D., Foster, R., Fruchter, A. and Taylor, J., *Nature* **331**, 50-54, (1988).

Kulkarni, S. R. and Narayan, R., *Astrophys. J.* **335**, 755-769, (1988).

Kulkarni, S. R. in *Physics of Neutron Stars and Black Holes*, Ed. Y. Tanaka, (Tokyo:Universal Press Academy), 37-54, (1988).

Kulkarni, S. R. and Heiles, C. E., in *Galactic and Extragalactic Radio Astronomy*, Eds. G. L. Verschuur and K. I. Kellerman, (Berlin:Springer-Verlag), 95-153, (1988).

Kulkarni, S. R. and Hester, J. J., *Nature* **335**, 801-803, (1988).

Kulkarni, S. R., Narayan, R. and Romani, R. W., *sub. to Astrophys. J.*, (1989a).

Kulkarni, S. R., Wolszczan, A., Middleditch, J. M. and Goss, W. M., *in prep.*, (1989b).

Kulkarni, S. R., Djorgovski, S. and Klemola, A., *in prep.*, (1989c).

Lyne, A. G., *Nature* **310**, 300-302, (1984).

Lyne, A. G., Manchester, R. N. and Taylor, J. H., *M. N. R. A. S.* **213**, 613-639, (1985).

Lyne, A. G., in *The Origin and Evolution of Neutron Stars*, Eds. D. J. Helfand and J.-H. Huang (Dordrecht:Reidel), 22-23, (1987).

Lyne, A. G., Brinklow, A., Middleditch, J., Kulkarni, S. R., Backer, D. C. and Clifton, T. R., *Nature* **328**, 399-401, (1987).

Lyne, A. G., Biggs, J. D., Brinklow, A., Ashworth, M., and Mckeena, J., *Nature* **332**, 45-47, (1988).

Lyne, A. G. and Manchester, R. N., *M. N. R. A. S.* **234**, 477-508, (1988).

Lyne, A. G. and Biggs, J. D., *pers. comm.*, (1989).

Mahoney, M. J. and Erickson, W. C., *Nature* **154**, 154-155, (1985).

McKenna, J. in *Timing Neutron Stars*, Eds. H. Ögelman and E. P. J. van den Heuvel (Dordrecht:Kluwer), 143-152 (1989).

Narayan, R, *Astrophys. J.* **319**, 162-179, (1987).

Narayan, R. and Ostriker, J. O., *sub. to Astrophys. J.*, (1989).

Narayan, R. and Schaudt, K. J., *Astrophys. J.* **325**, L43-L46, (1988).

Nousek, J. A., Cowie, L. L., Hu, E., Lindblad, C. J. and Garmire, G. P., *Astrophys. J.* **248**, 152-160, (1981).

Phinney, E. S., Evans, C. R., Blandford, R. D. and Kulkarni, S. R., *Nature* **333**, 832-834, (1988).

Rawley, L. A., Taylor, J. H. and Davis, M. M., *Astrophys. J.* **326**, 947-953, (1988).

Radhakrishnan, V., *Contemp. Phys.* **23**, 207-231, (1982).

Rappaport, S., Putney, A. and Verbunt, F., *sub. to Astrophys. J.*, (1989).

Reynolds, R. J., *Astrophys. J.* **339**, L29-L32, (1989).

Ruderman, M., Shaham, J. and Tavani, M., *Astrophys. J.* **336**, 507-518, (1989).

Romani, R. W., Kulkani, S. R. and Blandford, R. D., *Nature* **329**, 399-400, (1987).

Romani, R. W., in *Goa Workshop on SN and Stellar Evolution*, Ed. A. Ray (in press).

Sang, Y. and Chanmugam, G., *Astrophys. J.* **323**, L61-L64, (1987).

Segelstein, D. L., Rawley, L. A., Stinebring, D. R., Fruchter, A. S. and Taylor, J. H., *Nature* **322**, 714-717, (1986).

Seward, F. D. and Wang, Z. -R., *Astrophys. J.* **332**, 199-205, 1988.

Seward, F. D., *Space Sci. Rev.* **49**, 385-424, (1989).

Smith, A., Jones, L. R., Watson, M. G., Willingale, R., Wood, N. and Seward, F. D. , *M. N. R. A. S.* **217**, 99-104, (1985).

Smith, F. G. in *Timing Neutron Stars*, Eds. H. Ogelman and E. P. J. van den Heuvel (Dordrecht:Kluwer), 133-142, (1989).

Srinivasan, G., Bhattacharya, D. and Dwarkanath, K. S., *J. Astr. Astrophys.* 5, 403-424, (1984).

Stollman, G. M., *Astron. Astrophys.* 178, 143-152, (1987).

Stokes, G. H., Segelstein, D. J., Taylor, J. H. and Dewey, R. J., *Astrophys. J.* 311, 694-700, (1986).

Strom, R. G., *Astrophys. J.* 319, L103-L108, (1987).

Sweeney, M. A., *Astr. Astrophys.* 49, 375-385, (1976).

Taylor, J. H. and Stinebring, D. R., *Annu. Rev. Astr. Astrophys.* 24, 285-328, (1986).

van den Heuvel, E. P. J., van Paradijs, J. A., and Taam, R. E., *Nature* 322, 153-155, (1986).

van den Heuvel, E. P. J., in *The Origin and Evolution of Neutron Stars*, Eds. D. J. Helfand and J.-H. Huang (Dordrecht:Reidel), 393-406 (1987).

van den Heuvel, E. P. J., in *Timing Neutron Stars*, Eds. H. Ogelman and E. P. J. van den Heuvel (Dordrecht:Kluwer), 523-548 (1989).

Verbunt, F., van den Heuvel, E. P. J., van Paradijs, J. and Rappaport, S. A., *Nature* 329, 312-314, (1987).

Verbunt, F., in *Physics of Neutron Stars and Black Holes*, Ed. Y. Tanaka (Universal Press Academy:Tokyo 159-174, (1988a).

Verbunt, F., *Adv. Space Res.* 8, 529-538, (1988b).

Verbunt, F., Lewin, W. G. H. and van Paradijs, J., *sub. to M. N. R. A. S.*, (1989).

White, N. E. and Stella, L., *Nature* 332, 416-418, (1988).

Wolszczan, A., Cordes, J. M., Dewey, R. J. and Blaskiewicz, M., *I. A. U. Circ.* # 4694, (1988).

Wolszczan, A., Kulkarni, S. R., Middleditch, J. M., Backer, D. C., Fruchter, A. S. and Dewey, R. J., *Nature* 337, 531-533, (1989).

Wright, G. A. and Loh, E. D., *Nature* 324, 127-128, (1986).

# THE ASSOCIATION BETWEEN PULSARS AND SUPERNOVA REMNANTS

G. Srinivasan
Raman Research Institute
Bangalore - 560080, India

The true nature of the association between pulsars and SNRs has remained an intriguing and poorly understood problem even after all these years of research on them.

The idea that there should be an association at all dates back to Baade and Zwicky (1934) who advanced the view that supernovae represent the transition from ordinary stars into neutron stars. Following their suggestion, all theories of supernovae for many years afterwards involved a neutron star as an essential member of the cast and one capable in prinicple, of releasing upto $10^{53}$ erg of energy at the time of its formation. But the details of how a part of this energy could be coupled to the infalling envelope of the star to arrest its collapse, and to accelerate it outwards has remained a major problem. It is only in recent years that plausible scenarios are being advanced. Regardless of this technical difficulty, it is now generally accepted that the formation of a neutron star is indeed origin of Type II supernovae.

On the other hand, according to current wisdom, the mechanism of Type I supernovae is very different; the star completely disrupts and no stellar remnant is left behind.

Although no supernova has been sighted in the Galaxy since the time of Kepler, the historical observations suggest a frequency of about one in 30 years (Clark and Stephenson 1977). At any rate, supernovae do leave behind supernova remnants (SNRs), and one would expect neutron stars to be associated with atleast the remnants of Type II supernovae. Soon after pulsars were discovered, and when it was suspected that they must be spinning neutron stars, Woltjer (1968) predicted that they should be found in SNRs, and in particular that there should be one in the CRAB NEBULA. In the case of the Crab nebula, however, there was the famous prediction by Pacini (1967) made before the discovery of pulsars! It is worth recalling that this prediction was motivated by the need to have a central engine to explain the continuing activity of the Crab nebula. The discovery of a pulsar in the Vela SNR within months (Large et al. 1966) and one in the Crab soon after (Staelin and Reifenstein 1968) seemed to have answered several questions all at once.

The extraordinary thing is that the total number of associations remained at these two for more than a decade although the number of pulsars and SNRs increased in number. Just in the last few years two new associations have been found - pulsars have been found in MSH 15-52

*W. Kundt (ed.), Neutron Stars and Their Birth Events, 91–106.*
© *1990 Kluwer Academic Publishers.*

(Seward et al. 1982) and CTB 80 (Kulkarni et al. 1988). Even if neutron stars are associated only with Type II supernova events, this poor association is striking. This was reconciled by invoking various selection effects like pulsar beaming, dispersion, interstellar scattering, low flux etc. Such reasoning was implicit already in the paper by Woltjer (1968). A careful analysis suggested that, given these selection effects, the poor association is not inconsistent with a pulsar being associated with most SNRs.

But such a reasoning is not entirely satisfactory. If there are active pulsars in SNRs then one would expect them to energize a centrally-condensed synchrotron nebula (figure 1) which should be seen regardless of the viewing geometry (Radharkishnan and Srinivasan 1980). And what is more, the luminosity of such a pulsar produced nebula will be much larger than the pulsed luminosity, so that the chances of detecting it are much greater than detecting the central pulsar itself. It is worth recalling that there is such a nebulosity (radio or x-ray, or both) surrounding the pulsar in every known case of a pulsar associated with an SNR (Crab, Vela X, MSH 15-52, CTB 80 and 0540-69.3 in the LMC); indeed, in many cases, it is the presence of such a centrally filled nebulosity that made people look hard for the energizing pulsar!

Thus, a reliable signature of a pulsar - SNR association appears to be related to the morphology of the remnant. If there is no pulsar associated with an SNR, then one expects it to have a well defined shell with hollow interior. But is the converse true? The overwhelming majority of SNRs, although they may not be highly circularly symmetric, show pronounced limb brightening with hardly any central emission (figure 2). In principle, this may be reconciled in several different ways:

1. There are no neutron stars in them. In this case, since most supernovae leave behind supernova remnants, one will be forced to manufacture the majority of neutron stars in less spectacular events than supernovae.

2. For some reason the neutron stars in them do not function as pulsars (Radhakrishnan and Srinivasan 1980). A modest version of this would have the neutron stars "turn on" as pulsars after the SNR has faded away or lost its identity. Although this sounds attractive there are difficulties with this which will be elaborated later on.

3. The absence of detectable central emission in the majority of SNRs can also be reconciled if the PULSARS in them were born with rather long initial periods (Radhakrishnan and Srinivasan 1983). The nebulae produced by such pulsars will have very low surface brightness. There is mounting evidence in favour of this, and hence it is this alternative that will be closely examined in what follows.


## IS THE CRAB PULSAR A PROTOTYPE?

Even before pulsars were discovered it was conjectured that they would be spinning very rapidly at birth, and would be strongly magnetized. With the discovery of a pulsar with just these characteristics in the young Crab nebula, these expectations based on

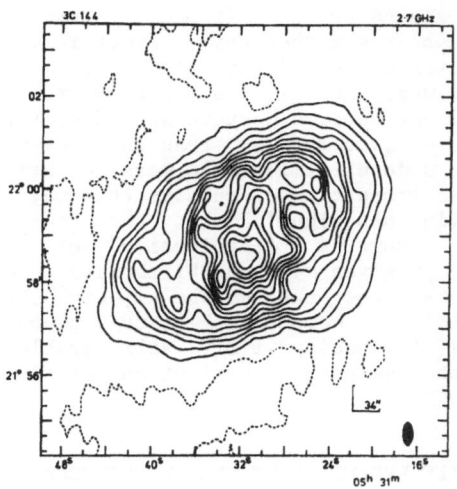

**Fig. 1.** *The CRAB NEBULA - an example of a synchrotron nebula produced by a pulsar (from Wilson, 1972).*

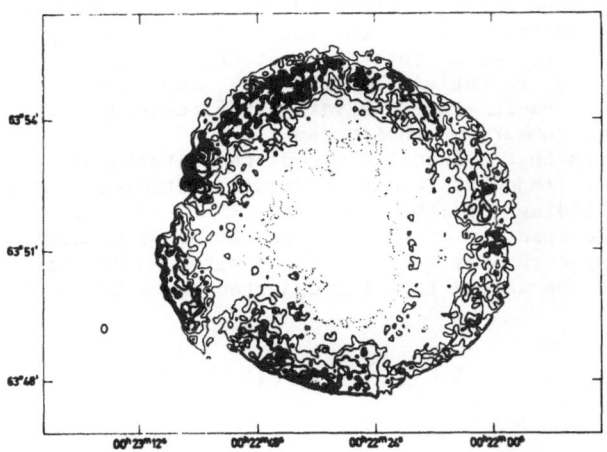

**Fig. 2.** *The remnant of TYCHO's supernova - an example of a shell SNR with a hollow interior (from Duin and Strom, 1975).*

general assumptions seemed to be confirmed. However, the supernova of 1054 AD must have been a rather unique event in many respects. It is not difficult to convince oneself that the Crab pulsar cannot be a prototype, for if it were then given a pulsar birth rate of one in 50 years (Narayan 1987) one would expect to see 40 or 50 pulsar-produced nebulae as luminous, or even more so, than the Crab nebula!

What follows is a descriptiion of a more quantitative attempt to get a handle on the properties of pulsars that may be present in SNRs. Since it is clear (in hindsight!) that the Crab pulsar is not a prototype, let us assume a "distribution" of initial periods and magnetic fields. Then, given a pulsar birth rate, one can estimate the number of pulsar-produced nebulae one expects to see above a certain chosen luminosity limit. A comparison of this number with the observed number above this luminosity limit will enable us to test the reasonableness of the assumed distribution of initial periods and magnetic fields.

## LUMINOSITY EVOLUTION OF PULSAR PRODUCED NEBULAE

In order to do this, one must have a theoretical framework to estimate the luminosity evolution of such nebulae. It is now well established that the relativistic particles and the magnetic field responsible for the synchrotron radiation from the Crab nebula are supplied by the central pulsar. The luminosity of the nebula at any instant is determined by the built up field in the cavity and the spectrum of the relativistic electrons. Since the walls of the cavity are expanding, one will have to take into account the adiabatic losses. For high energy particles, radiation losses are also important. In their classical work Pacini and Salvati (1973) developed such a formalism. Here is a brief summary of their result.

One assumes that a fraction $\epsilon_p$ of the rotational energy lost by the pulsar is injected as relativistic particles, and a fraction $\epsilon_m$ goes into building up the magnetic field in the nebula. The electromagnetic spectrum of the Crab pulsar and the nebula suggests that the injected spectrum may be a power law with one or more breaks. For simplicity let us assume that the injected spectrum is characterized by a single slope:

$$J(E,t) = K(t) E^{-\gamma}$$

The normlisation factor $K(t)$ is obtained from the condition that

$$\int_0^{E_m} E\, J(E,t)\, dE = \epsilon_p L_{psr}(t)$$

where $L_{PSR}(t)$ is the "mechanical luminosity" of the pulsar. According to the "standard model" the pulsar luminosity will evolve with time as follows:

$$L_{PSR}(t) = \frac{L_0}{(1 + t/\tau_0)^2}$$

where $L_0$ is the INITIAL luminosity and $\tau_0$ is the initial slow-down time scale. The evolution of the magnetic energy is obtained from the simple thermodynamic relation

$$\frac{dW_B}{dt} + \frac{1}{3} W_B \frac{1}{V} \frac{dV}{dt} = \epsilon_m L_{PSR}(t)$$

where $V(t)$ is the volume of the cavity and $W_B(t) = V \cdot B^2/8\pi$ is the magnetic energy.

To obtain the spectrum of the accumulated particles one has to allow for adiabatic and radiation losses of the particles. Formally this can be written as follows:

$$\frac{1}{ER} - \frac{1}{E_i R_i} = C_1 \int_{t_i}^{t} \frac{B^2(t')}{R(t')} dt' \quad .$$

In this expression $E_i$ is the energy of the particle injected at time $t_i$. The left hand side in the above expression represents adiabatic losses and the right hand side radiation losses ($C_1$ is the well known constant that appears in synchrotron radiation theory). Thus, given the injected spectrum and the law governing the evolution of the particle energy, one can calculate the spectrum of the particles present at any given time (freshly injected plus the accumulated particles). Knowing this and the magnetic field in the cavity one can calculate the spectral luminosity using the usual synchrotron formula.

Pacini and Salvati (1973) showed that most of the magnetic field gets built up during the initial characteristic slow-down time, $\tau_0$, of the pulsar (this is $\sim$ 300 years for the Crab pulsar). During this phase, the luminosity of the nebula will initially increase and then slowly decrease with time. After $t \sim \tau_0$, the pulsar luminosity starts to decrease, and consequently the nebular luminosity begins to decline rather rapidly.

The radio spectral luminosity of the nebula depends strongly on the initial period of the pulsar, its magnetic field and the expansion velocity. This may be seen in the following expression for the radio luminosity at times $t > \tau_0$:

$$L_{\nu}(t) \propto P_0^{2(\gamma-2)} \; B_*^{(3-5\gamma)/2} \; \nu^{-3(1+\gamma)/4} \; t^{-2\gamma} \; \nu^{(1-\gamma)/2} \, .$$

Here $P_0$ is the initial period, $B_*$ the magnetic field, v the expansion velocity of the nebular boundary, and $\gamma$ is the slope of the injected spectrum ($\gamma \sim 1.6$ for the Crab nebula).

Before we attempt to estimate the expected number of pulsar produced nebulae (or PLERIONS as they have come to be known) we must adopt a model for the expansion of the nebular boundary, and also fix a luminosity limit.

## PULSAR DRIVEN NEBULAE

One possibility is to model all such nebulae after the Crab nebula. It is well established that the kinetic energy of expansion of the filamentary shell in the Crab, as well as the acceleration experienced by it in the past, can be understood in terms of the energy being derived from the stored rotational energy of the pulsar. By analogy one can assume the same to be true for all plerions, namely, that the pressure of the relativistic wind from the pulsar accelerated the nebular boundary. One expects the pulsar to have a significant dynamical effect only during the first characteristic slow-down time during which it dumps half its stored rotational energy. The final velocity attained by the ejecta may be estimated from the simple relation

$$\tfrac{1}{2} M_{ej} \vartheta^2 \sim \tfrac{1}{2} E_R^o \, , \qquad E_R^o = \tfrac{1}{2} I \Omega_o^2$$

If one is interested in following the evolution of the nebula during the initial phase then one will have to take into account the accelerated expansion during that phase. But in the later free expansion phase, the velocity will scale as $v \propto 1/P_0$, where $P_0$ is the initial period of the pulsar. In the calculations presented here (Srinivasan et al. 1984; Bhattachrya 1987) it has been assumed for the sake of simplicity that in all cases the mass ejected is roughly same as in the Crab nebula.

## THE LUMINOSITY LIMIT

Let us recall what one is attempting to do. Under well defined assumptions about the birth rate of pulsars, and their initial periods and magnetic fields, one would like to estimate the number of plerions above a certain luminosity. Since one would eventually be comparing this with the observed sample, it makes sense to choose the luminosity limit such that one is unlikely to have missed many such remnants no matter where they are located in the galaxy.

The flux from the Crab nebula (at 1 GHZ) will be 10 Jy if placed at a distance of 20 kpc. It is reasonable to assume that not many sources with flux greater than 1 Jy would have been missed in surveys at frequencies around 1 GHz (Weiler, 1983). In view of this, one may be safe in taking 0.1 $L_{CRAB}$ as the luminosity limit; it should be born in mind that this limit corresponds to a flux of 1 Jy at the extreme distance of 20 kpc. This assumption will, of course, have to be modified if future surveys discover a large number of plerions above this luminosity limit.

## LUMINOSITY EVOLUTION

SENSITIVITY TO INITIAL PERIOD AND MAGNETIC FIELD: Figure 3 illustrates the importance of the initial characteristics of the pulsar in determining the luminosity. What is shown are contours of CONSTANT LUMINOSITY IN THE B-P plane. In the figure on the left, the different contours correspond to different luminosities but the SAME AGE, whereas in the figure on the right different contours correspond to different ages but the SAME LUMINOSITY. It is quite clear that it is not meaningful to assert that nebulae more luminous than a given one are necessarily younger. The case of MSH 15-52 is a good illustration of this. Even though the central pulsar is roughly of the same age as the Crab pulsar, there is hardly any central radio emission surrounding it!

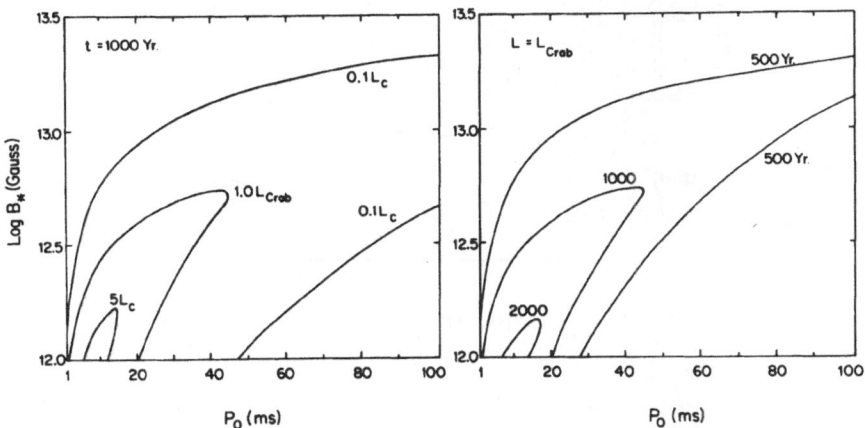

**Fig. 3.** Contours of constant luminosity for pulsar-driven SNRs are shown in $B_*$-$P_0$ plane; here $B_*$ is the surface magnetic field and $P_0$ the initial period of the pulsars. All pulsars born on a given contour will have the same luminosity at a specified age. (a) The three contours correspond to three different luminosities (measured in the units of the present luminosity of Crab) and an age of 1000 yr. (b) The contours correspond to different ages, but the same luminosity, viz., the present luminosity of Crab.

98

Let us now turn to an estimate of the expected number of plerions with luminosities greater than our chosen luminosity limit.

Figure 4 is a plot of several contours all corresponding to 0.1 $L_{CRAB}$. The labels represent the duration for which the nebulae will be more luminous than the specified value, or in other words, their lifetime. If $\tau$ is the mean interval between the birth of pulsars, then the number of nebulae that one expects to see above the threshold luminosity is given by

$$N(>) = \frac{1}{\tau} \int_0^\infty t\, f(t)\, dt$$

here f(t) is the probability that the nebula will have a lifetime between t and t+dt.

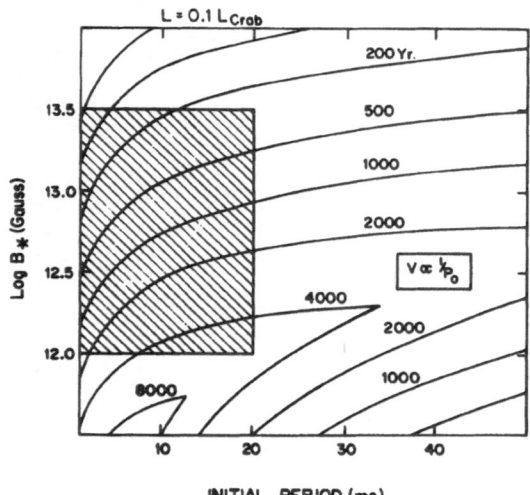

**Fig. 4.** *Pulsar-driven plerions. The contours of different ages for a luminosity of 0.1 $L_{Crab}$. In estimating the expected number of plerions with luminosities greater than the above-mentioned value we have assumed that pulsars are born anywhere inside the shaded regions.*

## THE DISTRIBUTION OF INITIAL PERIODS AND FIELDS

Before proceeding further let us choose a "reasonable" range of values for the initial periods and fields of pulsars. Although the distribution of the derived fields of pulsars extends from $10^{11}$ to $10^{13.5}$ gauss, there are strong reasons to think that the distribution of fields at birth may be narrower. At any rate, the majority of pulsars are observed to have fields greater than $10^{12}$ G. In view of this let us assume that the majority of pulsars have fields between $10^{12}$ and $10^{13.5}$ G at birth, with equal probability in equal logarithmic intervals. Regarding the initial periods, since it is our intention to question what is commonly believed, let us assume that pulsars are born spinning very rapidly with periods in the range 1 to 20 milliseconds.

The probability $f(t)$ that a nebula will have a lifetime between $t$ and $t+dt$ is related to the probability that the lifetime will be GREATER than $t$, namely $P(>t)$, through the relation

$$P(>t) = \int_t^\infty f(t')\,dt', \qquad f(t) = -\frac{dP(>t)}{dt}$$

Since we have restricted the region in 'phase space' where pulsars may be born, $P(>t)$ is simply given by

$$P(>t) = \frac{a(t)}{A}$$

where $a(t)$ is the area enclosed by the contour corresponding to an age $t$ and WITHIN the hatched area A.

### RESULTS

The procedure outlined above can be used to estimate the number of plerions. If pulsars are born in the hatched region in the B –P plane, then a birthrate of 1 in 40 years implies that there should be ~ 35 nebulae whose radio luminosities (at 1 GHZ) should be greater than 1/10th that of the Crab nebula, or in other words, whose fluxes should be greater than 1 Jy EVEN IF PLACED AT 20 kpc.

However, only four of the known sample of plerions, which number a dozen or so, have luminosities above this value. This glaring discrepancy between the expected and the observed number may, in principle, be reconciled in several ways. For example, it is possible that the assumed pulsar birthrate is too high. While it is quite likely that the estimates in the literature are in error, it seems unlikely that it is in error by a large factor ~ 10. Similarly, it is quite likely that the sample of plerions is incomplete. Although one can say with confidence that many new plerions will be found in the surveys that are planned, it is very unlikely that many of them will be as luminous as $0.1\ L_{CRAB}$, for if they were they would have been detected already.

There is one other possible resolution of this dilemma. We have assumed that all plerions are 'pulsar-driven' like the Crab nebula; this assumption may be wrong. One of the most remarkable things about the Crab nebula is its very low expansion velocity $\sim$ 1700 km/s. It may be that in most cases the pulsar does not have any dynamical effect on the ejecta. This leads us to explore the standard Type II supernova scenario in which the ejecta are accelerated to a high velocity $\sim 10^4$ km/s by a shock wave driven by the core-bounce during the formation of the neutron star. In this picture, unless the central neutron star is a millisecond pulsar AND has a strong field, it will not have any dynamical effect on the expanding shell. This is an attractive scenario to pursue for the following reason. Adiabatic losses will be more severe in a rapidly expanding cavity; the energy density of the particles and the magnetic field will also be smaller. As a consequence, one expects much weaker plerions around pulsars in a rapidly expanding cavity.

Unfortunately this doesn't solve the problem. A repetition of the procedure followed before, but now with an enpansion velocity of 10,000 km/s, suggests that there should be $\sim$ 16 standard 'limb brightened' SNRs with central plerionic components whose luminosities are greater than 0.1 $L_{CRAB}$. Unfortunately there is not a single such remnant known! Although a handful of SNRs with hybrid morphology are known (figure 5 is an example of a shell SNR with flat spectrum central emission) there is none with such pronounced plerion. The situation gets worse, of course, if one takes into account the expected deceleration of the ejecta as it sweeps up interstellar matter.

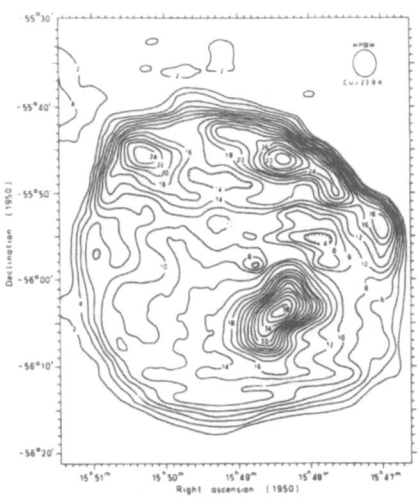

**Fig. 5.** *G326.3-1.8, the prototype of a hybrid SNR (from Clark et al. 1975)*

So far we have considered two mutually exclusive scenarios, one in which the pulsar decides the dynamics of the expanding shell, and the other in which it is a passive observer. Realistic situation may be in between. It has been suggested (Chevalier 1976, Weaver and Woosley 1980) that in some Type II supernovae although some of the ejected mass might be initially moving with very high velocity $\sim 10^4$ km/s, there might be (core) material moving much more slowly with velocities $\sim 300$ km/s. In such a situation, the pulsar wind will sweep up the core material into a thin shell. The spatial extent of the pulsar-produced nebula will then be determined by the inner boundary. Thus one is back to the 'pulsar driven' scenario.

Thus, in all the scenarios, faced with the gross discrepancy between the expected and the observed number of bright plerions, one is forced to question the assumed distribution of initial periods and fields. Clearly, there are three possibilities.

(1) The majority of pulsars are born with very high fields ($> 10^{13}$ G). Such pulsars will slow down very rapidly, so that by the time the SNR builds up, the pulsars will be too weak to energize a detectable nebula. While this sounds attractive there are difficulties. Although the nebulae produced by such pulsars will have a short lifetime, they will be extremely luminous in the early phase. Such ultra-luminous nebulae are not known. Secondly, if the majority are born with very high fields one will have difficulty explaining the fact the majority of pulsars are observed to have fields in the range $10^{12}-10^{13}$ G. Even the advocates of field decay would not like to entertain very short decay timescales!

2) The second possibility is that the majority of pulsars are born with fields much smaller than $10^{12}$ G. The difficulty with this suggestion is that with the exception of the recently discovered pulsar in CTB 80, no pulsar has been found with field less than that of the Crab or Vela pulsars and with period less than 150 ms (we exclude the "millisecond pulsars" from this discussion for there is mounting evidence that they are very old neutron stars which have been spun-up in binary systems. See, for example, Srinivasan and Bhattacharya 1987). In our opinion this cannot be a selection effect for two reasons. Firstly, assuming that they are born spinning rapidly but with low fields, their periods will lengthen much more slowly than their counterparts with high fields, thus enhancing the chance of detection. The paucity of short-period, low-field pulsars is unlikely to be due to luminosity selection effects either, since we see many pulsars with low fields AND long periods. In our opinion, it is the paucity of low-field short-period pulsars that provides the strongest evidence for field decay - pulsars with $B < 10^{12}$ G are presumably several millions of years old and their fields have decayed.

3) This leaves us with the last alternative which is that the majority of pulsars are born with periods well in excess of 20 ms. It turns out that only if the initial periods of the majority of pulsars is greater that 100 ms can one reconcile the observed and expected number of plerions. This conclusion appears to be unavoidable in all the three scenarios we have discussed.

Interestingly, this conclusion was arrived at by Vivekanand and Narayan (1981) by a very different reasoning. While attempting to estimate a reliable b.ithrate for pulsars, they came to the conclusion that the distribution of the pulsars in the P - Ṗ plane (or P - B plane) can only be understood if the majority of pulsars are INJECTED with periods in excess of 0.5 seconds. A similar conclusion has been arrived at by Chevalier and Emmering (1986) and Narayan (1987).

It is worth mentioning at this stage that recently there have been several surveys specifically designed to look for short period pulsars. Although several new pulsars with periods 100 ms have been found, only one with a shorter period has been found (the CTB 80 pulsar). In fact, in one of the papers describing the results of one such survey, Taylor et al. conclude that "the Galaxy does not contain a significant population of pulsars with periods less than ∿ 100 ms".

## X-RAYS FROM PLERIONS

So far we have confined our attention to radio properties of pulsar – produced nebulae. The main reason for this is that the sample of radio supernova remnants is much greater than, say, X-ray emitting SNRs. Also, the radio luminosity is much more sensitive to the initial luminosity of the pulsar than the X-ray luminosity. The reason for this easy to see. Since the electrons responsible for radio emission are of very low energy, their radiative lifetimes are very large( ≲ million years in the Crab nebula); therefore only adiabatic losses are important. As a consequence, the population of low energy electrons in, say, the Crab nebula consist not only of the recently injected electrons but all the relic electrons as well – with the latter contributing signifcantly to the observed radio luminosity. On the contrary, in the case of X-ray emission, for example, since radiative lifetimes are very short it is only the 'freshly injected particles' that matter. Thus the X-ray luminosity of a plerion more directly reflects the present energy loss rate of the central pulsar.

There were attempts to detect X-ray emission from the centres of several known supernova remnants with the Einstein observatory (Helfand and Becker, 1984). But none of the 36 remnants observed showed any central feature down to a luminosity limit $\sim 10^{34}$ erg/s. This, as was argued above, sets limits on the energy loss rates of the central pulsars in those SNRs. Thus if there are pulsars in these remnants, and if they have fields $> 10^{12}$ G, then they must have long periods.

## AN ALTERNATIVE APPROACH

Throughout our discussion we have adopted the point of view that there are functioning pulsars in all SNRs produced by Type II supernovae. As was mentioned in the beginning, the poor association between pulsars and supernova remnants can also be reconciled if the neutron stars in most remnants do not, for some reason or other, function as pulsars.

Following the original suggestion by Woltjer (1964), it is generally accepted that neutron stars are endowed with strong fields at birth, and will therefore function as pulsars right away. However, it has recently been suggested that the observed fields of pulsars may be built up AFTER their birth over a timescale $\sim 10^5$ years (Woodward, 1978, 1984; Blandford et al. 1983). In the mean time the supernova remnant would have disappeared, and the problem of the poor association neatly solved!

But there are several difficulties with this seemingly attractive suggestion. Since there will be a seminar devoted entirely to this question, I shall not dwell on them here (Bhattacharya, this volume). For the sake of completeness of the present discussion I might mention that a careful examination of this suggestion shows that the above conclusion, namely that the majority of pulsars must be born with relativelylong periods, survives REGARDLESS of whether their magnetic fields are amplified fossil fields or built up later. This follows from the paucity of low—field short-period pulsars like the one in CTB 80. If future surveys find a large number of such pulsars then not only will one have to abandon field decay, but one will have to seriously entertain field growth!

## SUMMARY AND DISCUSSION

Our main conclusion is the following. There is general agreement that Type II supernovae are the result of the formation of neutron stars. Although there is as yet no consensus on this point, the progenitors of Type Ib supernovae may also be massive stars. If so, the possibility of a neutron star remnant is not ruled out. The recent analysis of van den Bergh et al. (1987) suggests that the relative frequency of supernovae of Type Ia,Ib and II may be in the ratio 3:4:11. This would make Type IIs more frequent than was thought previously. Thus, one would expect neutron stars to be associated with the majority of supernova remnants. If these function as pulsars then one would be able to detect their presence through the synchrotron nebulae they would produce regardless of whether the pulsar is beaming towards us or not. Moreover, if the Crab pulsar is a prototype of young pulsars, as is commonly believed, then these pulsar-produced nebulae would be very bright, resulting in a hybrid or composite morphology for the remnant - a steep-spectrum shell with pronounced flat- spectrum central emission. The paucity of such objects strongly suggests that the pulsar in the Crab nebula is not a prototype of pulsars. If pulsars are, indeed, associated with most SNRs then the fact that most of them have hollow interiors suggests that most pulsars must be born with long initial periods. This conclusion is in agreement with a similar conlcusion arrived at through an analysis of the period distribution of the observed pulsars, as well as with the results of recent surveys designed to look for short—period pulsars.

If one takes this conclusion seriously then it may be worth speculating on some possible inferences one can draw. The expectation that newly formed neutron stars should be spinning close to their

limiting period was based on the assumption that the presupernova core
had large angular momentum. But this is not at all clear. Fricke and
Kippenhahn (1972) have argued, for example, that if the core is coupled
to the envelope upto the helium burning phase, and thereafter evolves
conserving angular momentum, the resulting neutron star will have a
small angular velocity. However, it is not clear if differential
rotation between the core and the envelope would be damped out. If the
main torque on the core is its magnetic coupling with the exterior, and
if significant magnetic field gets built up only during the carbon
burning phase (Ruderman and Sutherland 1973), then the presupernova core
would have large angular momentum. The reason being that the carbon
burning and subsequent phases do not last long enough for the torque to
brake the core. In this case, the resulting neutron star will be
spinning rapidly. If the binding energy released is not able to produce
a "prompt" explosion, then the rotation of the neutron star can be
slowed down by magnetic torque. Because of the high field and rapid
rotation, the timescale for extracting angular momentum and rotational
energy can be sufficiently short to have interesting consequences.

Let us digress a bit and examine this a little more closely.
Although the magnetic field may be a silent spectator during the
collapse, it may assume a major importance after the collapse. The
importance of winding up the magnetic field was first pointed out by
Kardashev (1965). This has been amplified by Bisnovatyi-Kogan (1971)
and Kundt (1976). Due to the large electrical conductivity of the
stellar matter the magnetic field of the neutron star will be anchored
to the matter surrounding it. Due to the large differential rotation
between the neutron star and the enevelope, a large toroidal field will
be built up. The energy in this field comes at the expense of the
rotational energy of the neutron star. The strength of the toroidal
field generated after $n$ cycles of differential rotation is $B_t \sim B_o \cdot n$,
where $B_o$ is the initial field. Thus, the magnetic energy grows as:

$$B_t^2 R^3 \sim B_o^2 n^2 R^3 \sim B_o^2 \Omega^2 t^2 R^3 \sim E_{mag}^o \Omega^2 t^2$$

where $\Omega$ is the angular speed of differential rotation. The energy in
this wound-up field will become comparable to the rotational energy in a
timescale

$$\tau \sim \left( \frac{E_{rot}}{E_{mag}^o} \right)^{1/2} \Omega^{-1}$$

If the neutron star is born spinning maximally (period millisecond)
then its rotational energy is comparable to the gravitational BINDING
ENERGY and the above timescale can be expressed as:

$$\tau \sim 10^2 B_{12}^{-1} \; sec$$

where $B_{12}$ is the strength of the magnetic field in units of $10^{12}$ G. Thus, in a timescale 3 to 100 seconds, the rotational energy of the newborn neutron star can be converted to magnetic energy. Bisnovatyi-Kogan (1971), Bisnovatyi-Kogan et al. (1976), Ardelyan et al. (1979) and Kundt (1976) have suggested that this conversion may be responsible for supernova explosions.

This interesting suggestion is worth taking seriously particularly since attempts to produce supernova explosions utilizing only the gravitational binding energy released have not been very successful. In the standard scenario, as the collapsing core reaches nuclear density the collapse is arrested, but the kinetic energy of the infall allows a certain overcompression of the core beyond its equilibrium size. Finally a "bounce" from this overcompressed state sends a shock wave through the surrounding matter. It has been the longstanding hope that this shock wave will produce the explosion. But, alas, extensive computatons show that the shock stalls. And hopes of reviving this shock by neutrino heating (Bethe and Wilson 1985) are fading. In view of this one should perhaps take seriously the role played by rotation and magnetic field. It is conceivable that the bounce shock, aided by the toroidal magnetic field, may do the trick.

As a distinguished colleague of mine said, Baade and Zwicky may be right after all, but with an added twist!!

## REFERENCES

Ardelyan, N.V., Bisnovatyi-Kogan, G.S., Popov, Yu.P., 1979, Sov. Astr., **23**,705.

Baade, W., Zwicky, F., 1934, Phys. Rev., 45, 138.

Bethe, H.A., Wilson, J.R., 1985, Astrophys. J., **295**, 14.

Bhattacharya, D., 1987, Ph. D. Thesis, Indian Institute of Science, Bangalore.

Bisnovatyi-Kogan, G.S., 1971, Sov. Astr., **14**,652.

Bisnovatyi-Kogan, G.S., Popov, Yu.P., Samochin, A.A., 1976, Astrophys. Sp. Sci., **41**, 287.

Blandford, R.D., Applegate, J.H., Hernquist, L., 1983, Mon. Not. R. astr. Soc., **204**, 1025.

Chevalier, R.A., 1976, Astrophys. J., **207**, 872.

Chevalier, R.A., Emmering, R.T., 1986, Astrophys. J., **304**, 140.

Clark, D.H., Green, A.J., Caswell, J.L., 1975, Aust. J. Phys. Astrophys. Suppl., No. 37, 75.

Clark, D.H., Stephenson, F.R., 1977b, Historical Supernovae, Pergamon
     Press, Oxford.

Duin, R.M., Strom, R.G., 1975, Astr. Astrophys., **39**, 33.

Fricke, K.J., Kippenhahn, R., 1972, Ann. Rev. Astr. Astrophys., **10**, 45.

Kardashev, N.S., 1965, Sov. Astr., **8**, 643.

Kulkarni, S.R., Clifton, T.C., Backer, D.C., Foster, R.S., Fruchter,
     A.S., Taylor, J.H., 1988, Nature, **331**, 50.

Kundt, W., 1976, Nature, **261**, 673.

Large, M.I., Vaughan, A.E., Mills, B.Y., 1968, Nature, **307**, 215.

Pacini, F., 1967, Nature, **216**, 567.

Pacini, F., Salvati, M., 1973, Astrophys. J., **186**, 249.

Radhakrishnan, V., Srinivasan, G., 1980, J. Astrophys. Astr., 1, 25.

Radhakrishnan, V., Srinivasan, G., 1983, in: IAU Symp. No. 101:
     Supernova Remnants and their X-ray Emission, Eds. J. Danziger and
     P. Gorenstein, D. Reidel, Dordrecht, p. 487.

Ruderman, M.A., Sutherland, P.G., 1973, Nature Phys. Sci., **246**, 93.

Seward, F.D., Harnden, F.R., 1982, Astrophys. J., **256**, L45.

Srinivasan, G., Bhattacharya, D., Dwarakanath, K.S., 1984, J. Astrophys.
     Astr., **5**, 403.

Staelin, D.H., Reifenstein, E.C., 1968, IAU Circ. No. 2110.

Van den Bergh, S., McClure, R.D., 1987, Astrophys. J., **323, 44.**

**Vivekanand, M., Narayan, R., 1981, J. Astrophys. Astr., 2, 315.**

**Weaver, T.A., Woosley, S.E., 1980, Ann. N.Y. Acad. Sci., 336, 335.**

Wilson, A.S., 1972, Mon. Not. R. astr. Soc., **157**, 229.

Woltjer, L., 1964, Astrophys. J., **140**, 1309.

Woltjer, L., 1968, Astrophys. J., **152**, L179.

Woodward, J.F., 1978, Astrophys. J., **225**, 574.

Woodward, J.F., 1984, Astrophys. J., **279**, 803.

# MILLISECOND PULSARS: A NEW POPULATION OF GAMMA RAY SOURCES ?

G. Srinivasan
Raman Research Institute
Bangalore - 560080, India

## 1. INTRODUCTION

Pulsars have a pride of place in gamma ray astronomy. This is as it should be! For, after all, the Crab and Vela pulsars are the best studied gamma ray sources. These observations, in turn, have sparked renewed activity in trying to understand how pulsars work. The early attempts concentrated on the generation of the coherent radio radiation through which they were first detected. The energy in this radiation is, however, an extremely small fraction of the rotational energy loss of the pulsar most of which is believed to be carried away by an ultrarelativistic electron-positron wind with a frozen-in magnetic field. But there is no consensus as yet regarding HOW these particles are accelerated, or WHERE the accelerator is located. This is where the importance of the detailed gamma ray observations of pulsars lies. There are reasons to think that the electron-positron pair production and their acceleration, and the observed gamma ray emission may be from the same region, or nearby regions of the magnetosphere. Thus, in an indirect way, the pulsed gamma rays are telling us something about the location of the particle accelerator.

Triggered by the need to explain the detailed observations of the Crab and Vela pulsars, a variety of models have now been developed /1-3/ and (unfortunately!) most of them can be made to fit the gross features of the observations. But there are important differences in the predictions like, for example, the shape of the spectrum, the high energy cut off, the dependence on the magnetic field and the period of the pulsar, the gamma ray efficiency i.e. the ratio of the gamma ray luminosity to the mechanical luminosity etc.. Unfortunately, with only two pulsars observed in the gamma ray window it has not been possible to constrain these theoretical models in a significant way. From the point of view of understanding the electrodynamics of pulsars, it would be highly desirable to observe some more pulsars with different periods and magnetic fields, and this may be possible in the future.

There have been suggestions in the literature that some of the unidentified COS B gamma ray sources may, indeed, be pulsars /4/. There have also been estimates of the pulsar contribution to the gamma ray luminosity of the Galaxy /5/. In such discussions attention is usually

107

*W. Kundt (ed.), Neutron Stars and Their Birth Events, 107–119.*
*© 1990 Kluwer Academic Publishers.*

focussed on "young pulsars", with the Crab pulsar presumed to be the prototype. In this talk, I wish to concentrate on a different population of very much older pulsars with very low magnetic fields, but whose energy loss rate is comparable to that of the Vela pulsar and, therefore, of great interest in the present context. I am referring, of course, to the MILLISECOND PULSARS. Soon after the discovery of the first millisecond pulsar it was pointed out by several people, in particular by Usov /6/, that it might be a strong emitter at very high energies, $10^{11}$ eV. The basic point is that because these old neutron stars have very low magnetic fields, high-energy photons will be able to get out even if they are produced near the polar cap. Indeed, there have been reports of possible detection of high-energy gamma rays from the 1.5 ms and the 6 ms pulsars /7,8/. Although these observations are somewhat marginal at present, there is every reason to think that future observations of these millisecond pulsars will be able to settle whether or not they are gamma ray emitters. If they are, then their luminosities and the spectrum of radiation will greatly improve our understanding of pulsars.

It is to a discussion of the origin of these millisecond pulsars that I wish to devote the rest of this talk. Through this discussion I hope to convince you that the handful of millisecond pulsars discovereed so far may be just the 'tip of the iceberg', and that their total number in the Galaxy may be as large as 10,000.

## 2. THE EVOLUTION OF ORDINARY PULSARS

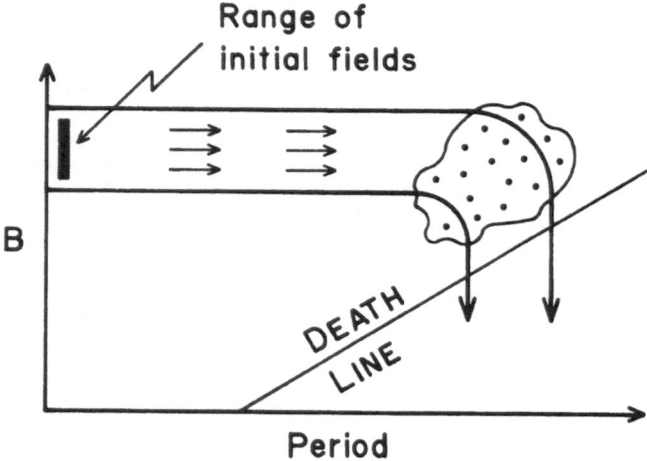

Fig. 1. The evolution of ordinary pulsars in the field-period plane. The 'island' with dots indicates the region where most of the observed pulsars lie.

Let us first briefly recall the evolutionary scenario for ordinary pulsars. A convenient representation to discuss this is the magnetic field – period diagram (fig. 1). As the pulsar ages its period will lengthen and, as long as the field remains constant, it will move horizontally in this diagram. However, there is strong observational evidence which suggests that the magnetic fields of neutron stars decay over a time scale of a few million years. When this becomes important the trajectory of the pulsar will swing downwards and eventually become vertical.

And for how long will a neutron star function as a pulsar? Again, observations seem to suggest that the lifetime of a neutron star as a pulsar is $\sim$ ten million years. Thus as the period lengthens and the field decays the neutron star ceases to function as a pulsar. In some theoretical models this happens rather suddenly, and the combination of B and P when this happens is shown in the diagram as the DEATH LINE. The region below this death line is the GRAVEYARD of pulsars. This is what one believes to be the life history of a pulsar. It will be seen from fig. 2 which shows the distribution of the observed pulsars that there is indeed a well defined boundary below which no pulsars are found /9/.

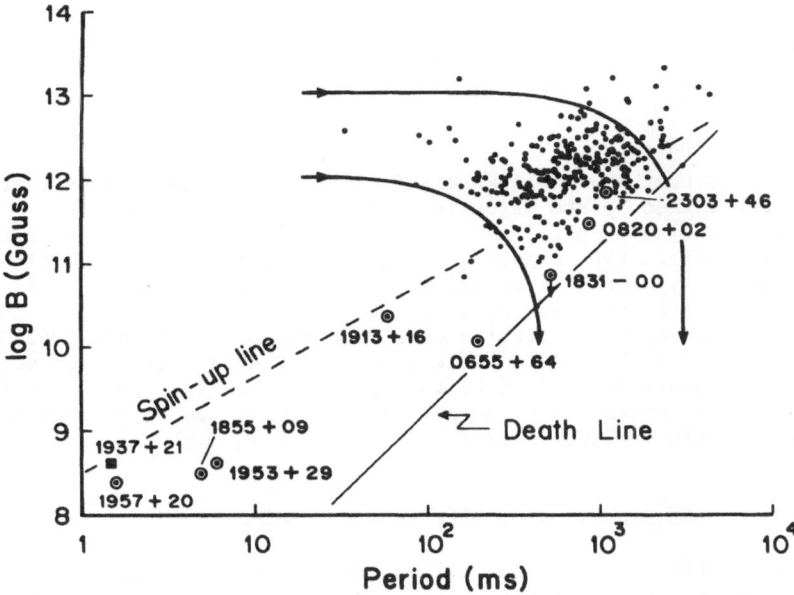

Fig. 2. The periods and derived magnetic fields of about 300 pulsars are shown. Dots with open circles are pulsars in binaries.

## 3. RECYCLED PULSARS

Under certain special circumstances a dead pulsar may be resurrected from the graveyard and given a second lease of life. This will happen if the neutron star in question is a member of a close binary, and accretes mass from its companion. The basic reason for this is that a neutron star can be spun-up during an accretion phase since the infalling matter has specific angular momentum. Although the last word has not been said on the subject, there is mounting evidence that the population of millisecond pulsars are such recycled pulsars /10,11/. What is even more interesting is that although they died their first death in a few million years, in their reincarnation they may live forever! To explain this let us outline the recycling scenario. As is well known, several of the accreting neutron stars are observed to be spinning up. The question is : can one say something about the period to which they will be spun up? The answer is 'yes'. It can be shown that an accreting neutron star will adjust its rotation period to an EQUILIBRIUM value at which the period of rotation is equal to the Keplerian period at the inner edge of the accretion disc where the infall has been arrested by the magnetic field. This equilibrium period is obtained by equating the ALFVEN RADIUS and the COROTATION RADIUS, and is given by:

$$ P_{eq} \approx 1.9\,ms \; R_6^{18/7} \left( \frac{M}{1.4 M_\odot} \right)^{-5/7} B_9^{6/7} \left( \frac{\dot{M}}{\dot{M}_{Edd}} \right)^{-3/7} . $$

It will be seen that given the mass and radius of the neutron star, this period is determined by the magnetic field and the accretion rate. This is shown as a series of parallel lines in fig. 3. For a given accretion rate, the smaller the magnetic field, the shorter is the equilibrium period. Similarly, for a given value of the magnetic field, the larger the accretion rate, the shorter is the equilibrium period. Since there is an upper limit to the accretion rate - corresponding to the EDDINGTON LUMINOSITY LIMIT - there is a minimum value for this equilibrium period and this is uniquely determined by the magnetic field. One therefore expects neutron stars accreting at the limiting value to be spun up to this critical spin-up line /12,13/.

## 4. THE ORIGIN AND EVOLUTION OF MILLISECOND PULSARS

Soon after the first millisecond pulsar (PSR 1937+21) was discovered it was clear that its magnetic field must be small. This led several people to simultaneously suggest that this might be such a recycled pulsar, but somehow the binary got disrupted /10,14-16/. If one took this hypothesis seriously then the observed period implied that the magnetic field of this pulsar ought to be close to $5 \times 10^8$ gauss, which is about four orders of magnitude smaller than that of the Crab pulsar.

When the period derivative was eventually measured the derived magnetic field turned out to be 4.7 x $10^8$ gauss. The remarkable proximity of this pulsar to the spin-up line (see fig. 2) seemed to rule out the possibility that this was just one of those odd pulsars born with a low field.

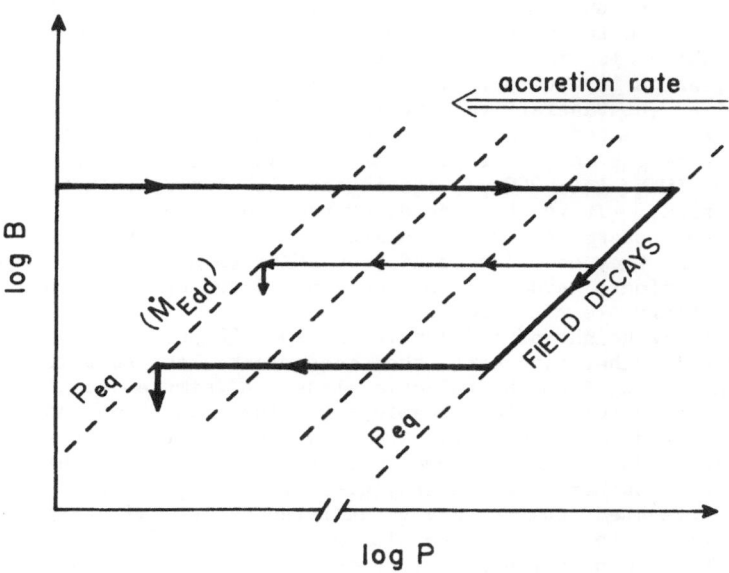

Fig. 3. The evolution of the spin period of an accreting neutron star. The EQUILIBRIUM PERIOD LINE corresponding to the Eddington accretion rate is shown at the extreme left.

Although the 1.5 ms pulsar has no companion at present, one can guess what kind of a progenitor it must have had. The reasoning is simple. To spin up a neutron star to a period $\sim$ millisecond it must accrete about 0.1 solar mass of material, which means that even if it accretes at the Eddington rate the mass transfer must be sustained for $> 10^7$ years. This, in turn, means that the companion could not have been a massive star, but must have been a low-mass star. In other words, the progenitor of the millisecond pulsar must have been a LOW MASS X-RAY BINARY (LMXB).

The confidence in this scenario has grown considerably because the next three "millisecond pulsars" to be discovered so far (outside of globular clusters) are all in binaries, with very circular orbits and low-mass companions - just as one would expect /17,18/. In addition, the latest member of the family - the 1.6 ms eclipsing pulsar - may be providing us an important clue as to how the solitary millisecond pulsar

got rid of its companion after being resurrected from the graveyard
/18/!

One of the really puzzling things about the first millisecond
pulsar was its location close to the galactic plane; considering that it
is probably a very old object this was disturbing. The discovery of the
next two ms-psrs underscored this puzzle for it turns out that they,
too, are very close to the plane. Normal pulsars have a scale height
$\sim 400$ pc, which is much larger than the scale height of their
progenitors, which is $\sim 60$ pc. This is easily understood if neutron
stars acquire substantial velocities at birth. Even if millisecond
pulsars are not created with such velocities one would expect them to
have a scale height comparable to that of their progenitors. The scale
height of the LMXBs is $\sim 300$ pc, consistent with their belonging to the old
disc population. It is, therefore, extraordinary that the first three
millisecond pulsars to be discovered are very close to the plane!! If
one accepts that they have descended from LMXBs then this must merely be
a selection effect, and by implication there must be many more
potentially observable millisecond pulsars /19/.

If the scale height of these pulsars is $\sim 300$ pc, like that of the
LMXBs from which they have come, then one expects less than 10% of them
to be within $\sim 30$ pc from the galactic plane. A simple scaling which
takes into account this factor, beaming of pulsar radiation and the fact
that the pulsars discovered so far are in a small sector of the galaxy,
suggests that there should be $\gtrsim 300$ 'millisecond' pulsars within $\sim 4$ kpc,
implying a total number in the Galaxy greater than two or three thousand
/19/. The expected number may be as large as $10^6$ if one takes into
account additional factors such as luminosity selection effects etc.
/20,21/. These expectations are consistent with the findings of the
recent Princeton-Arecibo survey /22/ which suggests that 'millisecond
pulsars' may constitute up to 10% of the population of pulsars. This
would imply a population as large as 10,000.

This expected number of millisecond pulsars is very much larger
than the number of LMXBs in the galaxy. Presently about 30 LMXBs are
known, and it is unlikely that there are many more than $\sim 100$. This has
many important implications.

If our basic premise, namely that millisecond pulsars evolve from
LMXBs, is correct then the lifetimes of these pulsars must be 20-30
times longer than the X-ray phase of these systems, which are believed
to last for $> 10^8$ years. In other words, millisecond pulsars must live
for $> 10^9$ years, or more than hundred times longer than normal pulsars
/19,23/.

As mentioned above, the statistics of the 500 or so pulsars
discovered so far suggests that their magnetic fields decay over a
characteristic time scale of a few million years. Clearly, if
millisecond pulsars are to live for $10^9$ years, this decay of the field
must either stop or after an initial rapid decay the time constant
should become comparable to the Hubble time /19,23,24/! The key
question at this stage is the following : Is the limiting value of the
field the same for all neutron stars? The answer is not clear at
present because one doesn't have a good picture of the nature of the

magnetic fields of neutron stars, let alone why and how the field decays. But the fact that the 1.5 ms, the 5 ms and the 6 ms pulsars have almost identical fields suggests that the limiting field may be around $4 \times 10^8$ gauss /19/.

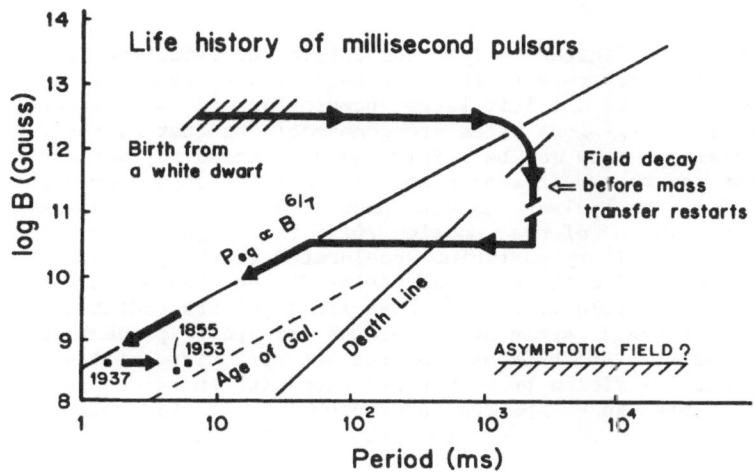

Fig. 4. The origin and evolution of millisecond pulsars

Figure 4 summarizes what one now believes to be the evolutionary history of these ultrafast pulsars /19/. Let us begin with the birth of the neutron star in a low-mass binary. Presumably it was spinning rapidly enough, and had strong enough field to function as a pulsar, and died eventually due to a combination of the lengthening of the period and field decay. With the onset of mass transfer it is spun up to its equilibrium period. Since in these low-mass systems mass transfer from the companion can last for $10^8$ years or more, the field will continue to decay and the neutron star will 'dribble down' the spin-up line till the field reaches its asymptotic value $\sim 5 \times 10^8$ G (fig. 4). By that time it would have been spun up to a millisecond. After the mass transfer stops, and the debris clears away, the recycled neutron star will once again start to function as a pulsar. Since the field doesn't decay any longer, it will move horizontally. Interestingly, given the age of the Galaxy the maximum spin period reached would only be $\sim 10$ ms. Thus, in their reincarnation they will live forever!

The very large population of millisecond pulsars implied by the number observed so far may be understood in these terms. The fact that all the three millisecond pulsars have almost identical fields (fig. 2), and the fact that the white-dwarf companion of the 5 ms pulsar has an estimated age of $> 10^9$ years /25/, implying that the pulsar must be even older, lend strong support to this scenario.

## 5. MILLISECOND PULSARS AS GAMMA RAY SOURCES

After this long digression into the origin and evolution of millisecond pulsars, let me repeat the main point I wish to make. I have argued that there must be a fairly large population of millisecond pulsars. Since their energy-loss rates are comparable to that of the Vela pulsar, and by analogy, they may be fairly luminous gamma ray sources.

What can we expect from them? The details depend upon the assumed model for the pulsar magnetosphere, and the gamma – ray emission mechanism. In one of the models /6/, the high – energy photons are produced by primary particles accelerated in the polar-cap gap and the radiation mechanism is CURVATURE RADIATION as these particles stream along the open field lines. In high-field pulsars, radiation above $\sim$ Gev will be strongly attenuated because of pair production in the magnetosphere. As USOV has pointed out /6/, since millisecond pulsars have rather low fields primary gamma rays with energies up to $10^{11}$ eV will be able to escape the magnetosphere, and the expected flux above $\sim 10^{11}$ eV will be comparable to that from the Crab pulsar in the same energy range.

In the "outer gap" model developed in detail by Cheng, Ho and Ruderman /3,26/, the outermagnetosphere current and the voltage drop of the millisecond pulsar should be about the same as that of the Vela pulsar since their energy – loss rates are roughly the same. Hence one may expect a similar gamma-ray spectrum and luminosity, i.e. an intensity spectrum roughly proportional to 1/E in the MeV to GeV range, with about 1% of the spindown power emitted as gamma rays.

If these expectations are correct, then millisecond pulsars may make a substantial contribution to the diffuse gamma-ray background of the galaxy. An estimate of this contribution can be made by integrating the gamma-ray emissivity of the galaxy due to millisecond pulsars along different lines of sight. This emissivity can be written as

$$\epsilon_\gamma (R,z) = \left[ \frac{1}{4\pi} \iint dP \, dB \, f(P,B) \, L_\gamma (P,B) \right] \rho(R,z)$$

where $\rho(R,z)$ is the number density of millisecond pulsars at a galactocentric radius R and at a height z from the galactic plane. $L_\gamma(P,B)$ is the gamma-ray luminosity of a millisecond pulsar with a spin period P and a magnetic field B, and f(P,B) is the normalized distribution of spin periods and magnetic fields of millisecond pulsars.

To proceed further we need to construct models for the functions f(P,B) and $\rho(R,Z)$. The gamma-ray luminsoity $L_\gamma(P,B)$ is predicted by the magnetospheric models /1,3/ mentioned above. To model f(P,B), we assume that all millisecond pulsars have the same magnetic field $\sim 4 \times 10^8$ gauss. We further assume that they are all born with a spin period $\sim 1$ ms and gradually slow down. This gives a period distribution $f(P,B) \propto P$. The space distribution $\varsigma(R,z)$ of these pulsars is expected to be similar to that of population II objects, and is modelled as an exponential distribution in R and z, with scale lengths $R_0 \sim 4$ kpc, and $z_0 \sim 300$ pc /20/. The only free parameter is the total number $N_{tot}$ of millisecond pulsars in the galaxy, which decides the overall scaling of $\varsigma(R,z)$.

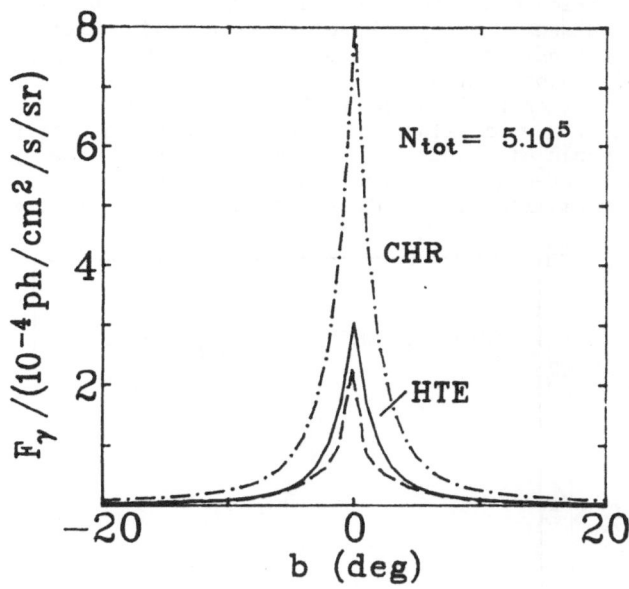

Fig. 5. Estimates of the specific intensity of gamma-rays ( >100 MeV) produced by millisecond pulsars plotted against the galactic latitude (for $l = 0°$). A total number of $5 \times 10^5$ millisecond pulsars in the Galaxy has been assumed. The solid line shows the results obtained using the curvature radiation model of Harding, Tademaru and Esposito (1978) and the dash-dotted line uses the outergap model of Cheng, Ho and Ruderman (1986). The dashed line at the bottom shows the contribution due to "normal" pulsars as estimated by Harding (1981).

The gamma-ray intensity at earth due to millisecond pulsars as a function of galactic longitude $l$ and galactic latitude b can now be computed by numerically integrating $\varepsilon_\gamma$ over the line of sight through the galaxy /5/. Fig. 5 shows the latitude profile at $l = 0°$ of the $\gamma$-rays expected from millisecond pulsars, for an assumed total number $N_{tot} = 5 \times 10^5$. The results for two emission models - namely, the outergap model due to Cheng, Ho and Rudermann (CHR) /3/ and the curvature radiation model due to Harding, Tademaru and Esposito (HTE) /1/ are shown, and a comparison is made with the $\gamma$-ray intensity from "normal" pulsars as estimated by Harding /5/.

There are two points to be noted from this diagram. First, the contribution of millisecond pulsars to the gamma-ray background may exceed that of the standard pulsars by a large factor. Second, the effective "width" of the latitude profile of the contribution due to millisecond pulsars is about twice that due to "normal" pulsars. This is in spite of the fact that normal pulsars have a scale height $\sim$ 400 pc, larger than that assumed for millisecond pulsars. The reason for this is that though normal pulsars have a large scale height, the younger short-period ones among them are expected to be very close to the galactic plane. Since these short-period pulsars dominate the gamma-ray production, the latitude profile is rather narrow. In the case of millisecond pulsars, even the "young" ones have a large scale height $\sim$ 300 pc, since this is the scale height of their progenitors. Further, even if a millisecond pulsar is born near the plane, it can migrate to a couple of hundred parsecs away from the plane before slowing down appreciably. Thus the latitude profile of the $\gamma$-ray intensity from millisecond pulsars merely reflects their tapering population.

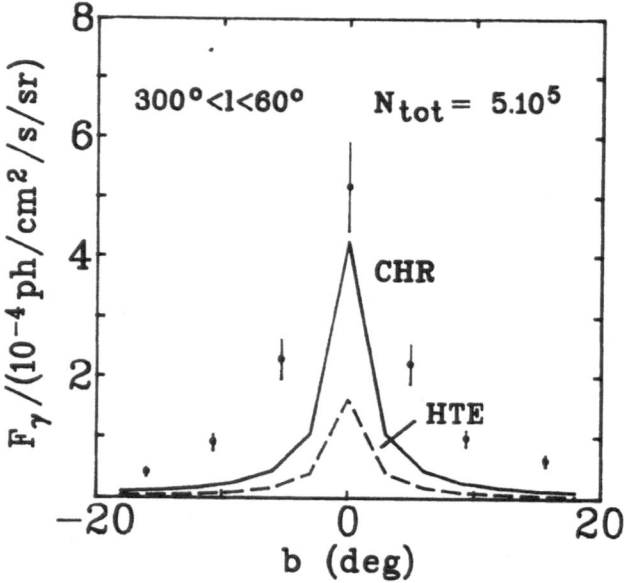

Fig. 6. The estimated contribution of millisecond pulsars to the diffuse gamma-ray background is compared with the intensities observed by SAS-2 at different galactic latitudes (Fichtel et al 1978). Results for two different magnetospheric models, as in fig. 5, are shown. Both the data and the theoretical estimates are averaged over $\pm 60°$ in galactic longitude around $l = 0°$.

Figure 6 compares these predictions with the diffuse $\gamma$-ray background intensities observed by SAS-2 /32/. The data points represent the diffuse gamma-ray specific intensity at different galactic latitudes, averaged over the galactic longitude range 300° through 0° to 60°. The expected contribution of millisecond pulsars averaged over the same longitude range is shown, for the two emission models CHR and HTE. As we can see, a good fraction of the background emission near the galactic plane may originate from millisecond pulsars. At higher latitudes, the millisecond pulsars can still account for 20-25% of the observed emission.

It is known, however, that much of the diffuse gamma rays near the galactic plane may originate due to the interaction of cosmic rays with the interstellar gas. If it is possible to obtain a firm lower limit to this contribution, then the argument may be turned around to place an useful upper limit on the number of millisecond pulsars in the galaxy.

I shall end with a speculation regarding the recently discovered 1.6 ms eclipsing binary pulsar. There have been reports that the Crab pulsar occasionally emits $10^{15}$ eV gamma rays. If this is true then it implies that nucleon beams with energy greater than this must be emitted. The reason is that even if such potential drops were to be generated in the pulsar magnetosphere, electrons can never be accelerated to such energies because of severe radiation losses. If the relative direction of the magnetic moment and the rotation axis is such that the gap-created electrons flow towards the stellar surface then protons (or positive ions) are expected to be pulled out of the surface. These will be accelerated to the full voltage drop since radiation losses are unimportant. Ruderman /29/ has estimated that the luminosity of such a $10^{15}$ eV nucleon beam will be $10^{35}$ erg/s in the case of Vela pulsar. If this nucleon beam were to be intercepted by some matter in the vicinity of the pulsar then there might be interesting consequences. The collision of this beam with nucleons will produce $\pi^o$ mesons, and the gammas produced by the decay of these $\pi^o$ mesons will have energy in excess of $10^{12}$ eV. Although this suggestion has been in the literature it has not attracted much attention. The reason is that one expects the gamma rays produced by such an interaction to be beamed in the forward direction, and this would require a very special geometry for them to be observed.

The recently discovered eclipsing millisecond pulsar may be offering us precisely such a geometry!! The eclipse suggests that we are viewing the pulsar from an angle very close to the plane of the orbit, and that at the middle of the eclipse the companion star is exactly in the line of sight to the pulsar. The very large duration of the eclipse ($\sim$ 50 minutes) implies that the size of the eclipsing object must be several times the Roche-lobe size of the $\sim$ 0.02 solar mass companion. This, and the enhanced dispersion at ingress and egress, strongly suggests that the radiation from the pulsar is slowly evaporating the companion /18,30,31/. The interaction of the pulsar wind with the evaporating companion results in a comet-like object. Given this geometry, if there is a nucleon beam from the pulsar then one would expect ultra high-energy gamma rays just before and after the radio blackout.

**ACKNOWLEDGEMENT**

The results presented in figures 5 and 6 are due to Dipankar
Bhattacharya of the Raman Research Institute. The details will be
published elsewhere.

**REFERENCES**

1. Harding, A.K., Tademaru, E., Esposito, L.W., Astrophys. J. **225**, 226
   (1978)

2. Lominadze, J.G., Machabeli, G.Z., Usov, V.V., Astrophys. Sp. Sci.
   **90**, 19 (1983)

3. Cheng, K.S., Ho, C., Ruderman, M., Astrophys. J. **300**, 500 (1986)

4. Harding, A.K., Astrophys. J. **245**, 267 (1981)

5. Harding, A.K., Astrophys. J. **247**, 639 (1981)

6. Usov, V.V., Nature **305**, 409 (1983)

7. Chadwick, P.M., Dipper, N.A., Kirkman, I.W., McComb, T.J.L., Orford,
   K.J., Turver, K.E., Turver, S.E., in: Very High Energy Gamma Ray
   Astronomy, ed. K.E. Turver, D. Reidel, Dordrecht, 1987, p. 159

8. Chadwick, P.M., Dowthwaite, J.C., Harrison, A.B., Kirkman, I.W.,
   McComb, T.J.L., Orford, K.J., Turver, K.E., Nature **317**, 236 (1985)

9. Radhakrishnan, V., Contemp. Phys. **23**, 207 (1982)

10. Radhakrishnan, V., Srinivasan, G., Curr. Sci. **51**, 1096 (1982)

11. Radhakrishnan, V., in : Highlights of Astronomy 7, ed. J.P. Swings,
    D. Reidel, Dordrecht, 1985, p. 3

12. Radhakrishnan, V., Srinivasan, G., in : Proceedings of the 2nd
    Asia-Pacific Regional Meeting in Astronomy, ed. B. Hidayat and
    M.W. Feast, Tira Pustaka, Jakarta, 1984, p.423

13. van den Heuvel, E.P.J., J. Astrophys. Astron. **5**, 209 (1984)

14. Alpar, M.A., Cheng, A.F., Ruderman, M.A., Shaham, J., Nature **300**,
    728 (1982)

15. Fabian, A.C., Pringle, J.E., Verbunt, F., Wade, R.A., Nature **301**,
    222 (1983)

16. Henrichs, H.F., van den Heuvel, E.P.J., Nature **303**, 213 (1983)

17. Taylor, J.H., Stinebring, D.R., Ann. Rev. Astr. Astrophys. **24**, 285 (1986)

18. Fruchter, A.S., Stinebring, D.R., Taylor, J.H., Nature **333**, 237 (1988)

19. Bhattacharya, D., Srinivasan, G., Curr. Sci. **55**, 327 (1986)

20. Kulkarni, S.R., Narayan, R., Caltech Astrophysics Preprint, #8, 1988

21. Cote, J., Pylyser, E.H.P., Preprint (1988)

22. Stokes, G.H., Segelstein, D.J., Taylor, J.H., Dewey, R.J., Astrophys. J. **311**, 694 (1986)

23. van den Heuvel, E.P.J., van Paradijs, J.A., Taam, R.E., Nature **322**, 153 (1986)

24. Kulkarni, S.R., Astrophys. J. **306**, L85 (1986)

25. Wright, G.A., Loh, E.D., Nature **324**, 127 (1986)

26. Cheng, K.S., Ho, C., Ruderman, M.A., Astrophys. J. **300**, 522 (1986)

27. Bignami, G.F., Hermsen, W., Ann. Rev. Astr. Astrophys. **21**, 67 (1983)

28. Srinivasan, G., Bhattacharya, D., Preprint (1988)

29. Ruderman, M.A., in : High Energy Phenomena Around Collapsed Stars, ed. F. Pacini, D. Reidel, Dordrecht, 1987, p. 145

30. Kluzniak, W., Ruderman, M., Shaham, J., Tavani, M., Nature, in press (1988)

31. Phinney, E.S., Evans, C.R., Blandford, R.D., Kulkarni, S.R., Nature **333**, 832 (1988)

32. Fichtel, C.E., Simpson, G.A., Thompson, D.J., Astrophys. J. **222**, 833 (1978)

# MILLISECOND PULSARS WERE BORN SPINNING FAST

Wilfred H. Sorrell
Astronomy Department
University of Wisconsin
Madison, Wisconsin 53706

Abstract. Occam's Razor Principle is invoked to explain the high spin rates and low surface magnetic field strengths of millisecond pulsars. The simplest explanation is that these objects were born as rapid rotators owing to conservation of angular momentum during core collapse of the presupernova star. It is shown that conservation of magnetic flux during core collapse leads to surface magnetic fields having the weak strengths ≈ $10^8$ G observed for the newlyborn pulsar in supernova 1987 A.

The outburst of neutrinos detected from SN 1987 A at the time of the explosion has been interpreted as the signature for the formation of a central neutron star, which is expected to become a rapidly rotating pulsar. There now appears to be direct optical evidence that a pulsar indeed formed ( Middleditch 1989 ; Kristian et al. 1989 ). The main surprises are that the object is spinning a lot faster ( ≈ 0.5 ms ) than expected, and the surface magnetic field strength ( ≈ $10^8$ G ) is weaker than canonical field strengths ≈ $10^{12}$ - $10^{13}$ G. The high spin rate could pose severe difficulties for current theories of nuclear matter at very high ( ≈ $10^{14}$ gm cm$^{-3}$ ) densities ( cf. Friedman et al. 1988 ), so another possibility is that the optical pulses are powered by radial oscillations of the star ( Wang et al. 1989 ). In this case, pulsations would produce a time-varying dipole moment, which causes the star to emit pulsational magnetic radiation . At present, it is still premature to decide on any particular explanation for the SN 1987 A pulsar. However, it is of some interest to invoke Occam's Razor Principle and ask whether the observations can be understood on the simplest hypothesis based on a rotation scenario. Another question of interest is how the sub-millisecond pulsar in SN 1987 A is related to millisecond pulsars as a class. To provide plausible answers to these questions, it is useful to first review some general characteristics of the pulsar population.

W. Kundt (ed.), Neutron Stars and Their Birth Events, 121–126.

There are roughly $\approx 500$ pulsars known at the present time, and statistics show that most of them are isolated objects rather than members of binary star systems. The spin periods P lie in the range from $\approx 0.5$ ms to 4 s. Once the slowing down rate $\dot{P}$ is determined, it is possible to estimate spin-down ages $\tau_{age} \approx P/\dot{P}$. The ages generally range from $\approx 10^3 - 10^9$ years. Taylor and Stinebring ( 1986 ) find the statistical distribution of Galactic pulsars suggests a birthrate $\approx 0.01 - 0.03$ year$^{-1}$, which is comparable to the Galactic supernova birthrate. This gives us some reason to believe that pulsars are born in supernova outbursts, perhaps immediately after collapse of the iron core of a massive presupernova star. If the Kristian et al. ( 1989 ) observation is confirmed for SN 1987 A, then we would have direct evidence for a supernova origin.

The surface magnetic field strengths of pulsars cannot be measured directly, so we must rely on theoretical models as a guide. In the canonical electrodynamic model, it is assumed that the magnetic field system has a dipole-like geometry ; and that the magnetic axis is non-aligned with the spin axis. It is further assumed that radiation observed from the supernova remnant is powered by magnetic dipole radiation from the pulsar light cylinder. The magnetic moment is then determined on the constraint that the magnetic dipole power never exceeds the observed luminosity of nebular emission. Once the magnetic moment is known, we can estimate the strength of the surface magnetic field by assuming the pulsar is a canonical neutron star of radius $\approx 10$ km .

It turns out that the Galactic pulsar population can be divided into two classes on the basis of their spin periods and surface magnetic field strengths. The Class I objects have spin periods $P \approx 33$ ms - 4 s and surface magnetic fields $B_s \approx 10^{12} - 10^{13}$ G. These objects spin down rapidly and they comprise the majority of the population, with less than one per cent being members of binary systems. The Class II pulsars have millisecond periods $P \approx 0.5 - 33$ ms and surface magnetic fields $B_s \approx 10^8 - 10^{10}$ G . At present, only nine millisecond pulsars are known ( including the optical pulsar in SN 1987 A ), with five of them being definite members of binary systems. Because the sample is so small, we can say nothing with certainty about their statistical properties. However, it is clear that millisecond pulsars have spin rates and magnetic field strengths that are quite different from those of ordinary pulsars. At the present time, a lively debate is focused on why the millisecond objects are so different.

The popular explanation is that a millisecond pulsar was once a slowly rotating ordinary pulsar in binary systems. In this scenario, mass transfer leads to accretion of high-angular-momentum material onto the neutron star. This causes the rapid spin-up.

The weak magnetic fields would arise from ordinary pulsars undergoing magnetic field decay. The net effect would then be a rapidly spinning object with low surface magnetic fields. Woosley and Chevalier ( 1989 ) discussed an alternate version of the accretion scenario to explain the sub-millisecond optical pulsar in SN 1987 A. In their version, angular momentum accretion occurred during the supernova event. The accreted angular momentum would originate from the mixed mantle and helium core of the supernova ejecta. It is expected that $\approx 0.1$ M$_\odot$ of ejected material would infall and spin up the newly formed neutron star.

A major requirement of the accretion scenario is that Ohmic dissipation would cause the magnetic fields of neutron stars to decay by factors $\approx 10$ -100 on timescales $\approx$ a few $10^6$ years. At present, however, all the models proposed for field decay are faced with severe difficulties ( Michel 1986 ; Kundt 1988 ). Detailed calculations of magnetic field decay in neutron stars have been reported by Sang and Chanmugan ( 1987 ). They find that even if field decay occurs, the time dependence of the decay does not proceed in the exponential fashion needed to explain pulsar observations. In light of this result, it might prove useful to abandon the accretion scenario and seek a solution for the millisecond pulsar problem elsewhere.

The simplest alternative point of view is that a millisecond pulsar is born as a weakly magnetized, rapidly rotating neutron star ( Pacini 1983 ; Arons 1983 ; Brecher and Chanmugam 1983 ; Kundt 1985 ). We may call this alternative picture the primordial scenario because it relates the initial spin and magnetic properties of pulsars to conditions during birth of the neutron star. Here the important question we must ask is whether the primordial scenario naturally explains the observed correlation between the magnetic field and spin period of pulsars in general.

To answer this question, consider the pre-supernova stage in which stellar magnetic fields remove spin angular momentum from the thermonuclear core before gravitational collapse and the supernova outburst occur. During the late stages of nucleosynthesis in massive stars, magnetic braking would provide a coupling between the stellar core and mantle. If the core of a massive star has radius $R_c$ and magnetic field strength $B_c$ , then its spin angular momentum $J_c$ can be transferred outwards across surface area $4\pi R_c^2$ by Maxwell stresses $\approx B_c^2 / 4\pi$ . The magnetic tensile force $\approx ( 4\pi R_c^2 ) ( B_c^2 / 4\pi )$ then gives a spin-down torque $\approx R_c ( 4\pi R_c^2 ) ( B_c^2 / 4\pi )$. It is possible that viscosity removes magnetic flux from the core while spin-down occurs, and thereby reduces the rate of magnetic braking. This process can happen by turbulent convection while the star evolves through successive stages of nuclear reactions from hydrogen, helium, oxygen, neon, to silicon burning in the core and ambient shell. Convection can also transport angular momentum out of the core. Because convective transport coefficients for both angular momentum and magnetic fields are unknown at present, we shall consider situations in which no magnetic flux loss from the core occurs at all.

The transport of angular momentum by Maxwell stresses occurs on the timescale $\tau_{mag} \approx J_c / R_c^3 B_c^2$. This time should be compared to the stellar evolution timescale $\tau_* \approx 10^{10} (M_*/M_\odot)^{-2}$ years, where $M_*$ is the initial total stellar mass. We expect efficient angular momentum transport for $\tau_{mag} \leq \tau_*$. This requires a core magnetic field strength $B_c \geq B_*$, where

$$B_* \approx 0.32 \ (J_c / 10^{49} \ \text{erg sec})^{1/2} \ (R_c/R_\odot)^{-3/2} \ (M_*/M_\odot) \ G \ . \qquad (1)$$

For field strengths $B_c < B_*$, Maxwell stresses are unable to transfer angular momentum out of the core during stellar lifetimes $\tau_* \approx 10^7 - 10^8$ years ($M_*/M_\odot \approx 5 - 30$). In such cases, the core will collapse with a high $J_c$ and a neutron star will form with a high spin rate. This would signal the birth of a Class II pulsar. In the ($B_c > B_*$) case, the core will collapse with a low $J_c$, leading to the formation of a slowing spinning neutron star and a Class I pulsar. The type of pulsar that forms depends on not only the initial core angular momentum, but also the initial mass of the pre-supernova star. The larger masses increase $B_*$ and thus favour the birth of millisecond pulsars. It follows that these objects should be common in the blue irregulars ( Magellanic Clouds ) and blue compact galaxies, which are enriched with high-mass OB stars. On the other hand, for galaxies like the Milky Way system, the steep decline of the stellar mass spectrum would cause millisecond pulsars to be rare. Although this point of view is consistent with present statistics of the Class II objects, it is emphasized that observational selection effects could also play an important part.

The typical B0 star on the main sequence has a rotation period $\approx 2$ days, which corresponds to a uniform spin rate $\approx 5.8 \times 10^{-6} \ s^{-1}$. Assuming a stellar mass $\approx 17 \ M_\odot$ and radius $\approx 8 \ R_\odot$, the star would have a total angular momentum $\approx 6 \times 10^{52}$ erg sec. Further assuming a core mass $M_c \approx 1.7 \ M_\odot$ and radius $R_c \approx R_\odot$, we find a typical core angular momentum $J_c \approx 9.4 \times 10^{49}$ erg sec for the same spin rate. In realistic situations, stars of a given spectral class will have a range of core values depending on the internal structure. We are unable to be more specific because the theory of stellar interiors with rapid rotation and magnetic fields is poorly understood at present. Thus, the above values are at best order-of-magnitude estimates. But using these estimates as a guide, we may still ask whether the spin and magnetic properties of pulsars can be understood from their birth conditions.

After silicon burning produces an inert iron core, the core would collapse to form a rotating neutron star on a dynamical timescale $t_{dyn} \approx (G \rho_{core})^{-1/2} \approx 1 - 100$ sec. In the primordial scenario, the collapse would proceed with an invariant angular momentum $J_0 \leq J_c$ and poloidal magnetic flux $F_0 \approx \pi R_c^2 B_c$. To make the picture definite, let us consider a canonical neutron star of radius $a \approx 10$ km and Chandrasekhar mass $M_{ch} \approx 1.4 M_\odot$. From angular momentum conservation, the initial spin rate is

$$\omega_0 \approx J_0 / M_{ch} a^2 \quad (J_{Birth} \approx J_0 \leq J_c) . \tag{2}$$

Assuming a dipole-like geometry, the magnetic flux threads the polar caps of the neutron star through an effective surface area $\pi (a \omega_0 / c) a^2$ where c is the light speed. Hence the neutron star would have a surface magnetic field $B_s$ and flux at birth

$$F_{Birth} \approx \pi (a \omega_0 / c) a^2 B_s . \tag{3}$$

Because the flux $F_0$ initially threading the core of the presupernova star now threads the polar caps of the neutron star ( $F_{Birth} \approx F_0$ ), we have the initial spin period

$$P_{spin} \approx 2\pi M_{ch} a^2 / J_0 \tag{4}$$

and surface magnetic field strength

$$B_s \approx (c F_0 / 2\pi^2 a^3) P_{spin} . \tag{5}$$

For presupernova stars with similar $F_0$, we find a field-period correlation for newly born pulsars in general. Table 1 lists values of $P_{spin}$ and $B_s$ for a canonical neutron star of radius $a \approx 10$ km, mass $M_{ch} \approx 1.4 M_\odot$, and magnetic flux $F_0 \approx 7 \times 10^{20}$ gauss cm$^2$.

Table 1. Spin periods and surface magnetic field strengths of newly born pulsars.

| $J_0$ ( erg sec ) | $P_{spin}$ (s) | $B_s$ (G) |
|---|---|---|
| $4.4 \times 10^{45}$ | 4.0 | $4.2 \times 10^{12}$ |
| $1.8 \times 10^{46}$ | 1.0 | $1.0 \times 10^{12}$ |
| $5.8 \times 10^{47}$ | 0.03 | $3.1 \times 10^{10}$ |
| $1.8 \times 10^{49}$ | 0.001 | $1.0 \times 10^{9}$ |
| $3.5 \times 10^{49}$ | 0.0005 | $5.3 \times 10^{8}$ |

It is seen that the millisecond objects have weak surface magnetic fields $\approx 10^8$ - $10^9$ G for angular momenta $\approx 10^{49}$ erg sec . Thus, in the primordial scenario, there is no reason to invoke magnetic-field decay. Both the spin and magnetic properties of pulsars are fixed by conditions at birth.

It is concluded that the primordial scenario can provide a reasonable explanation of pulsar observations. The millisecond Class II objects were born with high angular momentum when the stellar core collapsed. This situation arises because magnetic fields were too weak to transfer angular momentum out of the core during stellar evolution times. The weak-field conditions were simply carried into the newly formed neutron star as a result of flux conservation. For the ordinary Class I pulsars, core magnetic fields $B_c \approx 0.1$ G were strong enough to transfer angular momentum outwards. The strong-field conditions lead to the formation of a neutron star with low angular momentum ( long spin periods ) and surface magnetic fields $\approx 10^{12}$ G . It is emphasized that this scenario depends only on the structure and evolutionary history of the presupernova star itself. It is independent of whether the star is a member of a binary system.

References

Arons, J. , 1983. Nature, 302, 301.

Brecher, K. and Chanmugam,G., 1983. Nature, 302, 124.

Friedman, J.L., Imamura, J.N., Durisen, R.H., and Parker, L., 1988. Nature, 336, 560.

Kristian, J. et al. , 1989. Nature, 338, 234.

Kundt, W. , 1985. Bull. Astron. Soc. India. , 13, 12.

Kundt, W., 1988. Comments Astrophys., 12, 113.

Michel, C. F., 1986. Physics Today, 39, 9.

Middleditch, J. et al., 1989. IAU Circ. No. 4735.

Pacini, F., 1983. Astron. Astrophys., 126, L11.

Sang, Y. and Chanmugam, G., 1987. Astrophys. J., 323, L61.

Taylor, J. H. and Stinebring, D.R., 1986. Ann. Rev. Astron. Astrophys., 24, 285.

Wang, Q., Chen, K., Hamilton, T.T., Ruderman, M., and Shaham, J., 1989. Nature, 338, 319.

Woosley, S.E. and Chevalier, R.A., 1989. Nature, 338, 321.

# MAGNETIC FIELD DECAY IN COOLING NEUTRON STARS

YEMING SANG and G. CHANMUGAM
Department of Physics and Astronomy, Louisiana State University

SACHIKO TSURUTA
Department of Physics, Montana State University

ABSTRACT. We have carried out calculations of the Ohmic decay of dipolar magnetic fields which are created so that they are initially confined to the crust. The effects of the evolution of the neutron stars, which have not been previously taken into account, are included. The results confirm the earlier work of Sang and Chanmugam (1987) that the field does not decay exponentially as generally assumed, and that if it occupies the entire crust it decays by less than a factor of order 100 in the Hubble time. Fields may decay more quickly if they are confined to the outer crust, but again non-exponentially in conflict with observations.

## 1. Introduction

If pulsars spin down as a result of magnetic dipole radiation, their surface dipolar magnetic fields are typically $B_s \sim 10^{12}$ G (Ruderman 1986; Taylor and Stinebring 1986). Ostriker and Gunn (1969) have suggested that pulsars turn off when their magnetic fields decay on a time scale of about $10^6$ yr. However, detailed numerical calculations (Chanmugam and Gabriel 1971) taking account of the varying and high electrical conductivity in the interior of the star (Baym, Pethick and Pines 1969) showed that the time scale for the longest-living modes of decay, for a dipolar field which penetrates the interior of the star, $t_D > 10^{12}$ yr . Nevertheless, more recent analyses of pulsar spin-down statistics are consistent with the earlier analysis of Gunn and Ostriker (1970) that the magnetic fields decay *exponentially* on time scales of $(5 - 9) \times 10^6$ yr (Lyne, Manchester and Taylor 1985; Stollman 1987).

More recently, it has been proposed that the magnetic field is thermoelectrically generated in the crust after the formation of the neutron star (Urpin and Yakovlev 1980; Blandford, Applegate and Hernquist 1983; Urpin, Levshakov and Yakovlev

*W. Kundt (ed.), Neutron Stars and Their Birth Events, 127–131.*
© *1990 Kluwer Academic Publishers.*

1986). The field then decays on a time scale $t_D \sim 4\sigma_c L^2/\pi c^2 \sim 10^6$ yr, where $L$ is the thickness and $\sigma_c$ the value of the electrical conductivity in the crust. Other mechanisms for field decay involving various instabilities have also been proposed (Tayler 1973; Flowers and Ruderman 1977; Chanmugam 1984), if the magnetic field is frozen into the neutron star during its birth (Ginzburg 1964; Woltjer 1964). Here too the time scale of decay is essentially that for the decay of crustal currents or magnetic fields. Sang and Chanmugam (1987) made calculations of the evolution of magnetic fields which are initially confined to the crust for the case where the neutron star is isothermal and the temperature does not change with time. The electrical conductivity which depends critically on the temperature was therefore kept constant in time. They showed that the field does not decay in an exponential manner and may not decay by more than a factor of 100 in the Hubble time, thereby supporting the view that pulsar magnetic fields may not decay significantly (Kundt 1986; Beskin, Gurevich and Istomin 1983; Michel 1986). Here, we present calculations taking into account the cooling of the neutron star so that the conductivity varies with time.

## 2. Decay of Crustal Magnetic Field

### 2.1. EQUATIONS

The basic equation for the Ohmic decay of a magnetic field $\mathbf{B}$ is given, if internal motions are negligible, by

$$\frac{\partial \mathbf{B}}{\partial t} = -\nabla \times \left( \frac{c^2}{4\pi\sigma} \nabla \times \mathbf{B} \right) \tag{1}$$

where $\sigma(r,t)$ is the electrical conductivity of the material. If $\mathbf{B}$ is poloidal then $\mathbf{B} = \nabla \times \mathbf{A}$, where the vector potential $\mathbf{A} = (0, 0, A_\phi(r, \theta, t))$ in spherical polar coordinates $(r, \theta, \phi)$. By introducing the Stokes stream function $S(r, \theta, t) = -r \sin\theta \, A_\phi$ (Wendell, van Horn and Sargent 1987), this equation can be separated so that for a pure dipole $S = -g(x, t) \sin^2 \theta$, where

$$\frac{\partial^2 g}{\partial x^2} - \frac{2}{x^2} g = \frac{4\pi R^2 \sigma}{c^2} \frac{\partial g}{\partial t} \tag{2}$$

and $x = r/R$. The surface magnetic field at the pole is given by $B_s(t) = 2g(1,t)/R^2$. Equation (2), with appropriate boundary conditions, is solved numerically as an initial value problem.

## 2.2 CONDUCTIVITIES

The evolutionary model of the neutron star used in this work is the FP model of Nomoto and Tsuruta (1987). The mass of the neutron star is $1.4M_\odot$ and its radius $R = 10.9$ km. The electrical conductivity $\sigma(x, t)$ is calculated directly through the Wiedemann-Franz law from the thermal conductivity used in the neutron star model. In the thin surface layer of the neutron star where the radiative conductivity is dominant this conversion is no longer valid. But the decay time scale is not influenced by the surface layer conductivity since it is mainly determined by the highest conductivity to which the field penetrates. The neutron star evolution calculation was carried out up to a few million years. In this work we assume that the conductivity stays the same after the age of $1.84 \times 10^6$ years since the cooling is expected to slow down after that due to various heating effects (Pines and Alpar 1985). For the crust, the conductivity is essentially given by the lower of the values due to electron-phonon scattering and impurity scattering, the latter being independent of temperature. Since impurity scattering is unimportant at higher temperatures, it is not included in the models of Nomoto and Tsuruta (1987). But as the star cools down and the electron-phonon scattering becomes less important it becomes dominant (Flowers and Itoh 1976, 1981). However, the impurity conductivity is not well known at present due to uncertainties in the impurity concentration and charge fluctuations.

## 3. Discussions

For the initial field configuration $g(x, 0)$ we assumed that it is confined to the entire crust ($x > x_c = 0.885$) and satisfies the boundary conditions at the center and surface of the star. The variation of the surface polar field with time is shown in Figure 1. The field decays slowly and *non-exponentially*. The reason for this non-exponential behavior is that in addition to diffusing outwards there is also some inward diffusion of the field to regions where the conductivity is higher, so that the time scale of diffusion lengthens correspondingly, and the total decay in $10^{10}$ years is less than a factor of 100.

If $g(x, 0)$ is confined initially to the outer regions of the crust the field decays more quickly at first. The decay then slows down because of inward diffusion of the fields to regions of higher conductivity. The rate of evolution of the field depends on the location of the initial field and is again non-exponential.

Most of the analyses of pulsar statistics assume the fields decay exponentially with a time scale of few million years (e.g. Lyne, Manchester and Taylor 1986, Stollman 1987). Our calculations indicate that if the magnetic field initially occupies the entire crust it decays by less than a factor of order 100 in the Hubble time. Thus for initial surface fields of $10^{12}$ G the field does not decay down to $\sim 5 \times 10^8$

G, characteristic of millisecond pulsar fields, even within the Hubble time. It has been suggested by van den Heuvel, van Paradijs and Taam (1986) that the decay of pulsar magnetic fields slows down after about $10^8$ yr and that the residual field may be long lived (Kulkarni, 1986). Our results crudely support this view if the initial field is confined to an appropriate depth of the crust.

If the initial field is confined to the outer region of the crust(cf. Urpin, Levshakov and Yakovlev 1986) then sufficiently rapid field decay may be possible. However, the precise location of the field does not follow simply from these models. Furthermore, the decay would not be exponential, and hence conflict with observations, making such models difficult to justify at present.

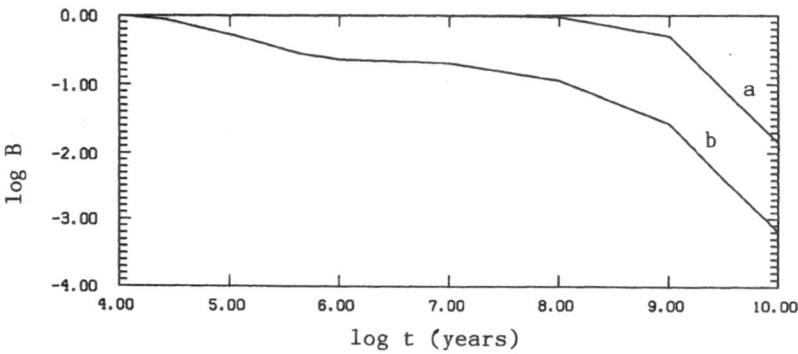

**Figure 1:** Surface magnetic field at the pole in arbitrary units as a function of time. Curve (a) is for the decay of a field initially occupying the entire crust; curve (b) is that of a field initially confined to the region outside the neutron-drip point ($x > 0.938$).

Mechanisms for field decay involving instabilities (Flowers and Ruderman 1977, Chanmugam 1984, Wang and Eichler 1988) require the Ohmic decay of fields or currents which occupy the entire crust, but such fields may not be able to decay exponentially.

To summarize, our calculations show that current models for field decay of neutron stars which have been proposed have difficulties. Our results, instead, tentatively supports the view that neutron star magnetic fields might not decay and other mechanisms may contribute to the death of radio pulsars.

**Acknowledgements**

Y. S. wishes to thank NATO and Professor Wolfgang Kundt for providing support in order to attend the School. This work was supported by NSF grant AST 87-00742.

# References

Baym, G., Pethick, C., and Pines, D. 1969, *Nature*, **224**, 674.
Beskin, V. S., Gurevitch, A. V., and Istomin, Ya. N. 1983, *Ap. Sp. Sci.*, **102**, 301.
Blandford, R., Applegate, J., and Hernquist, L. 1983, *M.N.R.A.S.*,**204**,1025.
Chanmugam, G. C. 1984, in *Birth and Evolution of Neutron Stars: Issues raised by Millisecond Pulsars*, ed. S. P. Reynolds and D. R. Stinebring (Charlottesville: N.R.A.O.), p. 213.
Chanmugam, G.,and Gabriel, M. 1971, *Astr.Ap.*, **16**, 149.
Flowers, E., and Itoh, N. 1976, *Ap.J.*, **206**, 218.
Flowers, E., and Itoh, N. 1981, *Ap.J.*, **250**, 750.
Flowers, E., and Ruderman, M. 1977, *Ap.J.*, **215**, 302.
Ginzburg, V. L. 1964, *Soviet Phys.* Doklady, **9**, 329.
Gunn, J. E., and Ostriker, J. P. 1970, *Ap.J.*, **160**, 979.
Kulkarni, S. R. 1986, *Ap.J(Letters)*, **306**, L85.
Kundt, W. 1986, in *The Evolution of Galactic X-ray Binaries*, eds. J. Truemper, W. H. G. Lewin and W. Brinkmann (Reidel, 1986), P. 263.
Lyne, A. G., Manchester, R. N., and Taylor, J. H. 1985, *M.N.R.A.S.*, **213**, 613.
Michel, C. F. 1986, *Physics Today*, **39**, No. 10, p9.
Nomoto, K., and Tsuruta, S. 1987, *Ap.J.*, **312**, 711.
Ostriker, J. P., and Gunn, J. E. 1969, *Ap.J.*, **157**, 1395.
Pines, D., and Alpar, M. A. 1985, *Nature*, **316**, 27.
Ruderman, M. A. 1986, in *Highlights in Modern Astrophysics*, S. L. Shapiro and S. A. Teukolsky (New York: Wiley), p 21.
Sang, Y. and Chanmugam, G. *Ap.J.(Letters)*, **323**, L61.
Stollmann, G. M. 1987, *Astr.Ap.*, **178**, 143.
Tayler, R. J. 1973, *M.N.R.A.S.*, **162**, 17.
Taylor, J. H. and Stinebring, D. R. 1986, *Ann. Rev. Astr.Ap.*, **24**, 285.
Urpin, V.A., Levshakov, S.A. and Yakovlev, D.G. 1986, *M.N.R.A.S.*, **219**, 703.
Urpin, V.A., Yakovlev, D.G. 1980, *Soviet Astr.*, **24**, 425.
van den Heuvel, E. P. J., van Paradijs, J. A., and Taam, R. E. *Nature*, **322**, 153.
Wang, Z., and Eichler, D. 1988, *Ap.J.* , **324**, 966.
Wendell, C. E., Van Horn, H. M., and Sargent, D. 1987, *Ap.J.*, **313**, 284.
Woltjer, L. 1964, *Ap.J.*, **140**, 1309.

# THERMAL FIELD GROWTH IN NEUTRON STARS: AN ALTERNATIVE TO "INJECTION"?

D. Bhattacharya
Raman Research Institute
Bangalore 560 080, INDIA

ABSTRACT: Analyses of the statistics of pulsars and of pulsar-produced nebulae have, over the past several years, led to the conclusion that most pulsars are born spinning rather slowly, with periods in excess of 100 ms. In this note we examine whether a slow build-up of neutron star magnetic fields from very low initial values ($\sim 10^8$ gauss) may provide a viable alternative to this, as has been suggested by some authors. We find that irrespective of whether the field grows or not, slow rotation of neutron stars at birth remains an inevitable conclusion.

## 1. INTRODUCTION

In the past few years several authors have suggested that most pulsars may be born spinning fairly slowly - with periods $P > 100$ ms (Vivekanand and Narayan 1981; Srinivasan et al. 1984; Chevalier and Emmering 1986; Stokes et al. 1986; Narayan 1987). The argument for this comes from two independent quarters. First, it was found by Vivekanand and Narayan (1981) that the evolutionary "current"[*] of observed pulsars rises to a peak at a spin period $\sim 0.5$ s, implying that many pulsars are being "injected" into the pulsar population at periods of this order. More recently, Narayan (1987) has reaffirmed this conclusion after taking into account many selection effects. Also, the results of the recent Princeton-Arecibo survey sensitive to detect short period pulsars strongly support this conclusion (Stokes et al. 1986; Dewey et al. 1988).

The second argument comes from the statistics of filled-centre supernova remnants (plerions) that young pulsars are expected to produce. The brightness of a plerion depends on the rate of supply of relativistic particles and magnetic fields into it, and hence on the spin period of the central pulsar. The absence of a large number of bright plerionic nebulae in our galaxy suggests that most pulsars are born spinning rather slowly, typically with periods $\gtrsim 50$-100 ms (Srinivasan et al. 1984).

---

[*] The "current" of pulsars, i.e., the number of pulsars moving per second from shorter to longer periods, is defined as $J(P)=N(P) \langle \dot{P} \rangle$, where $N(P)dP$ is the number of pulsars in our galaxy with spin periods between $P$ and $P+dP$, and $\langle \dot{P} \rangle$ is their average slowdown rate.

W. Kundt (ed.), Neutron Stars and Their Birth Events, 133–137.
© 1990 Kluwer Academic Publishers.

In drawing the above conclusions, it has been implicitly assumed that the magnetic field strengths of neutron stars are $\gtrsim 10^{12}$ gauss at birth. However, it has been suggested by Blandford, Applegate and Hernquist (1983) that most neutron stars may be born with very low magnetic fields $\sim 10^8$ gauss. A thermally driven battery effect then generates the observed $\sim 10^{12}$ gauss fields of pulsars in $\sim 10^5$ years. Meanwhile the supernova remnant around the pulsar would have dissipated, and the pulsar, whatever be its spin period, will produce a very weak plerion since the relativistic wind from it will be essentially unconfined. This seems to neatly overcome the plerion underpopulation problem, even if all pulsars are born with very short periods. Narayan (1987) suggests that such a field growth may also be the reason behind the apparent "injection" of pulsars at long periods: by the time the field grows enough to allow pulsar activity become possible, the star may already have slowed down to a fairly long period. In what follows we shall examine this possibility in some detail.

## 2. YOUNG PULSARS IN SUPERNOVA REMNANTS

If we grant that a slow growth of its magnetic field prevents a pulsar from becoming detectable during the lifetime of the associated supernova remnant, then the cases where we do see pulsar-SNR associations must be exceptional. At least in two known cases of such association, the objects are only $\sim 10^3$ years old (namely, the Crab Nebula and SNR 0540-69.3 in the LMC). And yet the pulsars in them have magnetic fields $>10^{12}$ gauss. Energetics of these nebulae require that the magnetic fields of these pulsars must have attained their present values, or very nearly so, within a few hundred years after their birth (see, e.g. Bhattacharya and Shukre 1985; Bhattacharya 1987). Blandford et al. (1983) suggest that such a quick generation of the magnetic field may be expected due to a dynamo action in the ocean of liquid metals at the surface of the newly born neutron star, if its rotation rate is very high. If so, then slow initial rotation for the majority of neutron stars becomes a necessity even in this model. Otherwise all fields will grow very fast and one will again end up predicting too many bright plerionic nebulae in the galaxy.

## 3. ABSENCE OF SHORT PERIOD PULSARS IN THE OBSERVED POPULATION

Let us, however, for the moment suppose that in most neutron stars the growth of the magnetic field occurs very slowly irrespective of the spin period, and regard pulsars like the Crab as representatives of a small population born with high fields. The question which we shall now ask is whether such a picture can explain the deficit of short period pulsars (P < 100 ms) in the observed population without requiring the newly born neutron stars to be spinning slowly.

The answer to this question depends on the evolution of the spin period of a neutron star during the field growth phase. Let us investigate this with a toy model in which the magnetic field B of the star grows exponentially from $10^8$ gauss to $10^{12.5}$ gauss in $10^5$ years (this corresponds to a growth timescale $\tau_m$ of 9650 yr: $B(t) = 10^8 \exp(t/\tau_m)$ gauss). Assuming a slowdown rate similar to that due to magnetic dipole radiation (i.e. $\dot{P} = 10^{-39} B^2(t)/P$), we can compute the spin period P as a function of time.

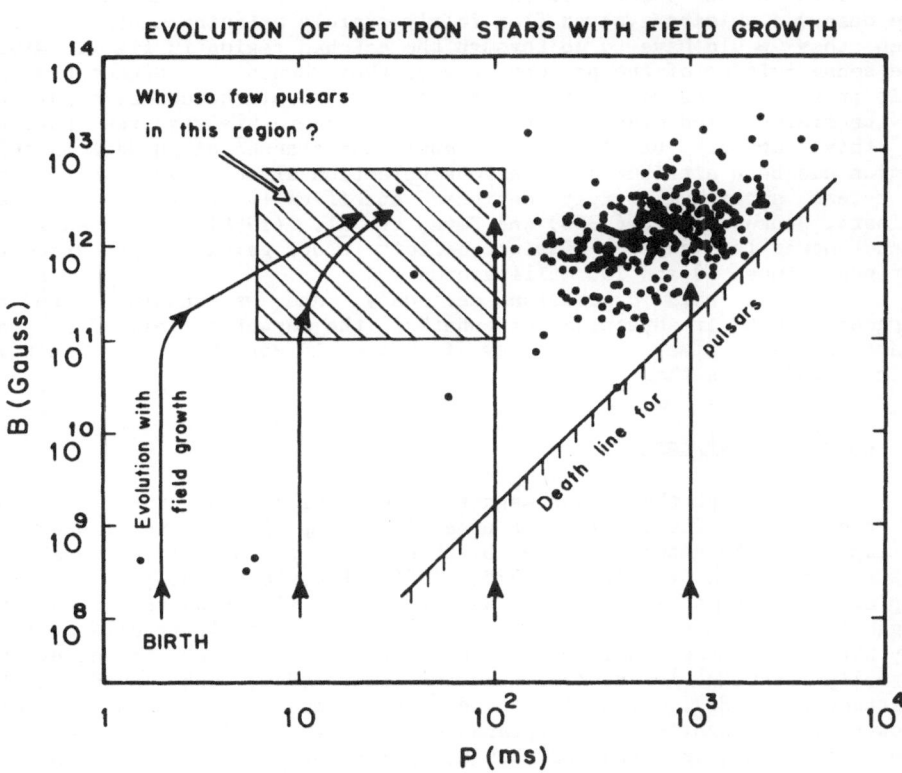

FIG. 1: *The evolution of neutron stars during an exponential growth of their magnetic fields. The e-folding timescale for the magnetic field has been assumed to be 9650 yr (see text). Trajectories corresponding to initial rotation periods of 2ms, 10ms, 100ms and 1s are shown. The periods and derived magnetic fields of observed pulsars are shown as dots. Neutron stars born with spin periods less than 100ms must pass through the hatched region as functioning pulsars, but very few pulsars have been detected in this region.*

In fig. 1 we have plotted the trajectories of pulsars in the field-period diagram for different values of the initial rotation period. One feature is immediately noticeable - namely, that most of the slowing down takes place at large fields. This is hardly surprising because the torque increases as the square of the magnetic field. We see that the slowdown is almost insignificant till a field strength of $\sim 10^{11}$ gauss is reached. A neutron star starting out with a spin period of $\sim 10$ ms slows down to at most $\sim 20$ ms when the field is fully grown, and one with a spin period of $\sim 100$ ms hardly slows down at all during the field growth phase. It is clear therefore that if the majority of the observed pulsars evolved from fairly rapidly spinning neutron stars, then they would have to go through the hatched region in fig.1. Given the sensitivities of the present surveys they should be detectable in this phase. But as we see from the distribution of the observed pulsars in the field-period diagram (shown as dots), there are very few pulsars in this region. Until quite recently the absence of pulsars in this region has been attributed to various selection effects. However, from a recent sensitive survey designed specifically to look for fast pulsars, Stokes et al. (1986) and Dewey et al. (1988) have concluded that there is indeed a distinct deficit in our galaxy of pulsars with periods between 10 and 100 milliseconds.

If we take this conclusion seriously, and we should, then it suggests that irrespective of whether the magnetic field of neutron stars grow with time, the majority of them could not have had initial spin periods less than $\sim 100$ ms.

## 4. CONCLUDING REMARKS

The idea that the strong magnetic fields of neutron stars may not be the fossil fields of their progenitors amplified in the process of collapse, but be generated after birth has been in the literature for quite some time (Woodward 1978, 1984; Blandford 1983; Blandford, Applegate and Hernquist 1983). The theoretical situation regarding the suggested mechanisms has never been very clear and it is difficult to say whether all the conditions necessary for such a growth to occur do actually obtain in a real neutron star (see, e.g. Kundt 1985; Bhattacharya 1987). Even so, it has been argued that such a field growth might provide an explanation for the deficit of detectable short-period pulsars and of bright pulsar-produced nebulae in our galaxy, without requiring most neutron stars to be born slow. The arguments presented above suggest that this is unlikely to be the case. Irrespective of whether the magnetic field grows or not, slow rotation at birth for most pulsars remains an inevitable conclusion.

## REFERENCES

Bhattacharya, D. 1987, Ph.D. Thesis, Indian Institute of Science, Bangalore.
Bhattacharya, D., Shukre, C.S. 1985, J. Astrophys. Astr. **6**, 233.
Blandford, R.D. 1983, Astr. J. **88**, 245.

Blandford, R.D., Applegate, J.H., Hernquist, L. 1983, Mon. Not. R. astr. Soc. **204**, 1025.
Chevalier, R.A., Emmering, R.T. 1986, Astrophys. J. **304**, 140.
Dewey, R.J., Taylor, J.H., Maguire, C.M., Stokes, G.H. 1988, Astrophys. J. **332**, 762.
Kundt, W. 1985, Bull. Astr. Soc. India **13**, 12.
Narayan, R. 1987, Astrophys. J. **319**, 162.
Srinivasan, G., Bhattacharya, D., Dwarakanath, K.S. 1984, J. Astrophys. Astr. **5**, 403.
Stokes, G.H., Segelstein, D.J., Taylor, J.H., Dewey, R.J. 1986, Astrophys. J. **311**, 694.
Vivekanand, M., Narayan, R. 1981, J. Astrophys. Astr. **2**, 315.
Woodward, J.F. 1978, Astrophys. J. **225**, 574.
Woodward, J.F. 1984, Astrophys. J. **279**, 803.

# 2D RELATIVISTIC PULSAR WINDS FROM RAPIDLY ROTATING NEUTRON STARS

M. CAMENZIND
*Landessternwarte Königstuhl*
*D-6900 Heidelberg 1*
*Federal Republic of Germany*

ABSTRACT. Pulsars are rotating neutron stars surrounded by strong magnetic fields and energetic particles. The structure of both the interior and exterior magnetosphere depends on the strength of the plasma source near the surface of the star. We start our analysis with magnetospheric models including pair-production as proposed by Sturrock, Ruderman and Sutherland, and Arons. In the light of these models, all neutron stars observed as pulsars must have relativistic magnetohydrodynamic wind exterior magnetospheres. Explicit solutions of the pulsar wind equation have so far only been obtained for monopole geometries. We present the most general axisymmetric theory for adiabatic pulsar winds including the gravitational background of the neutron star, the structure of the magnetic flux tube and finite pressure effects. These MHD equations can be reduced to a tractable set of equations consisting of the hot wind equation for the pair plasma and the relativistic Grad-Schlüter-Shafranov (GSS) equation for the magnetic flux function. The structure of the solutions for the hot pair wind equation is discussed. We also show how self-consistent axisymmetric MHD magnetospheres can be constructed by simultaneous solutions of this pair wind equation and solutions of the GSS equation. In the case of young pulsars and of the millisecond pulsars, the escaping pair wind is terminated by a strong shock. Downstream of the shock, the flow decelerates and increases its pressure in order to match the boundary conditions imposed by the outer edge of the nebula. We discuss the jump conditions for those types of relativistic MHD shocks, which consist of a family of nested toroids as shown by the hard X-ray observations on the Crab nebula. This is in conflict with standard spherically symmetric models. We also apply these pulsar winds to millisecond pulsars in low mass binary systems, such as PSR 1957 + 20, where the pulsar wind is expected to form a shock front around the evaporating low mass companion star.

## 1. Introduction

Pulsars were discovered 20 years ago. Looking back over 20 years in pulsar physics, it seems to be extremely difficult to present a globally accepted physical picture about a working active pulsar. In these lectures, various aspects of the production and acceleration of pulsar winds are discussed. We also show the importance of pulsar winds for the understanding of the Crab Nebula and of similar nebulae around millisecond pulsars. Since we are not following the historical order of events in these lectures, some cornerstones in the evolution of the pulsar theory will be

139

*W. Kundt (ed.), Neutron Stars and Their Birth Events, 139–177.*
© *1990 Kluwer Academic Publishers.*

described in this section. More details on particular pulsar models can be found in the reviews by Michel (1982) and Kundt (1985). Major developments taking place in pulsar radio observations are summarized by Taylor and Stinebring (1986), and Kulkarni (this Summer School).

In the next section, various arguments are presented for the existence of a dense pair plasma which escapes from the polar region of rapidly rotating neutron stars as a strong wind carrying off energy and angular momentum. According to this picture, the coherent radio emission is due to instabilities in this pair plasma and dead neutron stars are no longer capable of producing a sufficiently dense pair wind. Solutions for the dead neutron star problem are beyond the scope of these lectures. Charge-separated solutions of the magnetospheric problem have been considered among others by Mestel and Wang (1982), Mestel et al. (1985), Fitzpatrick and Mestel (1988a,b) and Shibata (1988).

But everybody still agrees on the fact that pulsars are rotating neutron stars surrounded by strong magnetic fields and energetic particles. During the last years, extremely rapidly rotating neutron stars have been detected, the pulsars PSR 1937+21 and PSR 1957+20 have periods of 1.6 milliseconds close to the limiting period of 0.5 milliseconds, or $\Omega_{\rm lim} \simeq \sqrt{GM/R^3} \simeq 10^4$ radian per second. This rapid rotation means that not only the interior of the neutron star must be based on general relativity, but also the exterior gravitational field has to be given in a general relativistic manner. Since the structure of the gravitational field is no longer a secondary question for wind theories, we discuss in the beginning of Sect. 3 the general form of this gravitational field for rapidly rotating neutron stars. In this context, we also discuss the question of the exact limiting period $P_{\rm lim} = 2\pi/\Omega_{\rm lim}$ for neutron star matter, which turns out to be closer to 1.5 milliseconds than to 0.5 milliseconds depending somewhat on the mass and the particular equation of state for neutron star matter.

A real pulsar is a 3-dimensional object and so is its wind. This 3D problem is one of the outstanding unsolved problems in relativistic astrophysics. Until a few years ago, in fact only 1-dimensional wind models were available. The relativistic magnetic monopole wind generated by an axisymmetric aligned rotator was first considered by Michel (1969). His treatment is a generalisation of the classical Weber-Davis model for solar winds. Kennel et al. (1983) extended this 1D model to include the finite pressure of the pair plasma, without considering the gravitational effects from the central source. Okamoto realized in 1978 how one can get rid of the unphysical monopole structure. He found a special relativistic formulation of the cold wind problem in terms of axisymmetric magnetic surfaces, gravity and pressure were still missing in this formalism.

In the last few years, we found a formulation of 2D relativistic axisymmetric MHD winds, which includes the gravitational effects from the central source, finite pressure and the structure of the magnetic surfaces (Camenzind 1986a,b, 1987; Camenzind and Endler 1989). This theory is the basic starting point for the construction of relativistic MHD-jets (Camenzind 1988, 1989a), but it can be similarly applied to the problem of relativistic pair winds (Camenzind 1989b). In Sect. 3 we present therefore the most general 2D axisymmetric pulsar wind theory which includes the general relativistic gravitational effects from the rapidly rotating compact source, the detailed structure of the magnetic surface which guides the outflowing plasma, and the effects from the finite pressure of the pair plasma. Radiation effects are however neglected, since they are of secondary importance.

According to the basic assumption in ideal MHD, the plasma is forced to stream along the magnetic surface which connects the polar cap region with the asymptotic region. The concept of magnetic surfaces is therefore crucial in our treatment of the problem. Its detailed form determines basically the overall acceleration of the pair wind. And here, we find indeed the biggest difference between 1D and 2D models for pulsar winds. In a 1D model, the boundary conditions implied at the Alfvén surface for the structure of the magnetic surface cannot be satisfied, and, therefore, the relationship between Michel's magnetisation parameter $\sigma$ (Michel 1969) and the terminal speed of the plasma found in a 1D model must be incorrect. We found that for a given $\sigma$ the final velocity is much higher in a 2D model than in the 1D case, the plasma is post-accelerated at the light cylinder by cross-field currents (Camenzind and Endler 1989). This helps to solve a long-standing puzzle for pulsar winds. A stronger acceleration means that more magnetic energy is converted into kinetic energy, the Poynting flux at infinity is drastically reduced in a 2D model as compared with the 1D case. In a fully 3D treatment, this Poynting flux could be reduced once more through reconnection processes (see e.g. Arons 1981). But this question is completely open.

In Sect. 4 we use therefore the asymptotic properties of the pulsar wind to couple the pulsar with its ambient medium, which is a supernova remnant in the case of young pulsars. In the case of millisecond pulsars, the ambient medium is just the interstellar medium. We show that this coupling is a real electrodynamic coupling in the sense that the rotating pulsar must drive a current system which can only be closed over the inner shocked region in pulsar nebulae. When the ram pressure of the pulsar wind equals the pressure of this surrounding medium, a shock must form producing an amplification of the toroidal magnetic field and a randomisation of the directed plasma energy flow. This type of consideration goes back to Rees and Gunn (1974). In the literature, only 1D models have been discussed for this pulsar nebula. Since the pulsar wind is essentially 2D, the structure of the nebula is also at least 2-dimensional. The basic properties of 2D axisymmetric pulsar nebulae will be discussed, and we also strengthen their similarity with hot spot regions in extragalactic radio sources.

Pulsar winds have also a certain importance for close binary systems consisting of a normal star in orbit around a weakly magnetized neutron star. In these systems, normal plasma from the accretion disk could be injected near the light cylinder and driven away as a marginally relativistic pulsar wind. The physics of this process depends crucially on the interaction of the magnetosphere with the accretion disk.

## 2. Relativistic Particle Winds from Magnetized Rotators

### 2.1. Pulsar Magnetospheres

The possibility that pulsars emit charged particles inspired essentially two directions of research aimed at delineating the exterior magnetosphere, which couples the neutron star with the surrounding nebula. The first approach was based on the vacuum Deutsch-Pacini magnetic dipole wave (Deutsch 1955), making it a nonlinear plasma wave whose amplitude is so large that it drives particles to relativistic energies. If pair production is efficient near the surface of the neutron star, this wave is quenched within a few wavelengths (Asséo et al. 1978, 1980). In addition, there is still not much known about such exotic waves and their interaction with the pulsar environment (Rees and Gunn 1974; Kundt and Krotschek 1980; Asséo, Llobet and Pellat 1984). The second direction was to consider relativistic winds, by analogy with stellar winds supposing a sufficiently high plasma density to justify the use of the magnetohydrodynamic approximation and also magnetohydrodynamic concepts. Therefore, the fundamental question of pulsar physics is really: *what is the density in the inner magnetosphere ?*

Since rapidly rotating pulsars are the most likely to have high densities, these pulsars will have relativistic winds. There exists, however, no 3D solution for an exterior wind which couples to an interior dipolar, or higher multipolar field. In addition, the 3D equations used are dissipation free, while we expect that in a real 3D configuration magnetic field line reconnection is an essential input in the energetics of pulsar winds. For these reasons we present in the following the 2D axisymmetric pulsar magnetosphere filled with plasma. Here, one has to admit that in a real pulsar the rotation and dipole axes must be misaligned. We hope that a small inclination will not drastically alter the involved physics. However, a 90° inclination can only be treated as a real 3D problem.

## 2.2. Pair Winds from Rotating Neutron Stars

Typical radiopulsars have been observed only through their coherent microwave emission, but this only accounts for a very small fraction of their measured spin-down energy loss rate. There is a whole bunch of much more energetic and more intense electromagnetic and particle emission, in particular observed for pulsars with strong magnetic fields and short periods. These include

a) pulsed X-rays

b) $\gamma$-rays in the MeV-region from young pulsars

c) strong optical emission from fast pulsars

d) very intense $e^{\pm}$ winds inferred for the Crab pulsar by modeling the emission from the surrounding nebula, which must be supported by injection from the pulsar. Pulsars with spin down energy loss rates in excess of $10^{34}$ erg/s have surrounding X-ray halos.

Attempts to understand the observed synchrotron emission from these nebulae are strongly supporting the idea that these neutron stars are sources of strong pulsar winds. The synchrotron X-rays and the optical emission from the Crab nebula together with the growing nebulae expansion kinetic energy imply an injection rate of $e^{\pm}$ of $\simeq 10^{39}$ particles per second with a typical mean energy of $\simeq 10^{12}$ eV. These particles must form a more or less neutral plasma, since otherwise the electronic current magnetic field would even exceed the field of the star. Therefore, if electrons are injected and accelerated into the nebula by the central star, an almost equal number of positrons must flow with them. On the other hand, there is no mechanism for lifting a flux of $10^{39}$ ions per second from the neutron star surface to locally cancel the electron flux. It is therefore suggestive that pair production near the surface of the Crab pulsar is the source of the particles injected into the nebula. Models for the nebula emission proposed by Rees and Gunn (1974), Kundt and Krotschek (1980), Kennel and Coroniti (1984a,b) and Emmering and Chevalier (1987) just start with this assumption.

There is still another reason which supports the presence of a dense pair plasma near the neutron star surface. Either the surface of the neutron star or some current-carrying field lines connecting to the polar cap can be sources of charged particles. Goldreich and Julian (1969) argued that field emission will inject charges of one sign, depending on whether $\Omega_* \cdot \mathbf{B} > 0$ or $\Omega_* \cdot \mathbf{B} < 0$, from both polar caps into the exterior magnetosphere. Those particles emitted on closed field lines will accumulate until a critical number density $n_{GJ}$, sufficient to suppress the parallel electric field, is reached

$$n_{GJ} = \frac{\Omega_* \cdot \mathbf{B}}{2\pi ec} = 7 \cdot 10^{10}\,\mathrm{cm}^{-3}\,B_{*,12}\left(\frac{R_*}{r}\right)^3 P^{-1}\,. \tag{1}$$

This Goldreich-Julian density is a critical density in the sense that it approximately separates regimes of space-charge physics and quasi-neutral plasma physics. Since some field lines must be open, their footpoints at the star define the polar cap. Sturrock (1971) suggested for the first time that current-carrying field lines can be unstable to pair production discharge. The current density threading the polar cap is of order of $n_{GJ}ec$. Such a discharge would then produce a plasma whose density exceeds $n_{GJ}$, suggesting that the physics will be magnetohydrodynamic and not space-charge dominated, at least outside to the production region. If a Goldreich-Julian current flows along open field lines, we get a total particle loss rate from the polar cap (if pair production is not included)

$$\dot{N}_c = A_c\, n_{GJ}\, c \quad , \quad A_c = \pi R_*^2\, \frac{R_*}{R_L}\,. \tag{2}$$

For the Crab pulsar we then have $\dot{N}_c \simeq 9.0 \cdot 10^{34}\,s^{-1}\,B_{12.7}$ This would indicate for the Crab pulsar that at least $10^4$ pairs are created per primary beam particle in the wind. The observed microwave radiation follows then from this relativistic electron beam-plasma system, which could excite plasma oscillations near $\omega_p$, $\omega_p = \sqrt{4\pi n_p e/m} = 5.6 \cdot 10^4\, n_p^{1/2}\,s^{-1}$,

$$\nu_{microwave} \simeq \omega_p/2\pi \simeq 0.9\,\mathrm{GHz}\,n_{p,10}^{1/2}\,. \tag{3}$$

For energetic reasons we have roughly $\gamma_b\, n_b \simeq 2\gamma_p\, n_p$, $n_b \simeq n_{GJ}$, $\gamma_b \simeq 10^6$, and $\gamma_p \simeq 10^2$. These plasma oscillations occur somewhere between the polar cap surface and the light cylinder. Therefore, the presence of the microwave radiation could also be an indicator for a dense pair plasma near the neutron star surface. The physics of these beam-plasma systems is vastly unexplored, however, see Ulm and Davidson (1980) and Ulm (1982) for an analysis of a homogeneous beam-plasma system, and Usov (1987) for a recent discussion of the two-stream instability in pulsar magnetospheres. Similar considerations also hold for the slot-gap model proposed by Arons (1981, 1983) based on electrodynamic considerations for inclined rotators.

The accelerator for the pair plasma cannot be based on electric fields, since these would accelerate electrons and positrons in opposite directions. In fact, the parallel electric field $\mathbf{E}_{\parallel} \simeq 0$ in the slot-gap models (Arons, 1983). An equal acceleration of electrons and positrons in a way that no net electric current results, would be produced by copious pair injection into the rapidly rotating magnetic field of the pulsar. The presence of the pulsar's optical and X-ray radiation also supports the existence of a pair wind. Incoherent synchrotron or curvature radiation is not sufficient to explain the observed intensities unless the pair wind exceeds $10^{37}$ particles per second for the Crab and $10^{36}$ for Vela.

## 2.3. Pair Production

Production of $e^{\pm}$ pairs in the magnetosphere of a rapidly rotating neutron star is caused by the materialisation of high energy $\gamma$-rays. The dominant source of convertible photons is the curvature photons emitted by ultrarelativistic electrons moving along curved field lines; their energy is given by (Jackson, 1975)

$$h\nu \simeq h\nu_c = m_e c^2 \, \frac{3}{2} \, \frac{\Lambda_c}{\rho_c} \, \gamma_b^3(h) \tag{4}$$

where $\Lambda_c = \hbar/mc = 3.87 \cdot 10^{-11}$ cm is the Compton wavelength, $\rho_c$ the curvature radius of the field line and $\gamma_b(h)$ is the Lorentz factor of the beam particle at the height $h$ above the polar cap. These $\gamma-$rays have typically energies of a few GeV, are emitted parallel to $\mathbf{B}$ and travel along straight paths, so that the pitch angle $\psi$ between $\mathbf{B}$ and the photon momentum increases along its path. The dominant opacity source is then magnetic conversion (Erber 1966; Sturrock, 1970, 1971)

$$\alpha \simeq 0.23 \, \frac{\alpha_F}{\Lambda_c} \, \frac{B}{B_c} \, \sin\psi \, \exp\left\{ -\frac{8}{3} \, \frac{m_e c^2}{h\nu} \, \frac{B_c}{B \sin\psi} \right\} \quad . \tag{5}$$

$\alpha_F = 1/137$ is the fine structure constant, and $B_c = 4.4 \cdot 10^{13}$ Gauss the critical magnetic field. This opacity increases with $\psi$, it is therefore possible that the optical depth unity is reached before the decline of B along the outward propagating photon prevents a further increase of the optical depth (since for $r > R_*$ the magnetic field rapidly decays). The surface where $\tau = 1$ is reached is called pair formation front (PFF in Fig. 1). From these considerations one gets a minimum energy $h\nu_{\min}$ which can be absorbed anywhere above the polar cap region

$$h\nu_{\min} \simeq 70 \, \text{MeV} \left( \frac{5 \cdot 10^{12} \, \text{G}}{B_*} \right) \left( \frac{10 \, \text{km}}{R_*} \right)^{1/2} \left( \frac{P}{100 \, \text{ms}} \right)^{1/2} . \tag{6}$$

There are additional sources for hard $\gamma-$rays above the polar cap. Inverse Compton scattering of thermal photons into $\gamma$-rays can be shown to be negligible compared to curvature emission for typical polar cap temperatures $< 10^7$ K. Also pair creation through the interaction of a $\gamma$-ray with a thermal photon is not important, except perhaps for millisecond pulsars. Synchrotron emission from newly formed pairs created below and at the pair formation front is also a negligible source for convertible $\gamma$-rays compared to curvature emission. This latter emission mechanism is however an important contributor to the total number of pairs formed above the pair formation front. Finally, in models where the primary ultrarelativistic beam consists of positive ions, the neutral pion decay $\pi^o \mapsto \gamma + \gamma$ and similar hadronic processes could be important sources of $\gamma$-rays.

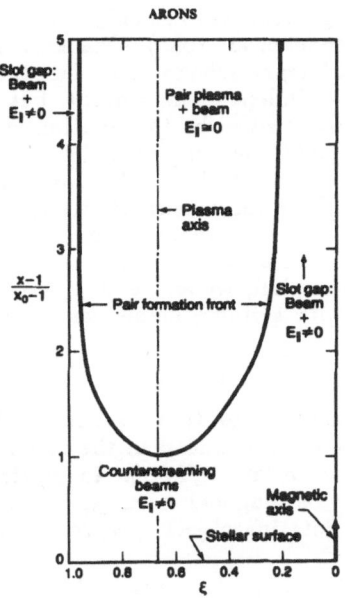

**Fig. 1.** Meridional cross section of the pair formation front in dipolar polar caps according to Arons (1983). $\xi$ is a measure for the distance from the magnetic axis, $\theta_c \xi r \sqrt{(r/R_*)}$, where $\theta_c \simeq 1.5 \cdot 10^{-2} \sqrt{(R_6/P)}$ is the opening angle of the polar cap, $x = \sqrt{r/R_*}$ so that the vertical distance is given as $s = R_*(x^2 - 1)$, $x_0 = \sqrt{1 + H_0/R_*}$.

Many authors carried out detailed calculations of pair-photon cascades initiated by high energy electrons above a pulsar polar cap (Sturrock 1971; Tademaru 1973; Daugherty and Harding 1982; Arons 1983; Jones 1983; Gurevich and Istomin 1985). In Fig. 1 we show a typical pair formation front in the slot gap models calculated by Arons (1983). Active pulsars always have strong pair winds, pair creation appears inevitable in short period pulsars, and one gets a maximum period for steady pair creation

$$P_{max} \simeq 0.2\,s \left(\frac{\mu_{30}}{R_{*,6}}\right)^{0.64} R_{*,6}^{-0.61} (\sin i)^{0.48} f_c^{0.89} f_p^{0.07} \left(\frac{100}{\ln \Lambda_a \ln \Lambda_p}\right)^{0.16}, \qquad (7)$$

which explains the cutoff line in pulsar diagrams (see lectures by Kulkarni). For the meaning of the various parameters in this formula, see Arons (1979, 1983).

## 3. The Quasi-Aligned Rotator

The general structure of hydromagnetic relativistic stellar winds whose source is within a rotating dipolar magnetosphere has only been outlined for monopole type geometries (Michel 1969; Kennel et al. 1983). Okamoto (1978) generalized the discussion of cold winds to an arbitrary axisymmetric flux tube. The structure of these flux tubes is, however, not an independent concept, but it follows from Ampère's equation with a toroidal current given by the solution of the equations of motion. This represents therefore a highly nonlinear problem. In order to understand the structure of the nestedness of this MHD problem, one should first study the corresponding Newtonian analog for self-consistent stellar winds (for this see Sakurai (1985)). In this section, we present a thorough reduction of the relativistic MHD equations for the stationary and axisymmetric case, in a fully general relativistic manner. At the end of this section, we arrive at a system of two nonlinear equations, which can be solved with suitable numerical methods for certain boundary conditions.

### 3.1. Rapidly Rotating Neutron Star Models

Due to the high compactness of the neutron star, $R_* \leq 3R_G$ for a mass of $1.4\,M_\odot$, general relativistic effects must be included in the formulation of the background space-time ($R_G = 2GM/c^2 = 3$ km $M/M_\odot$). In particular, the rotation of the central object introduces also off-diagonal terms $g_{t\phi}$ in the metric so that the space-time generated by a rapidly rotating object is represented by the metric

$$ds^2 = e^{2\nu}\,dt^2 - e^{2\psi}\,(d\phi - \omega dt)^2 - e^{2\mu}\,d\theta^2 - e^{2\lambda}\,dr^2\,, \qquad (8)$$

with the metric coefficients independent of $t$ and $\phi$. The coordinate $r$ is a Schwarzschild-like radial coordinate in the equatorial plane. This form of the metric is guaranteed by rigorous theorems, when the source of the gravitational field is only in toroidal motion (Carter 1973). The above metric has two Killing fields, $\mathbf{k} = \partial_t$ for stationarity, and $\mathbf{m} = \partial_\phi$ for axisymmetry. The 4-velocity $\mathbf{u}$ of the neutron star matter can then be written as

$$\mathbf{u} = u^t(\mathbf{k} + \Omega\mathbf{m})\,, \qquad (9)$$

where $\Omega$ is the angular velocity measured by an observer at infinity.

Relativistic models of slowly rotating neutron stars ($\Omega \ll \Omega_K(R_*)$) were first constructed by Hartle and Thorne (1968) based on a formalism developed by Hartle

(1967). An extensive study of the properties of these models can be found in the work by Ray and Datta (1984). In this approximation, the space-time generated by a rotating neutron star has the following metric coefficients

$$e^{2\nu} = \left\{1 + 2\Phi_G + \frac{2G^2 J_*^2}{c^6 r^4}\right\}\left\{1 + 2f_\nu(r, M, J_*, Q_*)P_2(\cos\theta)\right\}, \tag{10}$$

$$e^{2\lambda} = \left\{1 + 2\Phi_G + \frac{2G^2 J_*^2}{c^6 r^4}\right\}^{-1}\left\{1 - 2f_\lambda(r, M, J_*, Q_*)P_2(\cos\theta)\right\}, \tag{11}$$

$$e^{2\mu} = e^{2\psi}/\sin^2\theta = r^2\left\{1 + 2f_\mu(r, M, J_*, Q_*)P_2(\cos\theta)\right\}, \tag{12}$$

and involves 3 parameters (the total mass $M$, angular momentum $J_*$ and the quadrupole moment $Q_*$; for the functions $f_\nu$, $f_\lambda$ and $f_\mu$ see Hartle and Thorne (1968)). From these expressions we get the nonvanishing components of the metric tensor

$$g_{tt} = e^{2\nu} - \omega^2 e^{2\psi}, \tag{13}$$

$$g_{t\phi} = \omega e^{2\psi}, \tag{14}$$

$$g_{\phi\phi} = -e^{2\psi}, \tag{15}$$

$$g_{rr} = -e^{2\lambda}, \tag{16}$$

$$g_{\theta\theta} = -e^{2\mu}. \tag{17}$$

$\Phi_G$ is the Newtonian potential and $J_*$ the total angular momentum of the neutron star, $J_* = I_* \Omega_* = k^2 M R_*^2 \Omega_*$, $k^2 \simeq 0.4$ depending on the particular form of the equation of state for the neutron star matter. The metric coefficient $g_{t\phi}$ represents the frame dragging effect due to the rotation of the central gravitational force. The angular velocity $\omega(r)$ of the locally non-rotating frames with respect to distant stars is given by

$$\omega(r) = -g_{t\phi}/g_{\phi\phi} = \frac{2GJ_*}{c^2 r^3} = k^2 \Omega_* \frac{R_G}{r}\left(\frac{R_*}{r}\right)^2. \tag{18}$$

The quantity

$$\Omega_* g_{t\phi} = k^2 (R_G/r)(R_*/R_L)^2 \sin^2\theta \ll (R_G/r) \tag{19}$$

is of crucial importance for the following considerations. These models are certainly appropriate for a typical classical pulsar with a mean rotation period of 0.7 seconds.

For slowly rotating pulsars the gravitational binding energy plays the main role for the equilibrium structure. For millisecond pulsars, however, the rotational energy is already a significant fraction of the total energy of the system. According to Poincaré, the upper limit for the angular velocity to sustain stresses is

$\Omega_{max} = \sqrt{2\pi G \rho} \simeq \sqrt{1.5\,GM_*/R_*^3} \simeq 2 \times 10^4$ radian/s. A period of 1.6 milliseconds corresponds to $\Omega \simeq 0.4 \times 10^4$ radian/s. There are tighter limits on this period from secular instability theory. The rotation of a neutron star is limited by a gravitational instability to non-axisymmetric perturbations. They have a growth time determined by the rate at which energy is radiated. This gravitational radiation makes rotating perfect equilibria unstable to modes of the form $\exp(\imath m \phi)$ for sufficiently large $m$, where $\phi$ is the azimuthal coordinate (Friedman and Schutz 1978). In this process, the rotational energy of the star is converted into gravitational waves. In the absence of viscosity and heat conductivity, modes with the shortest wavelengths (i.e. with large $m$) become unstable at the lowest values of $\Omega$. But these short-wavelength modes also have the longest growth time. If neutron star interiors contain a large bulk viscosity, which damps out gravitational wave driven instabilities, a mode is stable for viscous damping times shorter than the gravitational time scale. It turns out that the viscosity rises rapidly with decreasing temperature, and it seems that for $T < 10^6$ K all modes are stable. Only for high temperatures (say $T > 10^8$ K), the non-axisymmetric instability sets the upper limit on the rotational frequency (for further details on this question see Friedman 1983; Wagoner 1984; Lindblom 1986; Cutler and Lindblom 1987). For temperatures in the range of $T \simeq 10^7$ K, which is typical for accreting sources, it is an open question whether rotation is limited by non-axisymmetric perturbations (with mode numbers $m = 3, 4$) or by Keplerian rotation of the equatorial region.

The structure of rapidly rotating neutron stars can only be computed numerically starting with Einstein's equations for the metric of the form of eq. (8) and using a particular equation of state for the neutron star matter. Friedman, Ipser and Parker (1986) calculated sequences of rotating stars with a baryon mass of approximately 1.4 $M_\odot$ using the 10 different equations of state discussed previously by Arnett and Bowers (1977). The numerical method follows the detailed discussion in Butterworth and Ipser (1976). It turns out that all the sequences end before the ratio $T/|W|$ of rotational energy to gravitational energy reaches 0.14, where the bar mode instability ($m = 2$) would set in. Rapidly rotating neutron stars are by large non-spherically symmetric, and the Keplerian frequency $\Omega_K$ at which a given sequence of neutron star models ends is substantially smaller than its value for a spherical model. For a neutron star of mass 1.4 $M_\odot$, a stiff equation of state would only allow for a minimum period of 1.6 milliseconds, while a soft equation of state (Reid potential) would allow for a minimal period of 1.0 milliseconds (Friedman et al. 1988). It is interesting that the two fastest known radio pulsars, PSR 1937 + 214 and PSR 1957 + 20 are close to this limiting period.

Millisecond pulsars are now thought to be neutron stars spun up by earlier accretion processes in low-mass X-ray binary systems (see the lectures given by Kulkarni). They could also be the remnants of an accretion-induced collapse of white dwarfs in cataclysmic variable systems (Chanmugam and Brecher 1987). In the LMXB scenario, a weakly magnetized neutron star which is initially slowly rotating is spun up by accretion of angular momentum from disk material. If the accretion disk extends all the way down to the neutron star's equator, a neutron star of 1.4 $M_\odot$ can be spun up to millisecond periods by the accretion of about 0.2 $M_\odot$ of disk material. The exact value depends on the moment of inertia of the neutron star and, therefore, on the equation of state. This would suggest that weakly magnetized neutron stars (with $B_* \leq 5 \cdot 10^8$ Gauss) in LMXBs would rotate at their critical frequency and could be used as test systems for the type of equation of state (Friedman et al., 1988).

The gravitational field of these critically rotating neutron stars is neither given by the above slow rotation approximation, nor by the Kerr metric, since this latter metric obeys strict relations between the mass multipole moments and the angular momentum multipole moments. The exterior gravitational field has still the form discussed in eq. (8), but the functions $\nu, \mu, \psi, \omega$ must be computed numerically. All these functions have angular expansions of the form

$$\nu(r, \theta) = \sum_{l=0}^{\infty} \nu_{2l}(r) \, P_{2l}(\cos \theta) \,, \tag{20}$$

where $P_l$ are Legendre polynomials. The functions $\nu_{2l}(r)$ have then an expansion in terms of multipole moments. The particular form of this expansion is not important for the following considerations. We just need the fact that the exterior gravitational field of rapidly rotating neutron stars has two Killing fields and can be written down in the form of equation (8).

## 3.2.  Relativistic MHD in the Gravitational Field of Rotating Stars

Whenever the local pair plasma density is higher than the Goldreich-Julian charge density, we can use the MHD condition in the sense that the electric field as seen in the plasma frame vanishes. The fundamental concepts of relativistic MHD can be found in the textbook written by Lichnérowicz (1967). The basic equations are particle number conservation

$$(n \, u^{\alpha})_{;\alpha} = 0 \quad , \tag{21}$$

magnetic flux conservation

$$(u^{\alpha} \, B^{'\beta} - u^{\beta} \, B^{'\alpha})_{;\alpha} = 0 \,, \tag{22}$$

and conservation of total energy-momentum

$$T^{\alpha\beta}_{;\beta} = 0 \,, \tag{23}$$

where $T^{\alpha\beta}$ is the total energy-momentum of the system

$$T^{\alpha\beta} = (\rho + P) \, u^{\alpha} \, u^{\beta} - (P + \frac{B^{'2}}{8\pi}) g^{\alpha\beta} - \frac{1}{4\pi} \, B^{'\alpha} \, B^{'\beta} \,. \tag{24}$$

$n$ is the proper particle density, $\rho$ the total energy density, $P$ the pressure, $u^{\alpha}$ the four-velocity of the plasma, and $B'$ the magnetic field as seen in the plasma frame

150

$$B'_\alpha = u^\beta * F_{\alpha\beta} \, . \tag{25}$$

The corresponding electric field $E'$ seen by the comoving plasma vanishes due to the MHD assumption $u^\alpha F_{\alpha\beta} = 0$. When written in this compact form, these equations look simple, but they are quite untractable when written down explicitly. We found in the last years a way to reduce this system of equations for *axisymmetric* configurations.

In the following we shortly describe the reduction of the above set of equations under axisymmetry. A full account for this procedure can be found in Camenzind (1986a, b, 1987a). The essential idea is based on the existence of two Killing fields $\mathbf{k} = \partial_t$ and $\mathbf{m} = \partial_\phi$ on stationary and axisymmetric space-times. The basic conservation laws, which follow from these Killing fields, have been written down for the first time by Bekenstein and Oron (1978).

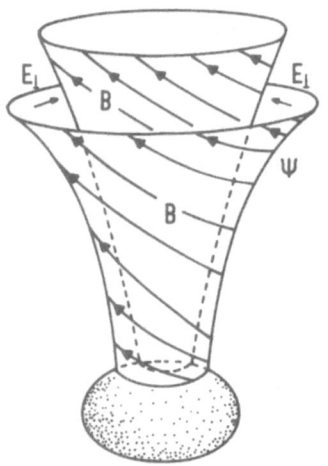

**Fig. 2.** The rotationally symmetric magnetic surfaces of an axisymmetric current-carrying magnetosphere. The magnetic surfaces are at the same time electric equipotential surfaces and plasma streaming surfaces. The magnetic field lines lie on rotating magnetic surfaces. The streaming plasma generates toroidal currents, which modify a vacuum solution.

Stationarity and axisymmetry of the flow provide the existence of a stream-function $\Phi = \Phi(x^A)$, following from particle number conservation

$$\sqrt{-g}\,nu^1 = -\partial_{x^2}\Phi \quad , \quad \sqrt{-g}\,nu^2 = \partial_{x^1}\Phi . \tag{26}$$

A conserved quantity is constant along the poloidal flow, $e = e(\Phi)$. The implied symmetries require $F_{t\phi} = 0$, and this condition together with the MHD condition

$E'_t = 0 = E'_\phi$ yields the relations

$$u^A F_{At} = 0 = u^A F_{A\phi} . \tag{27}$$

The second relation implies that the magnetic potential $\Psi = A_\phi = \Psi(\Phi)$ is conserved along the plasma stream. For practical purposes one uses therefore the magnetic potential $\Psi$ as the fundamental quantity instead of the stream line potential $\Phi$. The constants of motion $e(\Phi)$ are therefore considered as functions of $\Psi$. The above relation also tells us that the magnetic flux surfaces $\Psi = const$ are at the same time electric potential surfaces, $A_t = A_t(\Psi)$. Using the MHD condition and the symmetries of the Maxwell field $F$, we obtain further restrictions for the electromagnetic field

$$F_{tA} + \Omega^F(\Psi) F_{\phi A} = 0 \quad , \quad A = 1, 2 . \tag{28}$$

$\Omega^F(\Psi)$ is constant along flow lines and represents the angular frequency of the field lines. A second integration constant $\eta(\Psi)$ describes the particle-flux per unit flux tube

$$\sqrt{-g} \, n \, u^{1(2)} = -\eta(\Psi) \, F_{23(31)} , \tag{29}$$

$$\sqrt{-g} \, n \, u^t \, (\Omega - \Omega^F) = -\eta(\Psi) \, F_{12} . \tag{30}$$

$\Omega = u^\phi / u^t$ is the angular frequency of the plasma flow. In order to understand the physical meaning of these two constants of motion, it is useful to replace the Maxwell field $F$ by the magnetic and electric fields ($\mathbf{B}$ and $\mathbf{E}$) as seen by a stationary observer system on the rotating space time. From the experience with the Kerr metric we know that static observers would not be globally well defined. A stationary observer feels the frame dragging from the rotating central source. For these sources, the concept of fiducial observers has been introduced (the famous "FIDOs", see e.g. Thorne et al. 1986). Since the rotation of the central body drags all physical objects into orbital motion in the same direction as the body rotates, fiducial observers must also have some radius-dependent angular velocity $\omega$ as seen by asymptotic observers. The 4-velocity of the FIDOs is therefore given by

$$\mathbf{U}^{\mathrm{FIDO}} = \alpha^{-1} \, (\mathbf{k} + \omega \, \mathbf{m}). \tag{31}$$

Here, the redshift factor $\alpha$ follows from the lapse function of the metric in eq. (8), $\alpha = e^\nu$. In the case of the Kerr metric, this is the well known Bardeen observer. Physical processes on the background of a rotating body are then described in terms of the proper reference frame locally spanned by the orthonormal basis vectors (Camenzind 1975)

$$\mathbf{e}_t = \mathbf{U}^{\mathrm{FIDO}} , \; \mathbf{e}_r = e^{-\lambda} \partial_r , \; \mathbf{e}_\theta = e^{-\mu} \partial_\theta , \; \mathbf{e}_\phi = e^{-\psi} \partial_\phi . \tag{32}$$

In particular, the physical components of the magnetic field $\mathbf{B}$ follow from the defining equation (25), when the velocity field $\mathbf{u}$ is identified with the velocity field of the FIDOs, i.e.

$$B^\alpha = -\frac{1}{2} \eta^{\alpha\beta\gamma\delta} \, U_\beta^{\mathrm{FIDO}} \, F_{\gamma\delta} . \tag{33}$$

These fields have the following form

$$B^1 = -\frac{\alpha}{\sqrt{-g}} F_{23} \quad , \quad B^2 = -\frac{\alpha}{\sqrt{-g}} F_{31} \quad , \quad B^3 = -\frac{\alpha}{\sqrt{-g}} F_{12} . \tag{34}$$

Together with the defining equation for $\eta(\Psi)$ we get the well known relation between poloidal velocity and poloidal magnetic fields

$$u^A = \frac{\eta}{\alpha n} B^A \tag{35}$$

and similarly for the toroidal components

$$u^t(\Omega - \Omega^F) = \frac{\eta}{\alpha n} B^3 . \tag{36}$$

These relations are standard expressions in Newtonian axisymmetric MHD (see e.g. Sakurai 1985). In a similar way, we can also decompose the electric field **E** with respect to the FIDO system

$$\begin{aligned} \mathbf{E}_T &= 0, \\ \mathbf{E}_p &= -\mathbf{v}^F \wedge \mathbf{B}_p . \end{aligned} \tag{37}$$

$\mathbf{v}^F$ is the field line velocity with respect to the FIDO system $(R = e^\psi)$

$$\mathbf{v}^F = \frac{\Omega^F - \omega}{\alpha} R \, \mathbf{e}_\phi . \tag{38}$$

All these relations tell us that electrodynamics inside and outside of rapidly rotating neutron stars is a bit more complicated than in special relativity which neglects the gravitational effects. All treatments of magnetic fields for neutron stars have so far neglected a possible influence of the gravitational background. The lecturer of this topic felt a great response from the audience of this Summer School concerning the possible role of rapid rotation for the generation and decay of magnetic fields of rapidly rotating neutron stars. I leave this subject as a homework problem to the students, in particular resistive MHD should be formulated for rotating backgrounds. In the following I just outline one aspect of this general relativistic MHD which concerns the properties of the streaming pair plasma.

### 3.3. Energy, Angular Momentum and the Alfvén Cylinder

Stationarity and axisymmetry also require that the total energy $E$ and the total angular momentum $L$ are conserved along the flow, which involve the kinematic scalars $u_t$ (the redshift factor) and $l = -u_\phi/u_t$ (the specific angular momentum). The existence of the two Killing fields provides by standard methods two conserved currents

$$P^\alpha = T^\alpha_\beta k^\beta : \quad \text{energy current}, \tag{39}$$

$$J^\alpha = -T^\alpha_\beta m^\beta : \quad \text{angular momentum current}. \tag{40}$$

It is then a good exercise in MHD to show that in the ideal case these currents can be reduced to the form

$$P^\alpha = E(\Psi)\, N^\alpha \quad , \quad J^\alpha = L(\Psi)\, N^\alpha . \tag{41}$$

In special relativity, these two conserved quantities have the familiar form

$$E(\Psi) = \mu\, u_t - \frac{R\Omega^F B_\phi}{4\pi\eta} , \tag{42}$$

$$L(\Psi) = \mu l - \frac{RB_\phi}{4\pi\eta} . \tag{43}$$

$\mu = (\rho + P)/n$ is the specific enthalpy, and the first term on the right hand side represents the plasma contribution to total energy and angular momentum, while the second term involves the contribution from the magnetic fields. Similar expressions can be obtained for the rotating background so that we can solve these expressions for $u_t$ and $l$ using suitable expressions for the magnetic fields (Camenzind 1986b, Camenzind and Endler 1989)

$$u_t = \frac{E}{\mu}\, \frac{1 - M_g^2 - \Omega^F L/E}{1 - M_g^2 - \Omega^F l(\Omega^F)} , \tag{44}$$

$$l = \frac{l(\Omega^F)(1 - \Omega^F L/E) - M_g^2 L/E}{1 - M_g^2 - \Omega^F L/E} . \tag{45}$$

$M_g$ is a rescaled Mach number defined as

$$M_g^2 = \frac{M^2}{g_{tt} + \Omega^F g_{t\phi}} \tag{46}$$

for the Alfvén Mach number $M$, defined by

$$M^2 = \frac{4\pi\mu\eta^2}{n} = \frac{4\pi\alpha^2 \mu n u_p^2}{B_p^2} . \tag{47}$$

The quantity $l(\Omega^F)$ is the angular momentum of a corotating system with respect to the zero angular momentum frame defined by $\omega$

$$l(\Omega^F) = -\frac{g_{\phi\phi}\,(\Omega^F - \omega)}{g_{tt} + \Omega^F g_{t\phi}} . \tag{48}$$

It is instructive to compare the relativistic expression (45) with the Newtonian quantity for $j = R^2\,\Omega$

$$j = \frac{R^2\,\Omega^F - M^2\,L}{1 - M^2} , \tag{49}$$

with the Newtonian Mach number $M = v_p/v_A$. In view of this expression, the relativistic version contains three corrections. The Mach number $M$ is formed with the relativistic enthalpy $\mu = (\rho + P)/n$ and not with the rest mass. Secondly, there is a relativistic correction of the form $\Omega^F L/E$, which is important when magnetic fields carry an appreciable amount of energy. The third correction is of gravitational origin and involves the metric coefficients $g_{tt}$ and $g_{t\phi}$. The redshift factor $u_t$ contains as an additional correction in the denominator the term $\Omega^F l(\Omega^F)$, which is due to the existence of the light cylinder. In general, a "light cylinder" $\Omega^F l(\Omega^F) = 1$ means that the corotating field

$$\xi = \mathbf{k} + \Omega^F \mathbf{m} \tag{50}$$

spans a null surface defined by the condition $(\xi, \xi)_L = 0$, or

$$g_{tt}(X_L) + 2\Omega^F g_{t\phi}(X_L) + (\Omega^F)^2 g_{\phi\phi}(X_L) = 0. \tag{51}$$

In the absence of gravity ($g_{tt} = 1, g_{t\phi} = 0, g_{\phi\phi} = -R^2$), this is the well-known relation for the light cylinder in a flat space-time, $R_L = c/\Omega^F$. The presence of gravity will deform the "cylinder" somewhat. In addition, in a Kerr space-time e.g. we also find an inner light cylinder surface located close to the horizon of the black hole (Camenzind and Endler 1989). As an application, the reader is encouraged to calculate the form of the light cylinder for a slow rotator geometry, or a Kerr geometry.

The equation for the redshift factor $u_t$ would become singular at the cylinder $x = x_A$ defined by

$$M_{gA}^2 = 1 - \Omega^F l(\Omega^F) \quad , \tag{52}$$

unless we require at the same time

$$M_{gA}^2 = 1 - \Omega^F L/E \quad . \tag{53}$$

From these two relations we get the position $x_A$ in terms of

$$l_A(\Omega^F) = L/E \quad . \tag{54}$$

It is easy to show that the expression for the specific angular momentum becomes singular at the same point, but does not produce any additional constraints. The above expression reduces to $R_A^2 \Omega^F = L/E$ in Minkowski space and is well known in the Newtonian case, $R_A^2 \Omega^F = L/m$.

The expressions found above for the redshift factor $u_t$, the specific angular momentum $l$ and the location of the light cylinder and the Alfvén surface are completely general and valid for any stationary and axisymmetric space-time. They can be applied to any flow along a given magnetic flux tube anywhere in the meridional plane. This procedure generalizes the relations found by Okamoto (1978) for cold plasma flows on Minkowski space and the solutions found by Kennel et al. (1983) for the equatorial plane in a monopole geometry including pressure effects.

## 3.4. The Hot Wind Equation for Pair Plasmas

The poloidal velocity of the pair plasma when it streams along a given magnetic flux surface in the pulsar magnetosphere follows from the normalisation of the 4-velocity, $u^\alpha u_\alpha = 1$. Since the redshift factor $u_t$ and the specific angular momentum $l$ are determined by the conservation laws for the total energy and angular momentum, the solution of the poloidal velocity defined as

$$u_p^2 = -u^A u_A \quad , \quad A = 1,2 \quad , \tag{55}$$

follows from the expressions discussed in the last section

$$u_p^2 + 1 = \frac{u_t^2}{\Delta} \left( -g_{\phi\phi} - 2l\, g_{t\phi} - l^2\, g_{tt} \right) \quad . \tag{56}$$

$\Delta = g_{t\phi}^2 - g_{tt}\, g_{\phi\phi}$ is the square of the Weyl radius for axisymmetric metrics. The above equation for $u_p$ is still implicit, since the Mach number $M$ also contains the poloidal velocity. By using the definition of $\eta$, $\eta/n = \alpha\, u_p/B_p$, $B_p^2 = -B^A B_A$, $x = R/R_L$, we can, however, decompose this Mach number in a suitable way as

$$M^2 = 4\pi m \eta c^2 \frac{\mu}{mc^2} \frac{\eta}{n} = \frac{\mu}{mc^2} \alpha\, u_p\, x^2 \frac{\alpha_{in}\, f(R)}{\sigma_*(\Psi)} \tag{57}$$

with a similar decomposition for the specific enthalpy $\mu$

$$\mu = mc^2 \left\{ 1 + \mu_1 \left( \frac{\alpha_{in} u_{p,in} x_{in}^2}{\alpha u_p^2 x^2} \right)^{\Gamma-1} \right\}, \tag{58}$$

$$\mu_1 = \frac{\Gamma}{\Gamma - 1} \frac{P_{in}}{n_{in} mc^2} f^{\Gamma-1}, \tag{59}$$

$$f(R) = \frac{B_{p,in} R_{in}^2}{B_p R^2} \tag{60}$$

$$\sigma_*(\Psi) = \frac{B_{p,in} R_{in}^2}{4\pi \eta(\Psi) mc R_L^2} \quad . \tag{61}$$

With all these parameters, the above implicit equation for $u_p$ can be transformed to a polynomial of degree 14 for a polytropic index $\Gamma = 4/3$, when the quantity $z_p = u_p^{1/3}$ is used instead of $u_p$

$$\sum_{n=0}^{14} A_n \left( \frac{R}{R_L}, E, L, \frac{P_{in}}{n_{in} mc^2}; \frac{f}{\sigma_*}; g_{tt}, g_{t\phi}, g_{\phi\phi} \right) z_p^n = 0 \quad . \tag{62}$$

The detailed form of the coefficients $A_n$ for the hot pair wind can be found in Camenzind and Endler (1989) for any polytropic index $\Gamma = n/m$. The form of the solutions of this hot wind equation depends on the initial parameters $R_{in}$, $R_L$, $B_{p,in} R_{in}^2$, $u_{p,in}$, $P_{in}/n_{in} mc^2$, and the pair injection law $\eta(\Psi)$. Of central importance

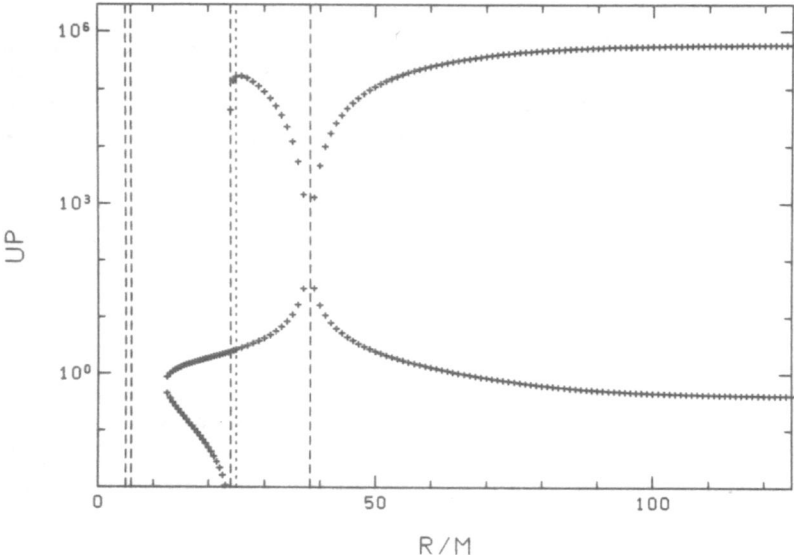

**Fig. 3.** The solutions of the hot pair wind equation for the following parameters: $P_{in}/n_{in}mc^2 = 0.45$, $\sigma_* = 1.0 \times 10^6$, with a critical energy of $E_c = 0.92312 \times 10^6\,mc^2$. The cylindrical radius $R$ is in units of $GM/c^2$, and the n* has a period of 1.6 ms, corresponding to a light cylinder $R_L = 5R_*$. The pair plasma is injected near the surface of the n* (inner dashed line), it passes through the Alfvén surface at the light cylinder and becomes supermagnetosonic at 1.53 $R_L$. The main acceleration occurs between the light cylinder and $\simeq 3$ light cylinder radii due to the deformation of the magnetosphere around the light cylinder.

is the form of the flux tube function $f$, defined by equation (60). $f$ is constant for a monopole geometry, but has a complicated form for an open flux tube in the pulsar magnetosphere. The parameters $L$ and $E$ are constrained by the requirement that the physical solutions pass through the Alfvén point and the fast magnetosonic point.

In Fig. 3 we show the various branches of solutions of the hot wind equation for a finite initial pressure in the background of the slow rotator with $P \simeq 1.6$ ms. The inner dashed line corresponds to the surface of the neutron star, the second one shows the position of the light cylinder, slightly deformed by the presence of gravity. The physical pair wind solution starts with finite velocity near the surface of the neutron star, crosses an unphysical branch at the Alfvén point and a second time at the fast magnetosonic point, and reaches finally a constant outflow velocity. Since the inertia of the plasma is included in our treatment, the light cylinder is no

157

critical point of the wind equation. The strong deformation of the magnetic surface around the light cylinder produces an enormous acceleration so that $u_{poo} \simeq \sigma_*$. This behaviour is completely different from monopole solutions, where $u_{poo} \simeq \sigma_*^{1/3}$. This drastic acceleration occurs at the cost of magnetic energy, the relative Poynting flux $\sigma$ as defined in eq. (87) is drastically reduced (see Fig. 4) and reaches a value of about 0.5 in the asymptotic region of the wind.

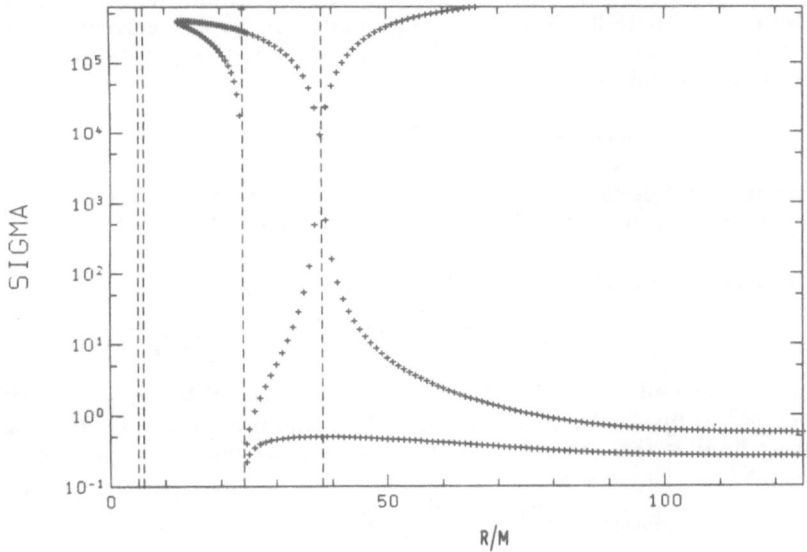

**Fig. 4.** The ratio between the Poynting flux and the total mass-energy flux of the solution shown in Fig. 3. This plot demonstrates how the magnetic energy is converted into particle energy in a very narrow zone around the light cylinder.

### 3.5. The relativistic Grad-Schlüter-Shafranov Equation

One of the essential points of the computation of self-consistent MHD pulsar winds is the particular form of the toroidal current $j_\phi$, which determines via Ampère's equation

$$\nabla \wedge \mathbf{B}_p = \frac{4\pi}{c} j_\phi \mathbf{e}_T \quad , \quad \mathbf{B}_p = \frac{1}{R}(\nabla \Psi \wedge \mathbf{e}_T) \tag{63}$$

the structure of the inner and outer magnetosphere. From here on we neglect the gravitational background, since this gives an unnecessary complication of the problem. With the introduction of the magnetic flux function $\Psi$, which we now use as our stream function, this equation can be transformed to a divergence equation of the form

$$R\nabla \cdot \left\{ \frac{1}{R^2} \nabla \Psi \right\} = -\frac{4\pi}{c} j_\phi \,. \tag{64}$$

It is well known that for non-rotating magnetic structures and static plasma equilibria, the current is simply given by two functions of $\Psi$

$$j_\phi = cR \, \partial_\Psi P(\Psi) + c\frac{B_\phi}{4\pi} \, \partial_\Psi(RB_\phi) \,. \tag{65}$$

Plasma confinement in the laboratory and in solar filaments is based on this equation, known as the Grad-Schlüter-Shafranov (GSS) equation. When plasma is moving along the magnetic flux surfaces, the general form for this current follows from a force balance perpendicular to the flux surfaces. For this purpose one considers the relativistic Lorentz force

$$n(\mathbf{u} \cdot \nabla)(\mu \mathbf{u}) + \nabla P = \rho_e \, \mathbf{E} + \frac{1}{c} (\mathbf{j} \wedge \mathbf{B}) \,. \tag{66}$$

The projection of this equation perpendicular to the magnetic flux surfaces provides then the current $j_\phi$ (a derivation is given in Camenzind 1987a)

$$\begin{aligned}
\frac{1}{c} B_p j_\phi (1 - M^2 - x^2) = &-\frac{RB_\phi}{4\pi R^2} \, \nabla_n(RB_\phi) + \rho_{\mathrm{GJ}} E_n \\
&- \mu\gamma n\{\nabla_n\gamma - \Omega\nabla_n l\} - \nabla_n P + \mu\eta B_p^2 \nabla_n(\eta/n) \,.
\end{aligned} \tag{67}$$

n is the normal unit vector for the magnetic surfaces. When $\eta = 0 = \Omega^F$ (and therefore $M^2 = 0$) we obtain the classical Newtonian limit of eq. (65). Another interesting limit is the *force-free limit*, where plasma inertia is neglected, $\nabla P = 0 = \nabla\gamma = \nabla l$

$$B_p j_\phi (1 - x^2) = -c\frac{RB_\phi}{4\pi R^2} \, \nabla_n(RB_\phi) + c\rho_{\mathrm{GJ}} E_n \,. \tag{68}$$

In this limit we also find from eq. (43)

$$RB_\phi = -T(\Psi) = -4\pi\eta(\Psi)L(\Psi) \,, \tag{69}$$

so that the current is simply given as

$$j_\phi (1 - x^2) = c\frac{T(\Psi)}{4\pi R} T'(\Psi) + c\rho_{\mathrm{GJ}} \frac{E_n}{B_p} \,. \tag{70}$$

It is then easy to show that the GSS-equation can be written in the suggestive form of

$$R\nabla \cdot \left\{ \frac{1 - x^2}{R^2} \nabla \Psi \right\} = -\frac{TT'}{R} \,, \tag{71}$$

which is the *classical pulsar equation* originally derived by Scharlemann and Wagoner (1973). From this equation we see that the poloidal current distribution $T(\Psi)$ also determines the effective source for the poloidal fields.

In this traditional approach to pulsar magnetospheres, one has to solve the nonlinear equation (71) for suitable inner boundary conditions (a dipole structure along the surface of the star) and strict boundary conditions on the light cylinder (which is identical with the Alfvén surface in the force-free limit) following from the regularity of the current in eq. (70) $(E_n = RB_p/R_L, \rho_e = Rj_\phi/(cR_L) - \rho_{GJ})$

$$\frac{\partial \Psi}{\partial x}\bigg|_{LC} = \frac{1}{2} TT' . \tag{72}$$

In the force-free approximation, it is assumed that the function $T(\Psi)$ can be arbitrarily prescribed except for some regularity conditions at $\Psi \mapsto 0$ and for the last closed field lines (see Scharlemann and Wagoner 1973). This assumption is incorrect, since the form of $T(\Psi)$ follows from particle injection $\eta(\Psi)$ and the distribution of total angular momentum $L(\Psi)$ carried away by the plasma. In fact, when the magnetosphere is filled up with plasma, eq. (69) has to be generalized

$$RB_\phi = -T(\Psi) I(x) \quad , \quad I(x) = \frac{x_A^2 - x^2}{1 - M^2 - x^2} , \tag{73}$$

where $x_A^2 = \Omega^F L/E$ gives the position of the Alfvén point. Now, in the force-free limit we have $x_A = 1$, $M^2 = 0$, and therefore $I(x) = 1$. When the force-free limit is slightly violated, drastic changes in $I$ occur around the light cylinder. This has the effect that cross-field currents appear around the light cylinder. The poloidal current density following from the expression for $RB_\phi$ has in general the form

$$\mathbf{j}_p = \frac{c}{4\pi R} \nabla(RB_\phi) \wedge \mathbf{e}_T = -\frac{cI}{4\pi} T' \mathbf{B}_p - \frac{cT}{4\pi R} (\nabla I \wedge \mathbf{e}_T). \tag{74}$$

The first term represents the usual current flowing along the magnetic surfaces, while the second term produces cross-field currents. They have two effects. First, they contribute to the closure of the global current system for rotating magnetospheres. In general, $T(\Psi)$ has such a form that a current flows along the equatorial plane outside of the light cylinder and flows back along the light cylinder from a region $z \gg R_L$. In between, we find a critical field line which carries no current, i.e. $T'(\Psi_c) = 0$ (Scharlemann and Wagoner 1973).

The second effect of cross-field currents consists in a post-acceleration of the plasma flow just outside of the light cylinder (Camenzind and Endler 1989). Cross-field currents produce a net Lorentz force along the plasma flow, which can drastically accelerate the plasma. As a result, we find that the force-free limit is a fruitful approximation for axisymmetric pulsar magnetospheres far inside and far outside the light cylinder, but not in the vicinity of the light cylinder itself. For this reason, one has to solve the GSS equation with the complete current $j_\phi$ in eq. (67). The force-free condition must, therefore, lead to complications beyond the Alfvén surface, where inertial effects of the plasma are important for the decoupling from the magnetic structure. All the troubles with the boundary conditions at the light cylinder (see e.g. Michel 1982) come from this assumption. There is a sizeable literature pointing up problems encountered near the light cylinder. The current density resulting from this general force-balance has various components, inertial currents, a current due to the toroidal field and a Goldreich-Julian current

$$j_\phi^{(GJ)} = c \frac{R}{R_L} \rho_{GJ}/(1 - M^2 - x^2) \quad . \tag{75}$$

This current and part of the inertial currents can be absorbed by the left hand side of Ampere's equation. In this way we obtain the relativistic version of the *Grad-Schlüter-Shafranov* equation

$$R\nabla \cdot \left\{ \frac{1 - M^2 - x^2}{R^2} \nabla\Psi \right\} = -\frac{4\pi}{c} j \quad . \tag{76}$$

For rigidly rotating magnetospheres, the remaining current $j$ has still two different components

$$j = j^{(n)} + j^{(0)} \quad , \tag{77}$$

the first one being of plasma origin, the second one due to the toroidal magnetic field. The above current and the toroidal magnetic field are regular at the Alfvén surface, which, however, also appears as a critical surface on the left hand side of the GSS equation (76). Since the Alfvén surface is a critical surface of the GSS equation, the current $j_\phi$ must also vanish there. This condition gives in general a quite complicated restriction for the vertical component $B_z$ of the magnetic field at the Alfvén surface. The relativistic GSS equation is elliptic between the surface of the slow magnetosonic points and the surface of the fast magnetosonic points. Outside of this latter surface, the equation becomes hyperbolic, similar to its Newtonian counterpart (Sakurai 1985).

From the above form of the current one can infer that a solution of the flow problem will provide us with the functions $E(\Psi)$, $L(\Psi)$, the poloidal velocity, Lorentz factor, Mach number and toroidal magnetic field. Consequently, the hot wind equation together with this GSS equation form a highly nonlinear system of equations which has to be solved simultaneously. In Camenzind (1987a) we also describe a numerical procedure based on the finite element method for partial differential equations to solve the system for adiabatic plasma flows. This method has been introduced to solve systems including the formation of relativistic jets in AGNs consisting of normal plasmas (Camenzind 1987a,b, 1988, 1989a). It is however of no principle difference to apply this technique also to pair plasmas, except that the magnetosphere outside the light cylinder couples with the surrounding medium. In particular the force-free limit can easily be solved with this technique. Calculations of self-consistent axisymmetric pulsar magnetospheres are in progress.

In Fig. 5 we show a self-consistent solution for a rotator surrounded by a disk. This solution shows the general features expected to occur for open pulsar magnetospheres. The inner magnetosphere is largely deformed due to the sharp boundary conditions required at the Alfvén surface (very near to the light cylinder) so that the dipole approximation breaks down already very near to the surface of the neutron star. The magnetic field lines cross the Alfvén surface without any cusp, and the exterior solution depends on the global boundary conditions at infinity.

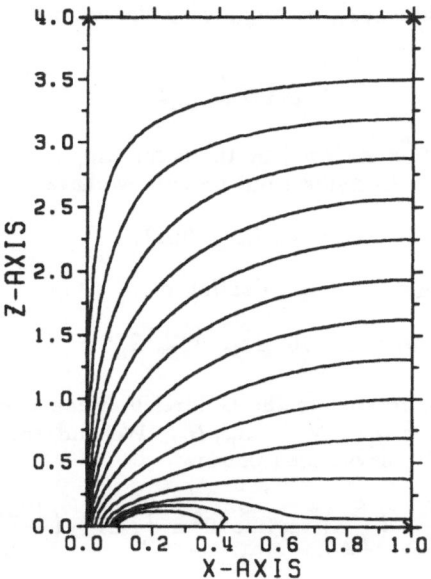

**Fig. 5.** The global structure of the inner magnetosphere sourrounding the neutron star, which could be surrounded by a neutral ionised disk (Michel 1983). The introduction of this boundary layer, which splits the dipolar field, provides a resistive load and a return current path to the star. This example demonstrates the ability of the code for the GSS equation to handle various boundary conditions.

When plasma is injected from the polar caps, all field lines from the polar cap are forced to cross the light cylinder - the region inside the light cylinder is evacuated in the vertical direction, $z \gg R_L$.

### 3.6. Essential Parameters for Pair Winds

The terminal Lorentz factor $\gamma_\infty$ for 2D pair winds is essentially given by 3 parameters. The first parameter concerns the initial temperature of the newly created pair plasma

$$\frac{P_*}{n_* m_e c^2} \leq 10^2 . \tag{78}$$

Synchrotron emission in the strong magnetic fields leads however to the instantaneous transverse cooling so that one should work with an anisotropic particle distribution. These effects are neglected by the present theory. The second parameter of central importance to magnetized winds is Michel's magnetization parameter

$\sigma_*(\Psi)$, as defined in equation (61), or roughly by the following relation

$$\sigma_* \simeq \frac{\Phi_c^2}{4\pi^2 \dot{N}_e m_e c R_L^2} , \qquad (79)$$

where $\Phi_c$ is the magnetic flux enclosed by the polar cap, $\Phi_c = B_* A_c$. For the light cylinder of a neutron star with millisecond periods we have

$$R_L = 4.8 R_* (P/\mathrm{ms}). \qquad (80)$$

For the Crab pulsar, this gives a magnetization of

$$\sigma_* \simeq 9 \times 10^5 \, \Phi_{c,23}^2 \, \dot{N}_{e,39}^{-1} \, R_{L,8}^{-2} . \qquad (81)$$

In general, pair creation is related to the Goldreich-Julian current $\dot{N}_c$ over a pair creation proliferation factor $\alpha_{\mathrm{pb}}$, $\dot{N}_e = \alpha_{\mathrm{pb}} \dot{N}_c$. We find therefore the following scaling relation for strongly magnetized pulsars

$$\sigma_* = \frac{eB_* R_*}{2m_e c^2 \alpha_{\mathrm{pb}}} \left(\frac{R_*}{R_L}\right)^2 \simeq 3 \times 10^6 \, \alpha_{\mathrm{pb},4}^{-1} \, B_{*,12} \left(\frac{100 R_*}{R_L}\right)^2 . \qquad (82a)$$

The pair proliferation factor also depends on the pulsar period $P$ and on the magnetic field strength. Gurevich and Istomin (1985) e.g. use the relation $\alpha_{pb} \simeq 10^4 \, P^{3/7} \, B_{*,12}^{-3/7}$. With this scaling relation we obtain the following expressions for the magnetization

$$\sigma_* = 1.1 \cdot 10^3 \, B_{*,12}^{10/7} \, (P/s)^{-17/7} \qquad , \quad \text{for classical pulsars} , \qquad (82b)$$

$$\sigma_* = 1.1 \cdot 10^6 \, B_{*,9}^{10/7} \, (P/\mathrm{ms})^{-17/7} \qquad , \quad \text{for millisecond pulsars} . \qquad (82c)$$

It is interesting to see that $\sigma_*$ has practically the same value for weakly magnetized millisecond pulsars as for young strongly magnetized objects, though the surface field strength differs by 3 orders of magnitude.

The third quantity, which has a great influence on the terminal Lorentz factor of the outflowing plasma is the form of the magnetic flux function $f$, defined in eq. (60). This function would be constant for spherically symmetric magnetic configurations. For a monopole type geometry one can derive the exact formula $\gamma_\infty = \sqrt{1 + \sigma_*^{2/3}} \simeq \sigma_*^{1/3}$ (for a $\Gamma = 5/3$). Since the magnetic surfaces must obey the boundary conditions of the type of eq. (72) at the Alfvén surface, this has the effect that post-acceleration beyond the Alfvén surface is essential for the terminal speed. From numerical experiments with the hot wind equation, we found the approximate relation

$$\gamma_\infty \leq \sigma_* \quad , \quad \sigma_* \gg 1. \qquad (83)$$

This means that the Poynting flux is strongly reduced in the real 2D problem when compared with the monopole result, the Poynting flux carries in general only a small fraction of the total energy. This terminal Lorentz factor $\gamma_\infty \simeq 5 \times 10^5$ nicely conforms with the observed value for the Crab Nebula, and the above relation predicts similar Lorentz factors for the pair plasma generated by the weakly magnetized millisecond pulsars, $\gamma_\infty \simeq 10^{5\pm1}$.

## 3.7. Open Problems

As a result of these investigations into the 2D pulsar wind problem, we now have a theory at our disposal which produces the correct values for the asymptotic wind properties when the pairs are generated near the polar cap surface. The proliferation factor $\alpha_{pb}$ is one of the quantities which should be analyzed more carefully. Each model for the pair creation process makes individual predictions for this factor.

The next major step forward would be the development of a 3D wind theory based on axisymmetric gravitational backgrounds. Advances in the understanding of the plasma physics for the polar beams are also urgently needed if the longstanding questions of the origin of the radio emission are to be solved. From the point of view of plasma physics, many attempts in this direction are in a rudimentary state. A consistent theory of excitation, stabilization and propagation of electromagnetic oscillations in a relativistic one-dimensional plasma flowing along curved magnetic field lines has been recently presented by Beskin et al. (1988). A presentation of this theory for the origin and the properties of the radio emission of pulsars is however beyond the scope of these lectures. This theory shows that polar cap theories including pair production can largely explain the observed radio structures. Polar cap theories also make some predictions about the gamma-ray emission from the recently detected millisecond pulsars (Usov 1983).

## 4. Pulsar Winds and the Crab Nebula

A rotating magnetosphere always produces a current system which has to be closed in some way, when the wind interacts with the ambient medium (Kennel et al. 1979; Camenzind 1986a; Benford 1987). Pulsar winds will interact either with a Supernova remnant or with the interstellar magnetic field.

## 4.1. MHD Flow Models of the Crab Nebula

Rees and Gunn (1974) described the basic features of how a pulsar's relativistic wind would interact with its surrounding Supernova remnant. They considered a hydromagnetic, spherically symmetric outflow of relativistic pairs and electromagnetic waves, with an unspecified ratio between the two energy forms. Since the equipartition pressure of the radio-emitting electrons exceeds the pressure of the ISM, the pulsar plasma must be confined by the ejected Supernova matter. This Supernova remnant expands with a speed of $\simeq 2000$ km/s. Since the pulsar wind is relativistic, a standing shock occurs at the inner side of the Supernova remnant which decelerates the flow to subsonic velocities. This shock occurs at a distance of $\simeq 3 \times 10^{17}$ cm from the location of the pulsar. This model has been extended by Kundt and Krotschek (1980) and various authors later on.

The Rees & Gunn model seems to conform with observations: we have an invisible relativistic wind flowing out from the pulsar. This wind is shocked, slows down and creates the synchrotron nebula, and then further decelerates as it expands, until the velocity of the filamentary structure is reached. Kennel and Coroniti (1984) have refined this analysis again under the assumption that shock heating causes the synchrotron emission. They also found that one needs a dense pair plasma with $\simeq 10^4$ pairs per charge unit for the resultant post-shock flow to conform to the nebular expansion.

In a pure MHD model the energy is carried away from the pulsar by the pair wind and a Poynting flux (here given for a flat space)

$$\mathbf{P}_p = \frac{c}{4\pi}(-B_\phi)\mathbf{B}_p \, \frac{R}{R_L} \, . \tag{84}$$

It is convenient to parametrize this Poynting flux in terms of the total particle energy flux

$$\sigma = \frac{P_p}{\gamma n u_p m c^2} = \frac{-B_\phi}{B_p} \, \frac{x B_p^2}{4\pi \mu n \gamma u_p} \, . \tag{85}$$

$\sigma$ is a flux-tube dependent quantity and varies along a given flux tube. Another quantity of interest is the ratio between the toroidal magnetic field and the poloidal field in the asymptotic regime of the wind

$$\frac{B_\phi}{x B_p} = \frac{4\pi \eta E}{B_p} \, \frac{x^2 - x_A^2}{1 - M^2 - x^2} \, . \tag{86}$$

When this expression and the solution for the Lorentz factor are used, $\sigma$ is just a function of the flow parameters

$$\sigma = \frac{x_A^2 - x^2}{1 - M^2 - x_A^2} \simeq \frac{mc^2}{\mu} \, \frac{\sigma_*}{u_p} \, \frac{1}{f} \, . \tag{87}$$

$\sigma$ is reduced with respect to the injection parameter $\sigma_*$ by the Lorentz factor, the internal energy and the flux tube function $f_\infty > 1$. This latter quantity is an essential ingredient with respect to the spherically symmetric wind discussed by Kennel and Coroniti (1983). When magnetic flux is annihilated along the chosen flux tube, the value of $\sigma$ is drastically reduced, though its initial value exceeds unity. The quantity $\sigma_\infty$ is the essential parameter for all MHD models of the Crab Nebula and of similar nebulae around rapidly spinning pulsars.

The total energy lost by a typical pulsar with a mean period of 0.7 seconds and a $\dot{P} \simeq 2.5 \times 10^{-15} \, \mathrm{s\,s^{-1}}$ is

$$-\dot{E}_{\rm rot} = 3.3 \times 10^{33} \, \mathrm{erg\,s^{-1}} \, I_{45} \left(\frac{P}{0.67\,\mathrm{s}}\right)^{-3} \frac{\dot{P}}{2.5 \times 10^{-15}} \, . \tag{88}$$

With this energy loss one can estimate the position of the shock $R_S$ of a typical pulsar wind expanding into the interstellar medium

$$R_S \simeq 0.9 \, \mathrm{lyr} \, \dot{E}_{\rm rot,35}^{1/2} \, B_{-6}^{-1} \, . \tag{89}$$

When the pair wind works against a Supernova remnant as in the case of the Crab Nebula, the pressure of the interstellar magnetic field must be replaced by the thermal pressure of the Supernova shell (for the Crab: $P_{\text{shell}} \simeq 10^{-8}$ dyn cm$^{-2}$, see Kundt and Krotschek 1980). The magnetic luminosity of the pair wind carried into the shock in a MHD model follows from the expression for the Poynting flux

$$L_{\text{mag}} = \frac{c}{4\pi} \frac{-R_S B_\phi^S}{R_L} \Phi_S, \tag{90}$$

with

$$\Phi_S = \int_S \mathbf{B}_p \cdot d\mathbf{S} = \int_{\text{polar cap}} \mathbf{B}_p \cdot d\mathbf{S} \simeq A_c B_*. \tag{91}$$

For this equation we assumed that no reconnection occurs along the magnetic surface, in general $\Phi_S < A_c B_*$. For the Crab pulsar we obtain

$$\Phi_* \simeq \pi R_*^2 \frac{R_*}{R_L} B_* \simeq 10^{23} \text{ Gauss cm}^2,$$
$$L_{\text{mag}} \simeq 2 \times 10^{38} \text{ erg s}^{-1} B_{-3.5}^{<S} \Phi_{S,23}. \tag{92}$$

$B^{<S}$ is the pre-shock magnetic field at the inner edge of the Crab Nebula. In terms of these relations, the magnetic luminosity of the Crab wind would be similar to the total particle luminosity, unless reconnection is an important process somewhere between the light cylinder and the shocked region (reconnection might occur e.g. in the     wisp region). This high Poynting flux disagrees with the predictions of spherically symmetric *ideal MHD shock models*, which require a value for $\sigma_S$ of at least two orders of magnitude lower in order to match the outer boundary conditions (Kundt and Krotschek 1980; Kennel and Coroniti 1984; Emmering and Chevalier 1987). It is therefore instructive to consider in more detail the physical processes at the shock.

A physically different picture has been discussed by Michel (1985) assuming a single particle description of a 3D wind. The plasma is assumed largely charge-separated. The deviation from spherical symmetry is an important input. We do not further discuss this model, since our fully axisymmetric treatment also contains deviations from spherical symmetry.

## 4.2. Collisionless Shocks in Pulsar Winds

The pair wind is decelerated by a fast magnetosonic shock from supermagnetosonic to submagnetosonic speeds. This shock produced by the relativistic pair plasma is not collision-dominated, but involves collisionless processes, and therefore has the width of a few electron gyroradii. Within this transition region, the incoming energy is dumped into particle heating and acceleration of a superrelativistic component. The downstream plasma of the shock consists therefore of three different components, the shocked thermal pair plasma, magnetic fields and ultrarelativistic particles. Heating and acceleration in collisionless shocks must occur over micro-turbulence. These processes are known for normal plasmas consisting of ions and electrons (Wu 1982; Winske et al. 1985), but they are completely unknown for pair plasmas.

Under the assumption that the pair plasma is thermalized by such processes, the jump conditions for relativistic MHD shocks can be locally analyzed. We have recently developed this theory for the hot spot region of relativistic jets (Appl and Camenzind 1988, 1989a). There is no intrinsic difference between shocks in electron-positron jets and pair winds from pulsars. The downstream state of the shocked pair plasma is then determined by the upstream state of the wind, which involves the Lorentz factor $\gamma_w$, the ratio of electromagnetic to material energy in the wind, $\sigma_w$, and the orientation of the magnetic field with respect to the shock. From the jump conditions one can then derive a complicated equation for the compression ratio $r = \gamma_2 n_2/\gamma_1 n_1$ (Appl and Camenzind 1988), which has simple asymptotic approximations for relativistic winds with $\Gamma_S = 4/3$ (already found by Kennel and Coroniti 1984):

$$r = 3(1 - 4\sigma_w) + O(\sigma_w^2) \quad , \quad \sigma_w \ll 1, \tag{93}$$

$$r = 1 + \frac{1}{2\sigma_w} + O(\sigma_w^{-2}) \quad , \quad \sigma_w \gg 1. \tag{94}$$

A strong Poynting flux considerably weakens the shock, and the shock would disappear already for $\sigma_w \simeq O(1)$. When the magnetic field is completely toroidal, this compression ratio will determine the post-shock pressure $P_S$

$$P_S = n_w mc^2 \beta_w^2 \gamma_w^2 \left\{ 1 - \frac{1}{r} + \frac{\sigma_w}{r}(r - 1) - \frac{\sigma_w}{2\beta_w^2}(r^2 - 1) \right\}. \tag{95}$$

The existence of prominent shock features at the inner edge of the Crab Nebula certainly limits the magnetic energy such that $\sigma_w < 1$ at the position of the shock. But there might exist pulsar winds with $\sigma_w \simeq 1$, which cannot evolve into prominent shock features.

The existence of a fast shock which terminates the wind zone also provides a nebular current flowing from the equatorial plane along the shock into the polar region where it closes the return current in the exterior pulsar magnetosphere (Camenzind 1986a; Kennel et al. 1979). This is a drift-current which must exist in order to close the global current system of rotating magnetospheres. Such drift currents have interesting microscopic properties suitable for particle acceleration. A thorough discussion of these microscopic aspects is beyond the scope of these lectures.

### 4.3. Axisymmetric MHD Nebular Models

Several characteristics of the Crab Nebula system indicate a general axial symmetry around a position angle of 150° - 160°. These features include the major axis of the radio and optical brightness distributions, as well as the E-vector direction of polarisation in the optical and low-energy X-rays. This conforms with the geometric

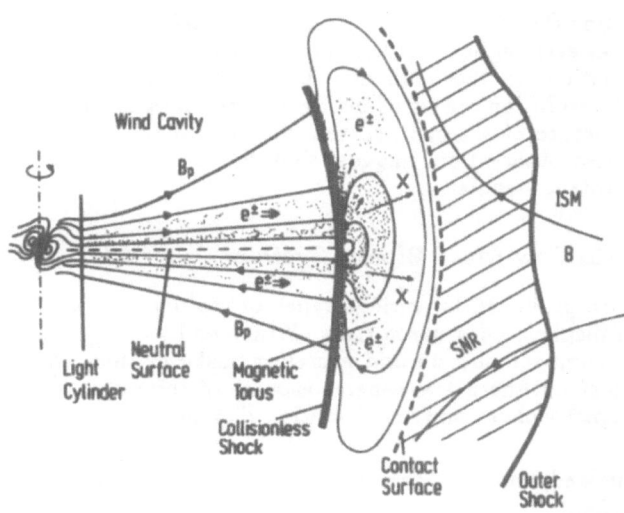

**Fig. 6.** The toroidal shock at the inner edge of the Crab Nebula. The pair wind forms a standing shock near the origin of the hard X-rays.

model derived from optical and X-ray observations by Aschenbach and Brinkmann (1975) (see also Brinkmann et al. 1985).

Instead of being spherically symmetric, the model for the Crab Nebula following from our previous considerations must be axisymmetric with a pair wind mostly concentrated towards the equatorial plane. Injection of the pair wind into the nebula occurs mainly along the equatorial plane. The shocked pair plasma forms a toroid-like structure around the rotational equatorial plane. The height of this toroid is essentially determined by the life time of the radiating electrons compared with their effective radial propagation time. The detailed emission structure of this toroidal shock including particle acceleration is completely unknown and is left to the audience as a homework problem.

Ricker et al. (1975) reported the first hard X-ray measurements of the size and location of the X-ray emitting region in the Crab Nebula using lunar occultation. They found that the centroid of this X-ray emitting region is offset 10"±4" to the northwest of the pulsar. Recently Pelling et al. (1987) have measured the size, shape and orientation of this hard X-ray source in the Crab Nebula. Their result is consistent with a toroidal emission zone tilted in a way that its near rim projects to the northwest of the pulsar (Fig. 6). These observations are consistent with the picture that the hard X-rays are emitted from the innermost toroidal shell with soft X-rays extending further out. The optical synchrotron emission originates from lower energy electrons, which are preferentially distributed in the direction perpendicular to the equatorial plane. The nearer side of the torus appears brighter due to Doppler boosting $((1+\beta_S \cos \theta)/(1-\beta_S \cos \theta))^3 \simeq 5$ for $\beta_S \simeq 1/3$ and $\theta \simeq 30^0$. Future observations at even higher energies and with better angular resolution could provide even information on the wind distribution and its interaction with the nebula. Recent high resolution observations by van den Bergh and Pritchet (1989) show the fibrous structure of the synchrotron nebula and the nature of the wisps near the pulsar. These observations suggest that the wisps are magnetic features in the fibrous synchrotron nebula.

## 5. Pulsar Winds in Low-Mass Binary Systems

Magnetized winds are produced by various types of rapidly rotating systems, and not only by isolated magnetized neutron stars. Winds and jets might also be driven by rapidly rotating compact objects and accretion disks in close binary systems. In this section we just outline a few general aspects of the interaction between a rapidly rotating magnetosphere and the accretion disk around weakly magnetized neutron stars.

### 5.1. Weakly Magnetized Neutron Stars with Millisecond Periods

The discovery of millisecond pulsars in the galactic bulge and in globular clusters has brought up a new trend in the interpretation of the evolution of low-mass X-ray binary systems. Various authors propose that the pulsar represents an evolutionary link between low-mass X-ray binaries and isolated pulsars with millisecond periods (Phinney et al. 1988; Kluzniak et al. 1988; van den Heuvel and van Paradijs 1988; lectures by Verbunt). The high rotation speeds observed for the millisecond pulsars are generally thought to result from a period of accretion earlier in the life of the neutron star. Many people now distinguish between *four phases* in the evolution according to the origin and characteristics of the secondary's mass-loss. We call them the *spin-up phase, equilibrium phase, wind phase* and the *pulsar phase*.

In *phase I*, systems with companion masses $M_c > 0.1\ \mathrm{M_\odot}$ evolve by standard Roche lobe overflow, which is most probably driven by gravitational radiation from the binary system with the typical time scale

$$t_G = 7.7 \cdot 10^7 \,\mathrm{yr} \left(\frac{P_b}{\mathrm{hr}}\right)^{8/3} \frac{(M_1 + M_2)^{1/3}}{M_1 \, M_2} \,. \tag{96}$$

Gas from this nearby companion accumulates in a disk around the neutron star. The inner edge of the disk is roughly located at the point $R_{\mathrm{in}}$ of equilibrium between disk pressure and magnetospheric pressure (see also lecture by Verbunt)

$$R_{\mathrm{in}} \simeq 3.8 \, R_* \, B_{8.7}^{4/7} \, R_{*6}^{4/7} \, \dot{M}_{18}^{-2/7} \, (R_G/R_*)^{1/7} \, (\alpha)^{2/7} \, (H/R)^{2/7} \,. \tag{97}$$

Here, we assume that these rapidly rotating neutron stars have magnetic fields similar to that of PSR 1937+21 with $B_* \simeq 5 \times 10^8$ Gauss. Another important quantity for these objects is the location of the corotation radius

$$R_{\mathrm{cor}} = 1.7 \, R_* \, (P/\mathrm{ms})^{2/3} \, (M_*/1.4 \, \mathrm{M_\odot})^{1/3} \,. \tag{98}$$

These two relations show that millisecond periods could be achieved for low surface magnetic fields. The neutron star will be spun up by the accretion process as long as $R_{in} \leq R_{cor}$. Since the accretion rate remains constant for a long time, the neutron star settles down asymptotically at the spin period

$$P_{eq} \simeq 1.3\,\mathrm{ms}\, B_{8.7}^{6/7}\, \dot{M}_{18}^{-3/7}\, M_*^{13/7} \left(\frac{R_*}{3R_G}\right)^{15/7} \left(\frac{\alpha\,H}{R}\right)^{3/7} \quad . \tag{99}$$

The mass $M_*$ of the neutron star is here always taken in the standard unit of 1.4 $M_\odot$. The exact value for this equilibrium spin period depends on the detailed structure of the interaction between the magnetosphere and the accretion disk. But it is encouraging that the rough value already follows from these simple pressure equilibrium arguments (Alpar et al. 1982; Camenzind 1982; van den Heuvel et al. 1986; White and Stella 1988). The time scale to attain this corotational equilibrium

$$t_{eq} = 1.2 \cdot 10^8\,\mathrm{yr}\, B_{8.7}^{-8/7}\, \dot{M}_{18}^{-3/7}\, M_*^{-22/7}\, I_{45} \left(\frac{R_*}{3R_G}\right)^{-20/7} \tag{100}$$

is comparable to the expected life-time for Roche lobe overflow in these systems. The light cylinder defined in eq. (80) is the third important radius. Please note that this light cylinder is extremely compact for millisecond pulsars, while it extends to the outer edge of the accretion disk in systems containing slowly rotating neutron stars, as e.g. in Her X-1. Accretion can only occur for $R_{in} < R_L$. The presence of this light cylinder completely destorts the inner part of the magnetosphere for rapidly rotating objects. This fact has long been ignored in the literature, since it was impossible to calculate the exact structure of rotating magnetospheres.

The interaction between the accreting neutron star and its feeding disk is certainly much more complicated than outlined above. The standard ideas on disk-magnetosphere interaction were invented to explain the behaviour of the rotation period $P$ and its derivative $\dot{P}$ for X-ray pulsators. As outlined above, accretion from the disk towards the neutron star can only occur when the corotation radius is at or outside the magnetopause radius. In the opposite case, the net gravity (= real gravity + centrifugal force) points away from the magnetopause and no accretion to the stellar surface could occur. Each bit of plasma that enters the magnetosphere from the inner edge of a Keplerian disk changes the specific angular momentum of the pulsar by the amount of $j_{in} = \sqrt{GM_* R_{in}}$. This gives a spin-up torque $T_+ = \dot{M}_{acc}\, j_{in}$. For strongly magnetized X-ray pulsars ($B_* \simeq 10^{12}$ Gauss) with periods greater than a few seconds, this theory seems to be qualitatively correct. In this respect, these X-ray pulsars are still in the *acceleration phase*.

However, in Her X-1, $\dot{P}/P \simeq 1/30$ of the value expected from the known strong magnetic field and the known accretion rate. This reduction cannot be obtained through a reduction of the magnetic moment - this is known from the observation of a cyclotron line in the spectrum of Her X-1. Therefore, one can conclude that some form of spin-down torque is operating in Her X-1 and in similar X-ray pulsators, which has to be almost in equilibrium with the spin-up due to accretion. You find different views on the nature of this spin-down torque in the literature. Usually, the disk-magnetosphere interaction is modelled under the assumption that the disk it-self does not carry any magnetic field and that ideal MHD is a good approximation (lecture given by Kundt). The disk plasma has however a finite conductivity. Some magnetic field lines will therefore cut through the disk, exerting in this way accel-erating and braking torques on the neutron star. In addition, the rapidly rotating magnetosphere must build up a current system driven by strong electric fields and the finite Goldreich-Julian charge density $\rho_{GJ}$. This current $I_c$ circulating in the polar cap region

$$I_c = \pi f_*(\chi) R_p^2 c \rho_{GJ} \simeq 4 \times 10^{13} B_{*,9} (P/\mathrm{ms})^{-2} \text{ Ampère} \tag{101}$$

will also exert braking torques on the surface of the neutron star

$$T_-^i \simeq \frac{i}{2c^3} f_*^2 B_*^2 R_*^6 \Omega^3 . \tag{102}$$

$\chi$ is the angle between the rotational axis of the star and its magnetic moment, $f_*$ measures the influence of the inclination on the torque, and $i \leq 1$ is a measure for the effective current flowing in the polar cap ($i = I_{\parallel}/I_c$). A similar braking torque would follow from the emission of strong dipole waves (Kundt 1985)

$$T_-^\mu = \frac{2}{3c^2} \mu_\perp^2 \Omega^3 \tag{103}$$

which might, however, be strongly depressed due to the presence of plasma flows.

When the spin-up by accretion is balanced by the above current-torquing, $T_+ = T_-^i$, we obtain a different expression for the equilibrium period for neutron stars in LMXBs

$$P_{eq}^i = 1.3 \text{ ms} \left(\frac{R_{cor}}{R_{in}}\right) i^{3/10} f_*^{3/5} B_{*,9}^{3/5} \dot{M}_{18}^{-3/10} R_{*,6.2}^{9/5} M_*^{-1/5} . \tag{104}$$

This true equilibrium period is quite close to the standard expression given in equation (99) for neutron stars in LMXBs indicating that the inner edge of the disk coincides practically with the corotation radius. For the parameters of Her X-1, however, we obtain $P_{eq}^i \simeq 0.5$ s meaning that the inner edge of the disk is here closer to the neutron star than the corotation radius, $R_{in} \simeq R_{cor}/2$. Torquing by circulating currents could be important for orthogonal rotators, since here the direct coupling between the magnetosphere and the disk is less efficient than in parallel rotators.

When the neutron star has reached its equilibrium period, it stays for a long time with the inner edge of the accretion disk located at the corotation radius. In this equilibrium phase, the angular momentum gain due to accretion must be balanced by some angular momentum loss over mass ejection from the system. LMXBs are then very similar to Her X-1, except for the weaker surface magnetic fields and perhaps the lower accretion rates in the system (Anzer and Börner 1983).

What happens, however, if the accretion rate suddenly drops due to the evolu-tion of the secondary towards a degenerate dwarf, when $M_c \simeq 0.08$ M$_\odot$ ? According to eq. (97), the radius of the inner edge of the disk increases then beyond the coro-tation radius so that accretion can no longer occur. Here, we must distinguish between two additional phases, *phase III and phase IV*. In *phase III*, the accre-tion rate stays in such a region that the inner edge remains between the corotation radius and the light cylinder, while in *phase IV*, the accretion rate drops that fast that now the inner edge stays outside of the light cylinder. Here, the pulsar turns on in a binary system and expels the accretion disk in the system. In phase III, the neutron star slows down from $P \simeq 1$ ms to the spin periods $\simeq 1.6$ ms found for the millisecond pulsars with the shortest periods. During phase III, also the mass of the secondary drops and the binary period increases from $\simeq 1$ hour to a few hours (in PSR 1957+20, the companion has a mass $\simeq 0.024$ $M_\odot$ and a binary period of 9.2 hours (Fruchter, Stinebring and Taylor 1988)).

The large size of the companion of PSR 1957+20 observed in the radio has been explained by a strong wind evaporating from the companion driven by a heating of the stellar atmosphere by the energy flow from the millisecond pulsar. Such a wind would suffice to block the radio emission from the pulsar to a distance many times the radius of the companion. The companion could be completely evaporated within $10^8$ years, which is much shorter than the pulsar spin down time scale (van den Heuvel and van Paradijs 1988). This evaporation process would drive neutron stars with millisecond periods in LMXBs into isolated millisecond pulsars.

## 5.2. Centrifugally Driven Winds from Neutron Stars

In ideal MHD, the magnetopause separating the disk and the magnetic field lines folded around is a tangential discontinuity - with no plasma transport across the field lines. In general, however, magnetosphere and disk plasma move at different speeds, with a toroidal velocity $v_{rel} = (\Omega^F - \Omega_K(R)) R$. Such a relative motion is unstable to the formation of breaking waves (Kelvin-Helmholtz instability). As a consequence, plasma from the disk is forced to mix into the magnetopause boundary (Anzer and Börner 1983; Arons 1987). This plasma in the magnetosphere is driven away by the centrifugal forces and forms a disk wind. The density $\rho_w$ in this mag-netospheric atmosphere can be estimated from pressure balance, $\rho_w \simeq B^2/8\pi c_S^2$. Because of X-ray heating, the sound speed in this atmosphere is quite high, $c_S \simeq 10^8$ cm/s. Even a low mass density in the plasma corotating in the magnetosphere at $R_{cor}$ exerts sufficient stress to burst the magnetosphere and convert closed field lines into open ones. From these considerations we obtain that even if only a small fraction of the disk plasma is scraped off at radii exceeding the corotation radius, the topology of the magnetosphere is completely altered due to the growth of the Kelvin-Helmholtz instability. The vertical extension of the atmosphere formed in this way is roughly given by the scale height. The plasma injected onto the field lines has sufficient inertial energy to break open the fields; the atmospheric plasma is flung away as a *centrifugal wind*. The resulting configuration is shown in Fig. 7.

As a consequence, the Kelvin-Helmholtz instability feeds plasma from the disk into the magnetosphere for radii where the field lines rotate faster than the un-derlying disk, at a rate sufficient to break open the field lines. This would be a nice simulation problem for time dependent resistive MHD. These winds also carry off angular momentum, since the plasma injected into the poloidal field lines is

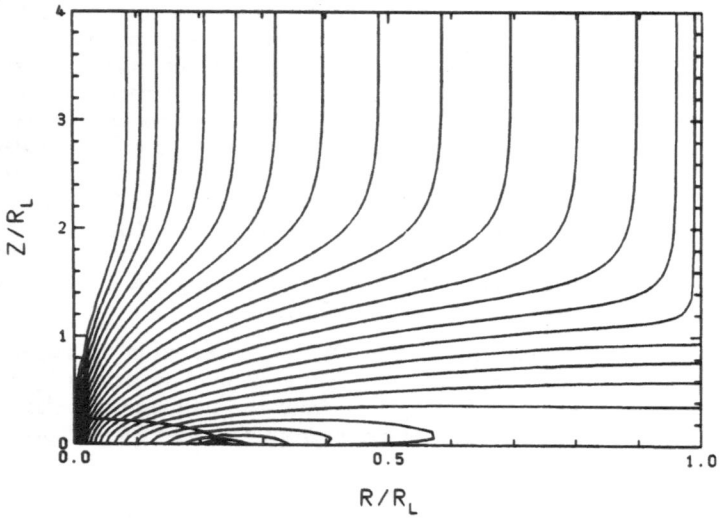

**Fig. 7.** Disk-magnetosphere interaction for a weakly magnetised millisecond n* surrounded by the accretion disk. Cylindrical coordinates are in units of light cylinder radii. The inner edge of the disk is at 0.4 $R_L$ with a disk of infinite conductivity. The isocontours show the poloidal field lines for an inner dipole as obtained by solving the GSS equation in the force-free limit. A small fraction of the magnetic flux is enclosed by the light cylinder, the open field lines must cross the Alfvén surface at the light cylinder restricted by the boundary conditions of eq. (72).

rotationally spun up by field lines swept backwards into a spiral. The outflowing plasma thus provides a braking torque on the neutron star.

This picture of the disk-magnetosphere interaction is however quite ideal, since the plasma in the accretion disk carries a finite conductivity, which could be related to the turbulent viscosity in the disk. Under this assumption, the magnetic field lines from the neutron star penetrate into the disk near the inner edge of the disk, but definitely not outside the light cylinder. Near, but inside the light cylinder the above discussed plasma feed is still operating.

In the *equilibrium phase* and in the *wind phase*, part of the rotating magnetosphere is filled up with normal plasma injected from the accretion disk between the inner edge of the disk and the light cylinder radius. Due to the extreme compactness of the light cylinder for neutron stars with millisecond periods, we are forced to use relativistic MHD for this plasma-magnetosphere problem. In addition, gravitational effects from the background geometry may also influence the structure of the magnetosphere and the solutions of the equations of motion. Since we already discussed the basic theory of relativistic MHD in the previous sections on pulsar winds, the only difference is in the nature of the plasma flowing in our rotating magnetosphere. Here, we are dealing with a normal plasma consisting of ions and electrons.

As in the case of pulsar winds, the equations of motion are integrable for axi-symmetric flows. The equation for the poloidal flow velocity $u_p$ of the plasma can be brought to a polynomial of degree 16 for $\Gamma = 5/3$, $z_p = u_p^{1/3}$ (see section 3.4)

$$\sum_{n=0}^{16} A_n\left(\frac{R}{R_L}, E, L, \frac{P_{in}}{n_{in}mc^2}; \frac{f}{\sigma_*}; g_{tt}, g_{t\phi}, g_{\phi\phi}\right) z_p^n = 0. \tag{105}$$

The detailed form of the coefficients $A_n$ of this hot wind equation can also be found in Camenzind and Endler (1989). In the cold limit, $P_{in}/n_{in}m_pc^2 \to 0$, the poloidal wind equation reduces to a fourth order polynomial in the velocity $u_p$ (Camenzind 1987a)

$$A_4\, u_p^4 + A_3\, u_p^3 + A_2\, u_p^2 + A_1\, u_p + A_0 = 0. \tag{106}$$

Of central importance is the form of the *flux tube function* $f(R)$ which is only known after a self-consistent calculation of the structure of the rotating magnetosphere, based on solutions of the GSS equation (76) with the corresponding current in eq. (77). In addition, there are three parameters which must be given at the base of the flux tube. These are $B_{p,in}R_{in}^2$, $P_{in}/n_{in}mc^2$ and the particular form of the particle injection function $\eta(\Psi)$. These parameters together with the light cylinder radius determine then Michel's magnetization parameter $\sigma_*(\Psi)$. When $\eta = const$, $\sigma_*$ just follows from the total mass flux $\dot{M}_w$ in the wind

$$\sigma_* = \frac{(B_{p,in}R_{in}^2)^2}{\dot{M}_w\, c R_L^2} \simeq 13.3\, B_{p,in,7}\, R_{in,6.5}^2\, \dot{M}_{w,15}^{-1}\, R_{L,6.7}^{-2}. \tag{107}$$

But note that $\eta(\Psi)$ has a complicated form, since particle injection only occurs in the immediate vicinity of the equatorial plane. $\eta$ vanishes at the rotational axis and reaches its maximum around the equatorial plane. A mass-loss rate from the disk of $\simeq 10^{15}$ g/s would correspond to an initial density of $\simeq 10^{18}$ cm$^{-3}$, and therefore to a scattering depth of order unity for lines of sight along the disk. The wind driven away from the disk would have semi-relativistic speeds with Lorentz factors $\gamma_w < \sigma_*$.

Such a disk wind could be the origin of the plasma flow in SS 433 with a $\sigma_* \simeq 0.1$, when the magnetic field at the injection point $B_{p,in} \simeq 10^9$ Gauss and the mass-loss rate in the jets is in the observed region, $\dot{M}_w \simeq 10^{19}$ g/s. The neutron star behind this system should be moderately rotating, $P \simeq 10$ ms, with a somewhat higher magnetic field than in typical millisecond pulsars, $B_* \simeq 10^{11}$ Gauss. We cannot give a self-consistent solution of the GSS equation for this parameter range, since our methods to solve the GSS equation only work for $\sigma_* > 1$. But we expect that the pinch forces exerted by the toroidal magnetic field on the plasma outflow are able to collimate the escaping wind on the typical scale of the binary system. This is work for the future.

## 6. Conclusions

Rapidly rotating magnetized neutron stars are electromagnetic engines and gravitational bugs. We tried to reconcile both aspects of pulsar theory in a quantitative way and therefore had to restrict ourselves to two dimensions. But two is between one and three, and three is left to the audience.

But even within two dimensions, many aspects of the link between the rapid rotation and the structure of the surrounding magnetosphere are still unclear. When people draw pictures e.g. about magnetospheres in close binary systems, they always use non-rotating magnetospheres. Using the theory and numerical procedures presented in these lectures, you can draw physically correct pictures, and these pictures look quite different   when compared with the non-rotating analogs. The physical information hidden behind these pictures is enormous and has to be worked out in the future. My personal view is therefore to stay with two for the next years and to grow slowly from two to three.

### Acknowledgements

The author would like to acknowledge enlightening and stimulating discussions with W. Kundt and all the lecturers and students of this Summer School as well as with S. Appl, M. Endler and H. Lesch in Heidelberg. Many of the results reported in these lecture notes originate from a research program on relativistic MHD jets in AGNs, supported by the Deutsche Forschungsgemeinschaft (SFB 328).

## 7. References

Alpar, M.A., Cheng, A.F., Ruderman, M., Shaham, J.: 1982, *Nature* **300**, 728

Anzer, U., Börner, G.: 1983, *Astron. Astrophys.* **122**, 73

Appl, S., Camenzind, M.: 1988, *Astron. Astrophys.* **206**, 258

Appl, S., Camenzind, M.: 1989, in *Hot Spots in Extragalactic Radio Sources*, ed. K. Meisenheimer, H.-J. Röser, Springer Proceedings (Springer-Verlag, Berlin)

Arnett, W.D., Bowers, R.L.: 1977, *Astrophys. J. Suppl.* **33**, 415

Arons, J.: 1979, *Space Science Rev.* **24**, 437

Arons, J.: 1981, in *Pulsars, IAU Symp.* **95**, ed. W. Sieber and R. Wielebinski (Reidel, Dordrecht), p. 69

Arons, J.: 1983, *Astrophys. J.* **266**, 215

Arons, J.: 1987, in *The Origin and Evolution of Neutron Stars, IAU Symp.* **125**, ed. D.J. Helfand and J.-H. Huang (Reidel, Dordrecht), p. 207

Aschenbach, B., Brinkmann, W.: 1975, *Astron. Astrophys.* **41**, 147

Asséo, E., Kennel, F.C., Pellat, R.: 1978, *Astron. Astrophys.* **65**, 401

Asséo, E., Llobet, X., Schmidt, G.: 1980, *Phys. Rev.* **A22**, 1293

Asséo, E., Llobet, X., Pellat, R.: 1984, *Astron. Astrophys.* **139**, 417

Bekenstein, J.D., Oron, E.: 1978, *Phys. Rev.* **D18**, 1809

Benford, G.: 1987, in *Astrophysical Jets and their Engines*, ed. W. Kundt (Reidel, Dordrecht), p. 197

Beskin, V.S., Gurevich, A.V., Istomin, Ya.N.: 1988, *Astrophys. Space Sci.* **146**, 205

Brinkmann, W., Aschenbach, B., Langmeier, A.: 1985, *Nature* **313**, 662

Butterworth, E.M., Ipser, J.R.: 1976, *Astrophys. J.* **204**, 200

Camenzind, M.: 1975, *J. Math. Phys.* **16**, 1023

Camenzind, M.: 1982, in *Accreting Neutron Stars*, ed. W. Brinkmann, J. Trümper (Max-Planck-Institut für Extraterrestrische Physik, Garching), p. 156

Camenzind, M.: 1986a, *Astron. Astrophys.* **156**, 137

Camenzind, M.: 1986b, *Astron . Astrophys.* **162**, 32

Camenzind, M.: 1987a, *Astron. Astrophys.* **184**, 341

Camenzind, M.: 1987b, in *Interstellar Magnetic Fields*, ed. R. Beck, R. Gräve, Springer Proceedings (Springer-Verlag, Berlin), p. 229

Camenzind, M.: 1988a, in *High Energy Astrophysics*, ed. G. Börner, Springer Proceedings (Springer-Verlag, Berlin)

Camenzind, M.: 1989a, in *Accretion Disks and Magnetic Fields in Astrophysics*, ed. G. Belvedere, Kluwer (Dordrecht), in press

Camenzind, M.: 1989b, in preperation

Camenzind, M., Endler, M.: 1989, to appear in *Astron. Astrophys.*

Carter, B.: 1973, in *Black Holes*, ed. C. DeWitt and B. DeWitt (Gordon and Breach, New York)

Chanmugam, G., Brecher, K.: 1987, *Nature* **329**, 696

Coroniti, F.V., Kennel, C.F.: 1985, in *The Crab Nebula and Related Supernova Remnants*, ed. M.C. Kafatos, R.B.C. Henry (Cambridge University Press, Cambridge), p. 25

Cutler, C., Lindblom, L.: 1987, *Astrophys. J.* **314**, 234

Daugherty, J.K., Harding, A.K.: 1982, *Astrophys. J* **252**, 337

Deutsch, A,J.: 1955, *Ann. d'Astrophys.* **18**, 1

Emmering, R.T., Chevalier, R.: 1987, *Astrophys. J.* **321**, 334

Fitzpatrick, R., Mestel, L.: 1988a, *MNRAS* **232**, 277

Fitzpatrick, R., Mestel, L.: 1988b, *MNRAS* **232**, 303

176

Friedman, J.L.: 1978, *Comm. Math. Phys.* **62**, 247

Friedman, J.L.: 1983, *Phys. Rev. Letters* **51**, 11

Friedman, J.L., Schutz, B.F.: 1978, *Astrophys. J.* **222**, 281

Friedman, J.L., Ipser, J.R., Parker, L.: 1986, *Astrophys. J.* **304**, 115

Friedman, J.L., Imamura, J.N., Durisen, R.H., Parker, L.: 1988, *Nature* **336**, 560

Fruchter, A.S., Stinebring, D.R., Taylor, J.H.: 1988, *Nature* **333**, 237

Goldreich, P., Julian, W.H.: 1969, *Astrophys. J.* **157**, 869

Gurevich, A.V., Istomin, Ya.N.: 1985, *Soviet Phys. JETP* **62**, 1

Hartle, J.B.: 1967, *Astrophys. J.* **150**, 1005

Hartle, J.B., Thorne, K.S.: 1968, *Astrophys. J.* **153**, 807

Jackson, J.D.: 1975, *Classical Electrodynamics* (J. Wiley Publ. Co., New York)

Jones, P.B.: 1983, *MNRAS* **204**, 9

Kennel, C.F., Fujimura, F.S., Pellat, R.: 1979, *Space Science Rev.* **24**, 407

Kennel, C.F., Fujimura, F.S., Okamoto, I.: 1983, *J. Astrophys. Geophys. Fluid Dyn.* **26**, 147

Kennel, C.F., Coroniti, F.V.: 1984a, *Astrophys. J.* **283**, 694

Kennel, C.F., Coroniti, F.V.: 1984b, *Astrophys. J.* **283**, 710

Kluźniak, W., Ruderman, M., Shaham, J., Tavani, M.: 1988, *Nature* **334**, 225

Kundt, W.: 1985, *Bull. Astr. Soc. India* **13**, 12

Kundt, W., Krotschek, E.: 1980, *Astron. Astrophys.* **83**, 1

Lichnérowicz, A.: 1967, *Relativistic Hydrodynamics and Magnetohydrodynamics* (Benjamin Press, New York)

Lindblom, L.: 1986, *Astrophys. J.* **303**, 146

Mestel, L., Wang, Y.-M.: 1982, *MNRAS* **198**, 405

Mestel, L., Robertson, J.A., Wang, Y.-M., Westfold, K.C.: 1985, *MNRAS* **217**, 443

Michel, F.C.: 1969, *Astrophys. J.* **157**, 1183

Michel, F.C.: 1982, *Rev. Mod. Phys.* **54**, 1

Michel, F.C.: 1983, *Astrophys. J.* **266**, 188

Michel, F.C.: 1985, in *The Crab Nebula and Related Supernova Remnants*, ed. M.C. Kafatos and R.B.C. Henry (Cambridge University Press, Cambridge), p. 55

Okamoto, I.: 1978, *Mon. Not. Roy. Astron. Soc.* **185**, 69

Pelling, R.M., Paciesas, W.S., Peterson, L.E., Makishima, K., Oda, M., Ogawara, Y., Miyamoto, S.: 1987, *Astrophys. J.* **319**, 416

Phinney, E.S., Evans, C.R., Blandford, R.D., Kulkarni, S.R.: 1988, *Nature* **333**, 832

Ray, A., Datta, B.: 1984, *Astrophys. J.* **282**, 542

Rees, M.J., Gunn, J.E.: 1974, *Mon. Not. Roy. Astron. Soc.* **167**, 1

Ricker, G.R., Scheepmaker, A., Ryckman, S.G., Ballantine, J.E., Doty, J.P., Downey, P.M., Lewin, W.H.G.: 1975, *Astrophys. J. Lett.* **197**, L87

Ruderman, M.A., Sutherland, P.G.: 1975, *Astrophys. J.* **196**, 51

Sakurai, T.: 1985, *Astron. Astrophys.* **152**, 121

Scharlemann, E.T., Wagoner, R.V.: 1973, *Astrophys. J.* **182**, 951

Shibata, S.: 1988, *MNRAS* **233**, 405

Sturrock, P.A.: 1970, *Nature* **227**, 465

Sturrock, P.A.: 1971, *Astrophys. J.* **164**, 529

Tademaru, E.: 1973, *Astrophys. J.* **183**, 625

Taylor, J.H., Stinebring, D.R.: 1986, Ann. Rev. Astron. Astrophys. **24**, 285

Thorne, K.S., Price, R.H., MacDonald, D.A.: 1986, *Black Holes: The Membrane Paradigm* (Yale University Press, New Haven)

Ulm: 1982, *Phys. Fluids* **25**, 1908

Ulm, Davidson, R.C.: 1980, *Phys. Fluids* **23**, 813

Usov, V.V.: 1983, *Nature* **305**, 409

Usov, V.V.: 1987, *Astrophys. J.* **320**, 333

van den Bergh, S., Pritchet, C.J.: 1989, *Astrophys. J. Lett.* **338**, L69

van den Heuvel, E.P.J., van Paradijs, J., Taam, R.E.: 1986, *Nature* **322**, 153

van den Heuvel, E.P.J., van Paradijs, J.: 1988, *Nature* **334**, 227

Wagoner, R.V.: 1984, *Astrophys. J.* **278**, 345

White, N.E., Stella, L.: 1988, *Nature* **333**, 708

Winske, D., Tanaka, M., Wu, C.S., Quest, K.B.: 1985, *J. Geophys. Res.* **90**, 123

Wu, C.S.: 1982, *Space Science Rev.* **32**, 83

THE ORIGIN AND EVOLUTION
OF
X-RAY BINARIES AND LOW-MAGNETIC-FIELD RADIO PULSARS

Frank Verbunt
Max Planck Institut für Extraterrestrische Physik
D-8046 Garching bei München
Federal Republic of Germany

**Abstract.** This paper gives an introduction to the theory of the formation and evolution of X-ray binaries and binary or low-magnetic-field radio pulsars. It deals with massive X-ray binaries, low-mass X-ray binaries, and X-ray binaries in globular clusters, as well as the transformation of some of these into millisecond radio pulsars. The paper is tutorial, the emphasis is on simple semi-analytical description.

## 1. Introduction

Many of the brightest X-ray sources in the sky are binaries in our own galaxy, in which a neutron star (or in a few cases a black hole) accretes matter from a companion donor star. The nature of these binaries, their formation, and their evolution have been gradually unravelled in several decades of intense observational and theoretical study. Our understanding is still far from complete.

Recently a number of low-magnetic-field pulsars have been discovered that are thought to have evolved from X-ray binaries.

I describe our current ideas on the formation and evolution of the X-ray binaries and low-magnetic-field pulsars. This paper is meant to be introductory, and the emphasis is on paedagogic exposition, rather than on completeness. This does not imply that the complete, detailed calculations aren't interesting! However, the main features of the origin and evolution of X-ray binaries can be described with a limited number of fairly simple equations. A brief overview in Section 2 of the observed classes of X-ray binaries and low-magnetic field radio pulsars precedes a description of these simple equations in Section 3. They are applied to the formation and evolution of massive X-ray binaries and related radio pulsars in Section 4, and to the formation and evolution of low-mass X-ray binaries and related radio pulsars in Section 5. The systems in globular clusters are formed via close stellar encounters, as is described in Section 6.

This paper relies heavily on earlier reviews, to which the reader is referred for more detail, and, occasionally, a different opinion. For reviews of observed properties one may consult Joss & Rappaport (1983, 1984), and Van den Heuvel

*W. Kundt (ed.), Neutron Stars and Their Birth Events, 179–218.*
© *1990 Kluwer Academic Publishers.*

& Rappaport (1987), for massive X-ray binaries; Mason (1986) for low-mass X-ray binaries; and Van Paradijs (1983) for optical observations of both classes. Bradt & McClintock (1983) give a compilation of references on individual X-ray sources and their optical counterparts. The evolution of massive binaries is well described in the research papers by Kippenhahn & Weigert (1967) and Paczyński (1967a). Their subsequent evolution into massive X-ray binaries is reviewed by Van den Heuvel (1978, 1983). For the formation and evolution of low-mass X-ray binaries one may consult the reviews by Savonije (1983), King (1988) and Verbunt (1988a), and for the systems in globular clusters Verbunt (1988b). The relation between X-ray sources and low-magnetic-field pulsars is reviewed by Van den Heuvel (1984) and Kulkarni (1988).

## 2. High-mass and low-mass systems

The X-ray binaries can be divided in two distinct groups, on the basis of the mass of the donor star. The systems with a massive ($\gtrsim 5M_\odot$) donor are often X-ray pulsars, and eclipses of the X-ray source are common in these binaries. None of the systems with O- or B-star donors shows X-ray bursts. The systems with a low-mass K or M star donor are — with a few exceptions — no pulsars, and they rarely show X-ray eclipses. Many of these sources show X-ray bursts. The X-ray spectra of the pulsars in both classes are hard (blackbody temperatures of $\sim 15$ keV). The X-ray spectra of the non-pulsating low-mass X-ray binaries are soft ($\sim 5$ keV).

O and B stars have short evolution time scales, and the binaries containing such stars must be very young, less than some $10^7$yr. In contrast, the binaries containing low-mass donor stars can be very old, up to $\sim 10^{10}$yr. This age difference is reflected in the distribution of the sources in the galaxy. The high-mass systems are all very close to the galactic plane, where young stars are formed, whereas the low-mass systems have high average distances to the plane (up to several hundred parsec), or are located in globular clusters.

The orbital periods of high-mass X-ray binaries range from 2 days to 20 days and longer. The orbits of most low-mass binaries are much more compact, with periods as short as 685 seconds, but more commonly in the hour range. A few examples of high-mass and low-mass systems are given in Table 1.

Radio pulsars are usually single, but some 11 are in a binary. In each case, the companion to the pulsar is a compact star, i.e., a white dwarf or a neutron star. In three of the binaries, the companion is thought to have a mass of $\gtrsim 1M_\odot$. Such massive white dwarfs or neutron stars must have relatively massive progenitors, $\gtrsim 5M_\odot$, say. These 3 systems are related to the high-mass X-ray binaries. In 7 binaries, the companion mass is between 0.17 and 0.4 $M_\odot$. These systems are related to the low-mass X-ray binaries.

One of the systems has a companion with an extremely low mass of $\lesssim 0.02M_\odot$. In this system, the radiopulsar appears to be ablating its companion. The binary radio pulsars and three related single radio pulsars with low magnetic field are listed in Table 1.

## 3. Equations for mass transfer and mass loss in a binary

### 3.1 Stellar time scales.

Three time scales associated with single stars are important for the study of binary evolution (Morton 1960). In order of increasing length these are:

1. the pulsational time scale. This is the time scale on which a star counteracts a perturbation of its hydrostatic equilibrium. It is given by the ratio of the radius of the star $R$ and the average sound velocity of the stellar matter $c_s$:

$$\tau_p = \frac{R}{c_s} \simeq 0.04 (\frac{M_\odot}{M})^{1/2} (\frac{R}{R_\odot})^{3/2} \text{day} \tag{1}$$

2. the thermal time scale. This is the time scale on which a star reacts when energy loss and energy production are no longer in equilibrium. It is given by the ratio of the thermal energy content of the star $E_{th}$ and the luminosity $L$:

$$\tau_{th} = \frac{E_{th}}{L} \simeq 3.1 \times 10^7 (\frac{M}{M_\odot})^2 (\frac{R_\odot}{R}) (\frac{L_\odot}{L}) \text{yr} \tag{2}$$

3. the nuclear time scale. This is the time scale on which a star uses its nuclear fuel. It is given by the product of the available fusable matter $M_{core}$ and the fusion energy $\epsilon_N$, divided by the stellar luminosity. For hydrogen fusion, this is:

$$\tau_N = \frac{M_{core}\epsilon_N}{L} \simeq 10^{10} \frac{M}{M_\odot} \frac{L_\odot}{L} \text{yr} \tag{3}$$

In the course of its evolution, a star fuses hydrogen in its core on the nuclear time scale. During this time, on the main sequence, the star does not change its radius very much. After exhaustion of the hydrogen in the core, the star starts expanding, on a thermal time scale.

### Table 2

*This Table gives approximate mass-radius and mass-luminosity relations for zero-age main-sequence stars in thermal equilibrium, for use in Eqs.(1-3). Cf. Allen (1976).*

|  | $M \gtrsim M_\odot$ | $M \lesssim M_\odot$ |
|---|---|---|
| mass-radius relation | $R/R_\odot \simeq (M/M_\odot)^{0.75}$ | $R/R_\odot \simeq (M/M_\odot)$ |
| mass-luminosity relation | $L/L_\odot \simeq (M/M_\odot)^{3.5}$ | $L/L_\odot \simeq (M/M_\odot)^3$ |

# Table 1

## massive X-ray binaries

| name | position | $P(s)$ | $P_{orb}(d)$ | $e$ | sp.type | $L_x$ | ref |
|------|----------|--------|--------------|-----|---------|-------|-----|
| LMC X-4 | $0532-66$ | 13.5 | 1.4 | 0.011 | O7III | $4 \times 10^{38}$ | 1 |
| LMC X-3 | $0538-64$ | - | 1.7 | $\sim 0$ | BIII-IV | $3 \times 10^{38}$ | 2 |
| Cen X-3 | $1119-60$ | 4.8 | 2.1 | 0.0007 | O6.5II | $8 \times 10^{37}$ | 3 |
| SMC X-1 | $0115-74$ | 0.7 | 3.9 | $< 0.0008$ | BOI | $6 \times 10^{38}$ | 4 |
| Cyg X-1 | $1956+35$ | - | 5.6 | $\sim 0$ | O9.7I | $2 \times 10^{37}$ | 5 |
| Vela X-1 | $0900-40$ | 283 | 9.0 | 0.092 | B0.5I | $6 \times 10^{36}$ | 6 |
| LMC tran | $0535-67$ | .069 | 16.7 | $\sim 0.7$ | B2IV | $1 \times 10^{39}$ T | 7 |
| - | $0115+63$ | 3.6 | 24.3 | 0.34 | Be | $8 \times 10^{36}$ T | 8 |
| V725 Tau | $0535+26$ | 104 | 111.0 | 0.3-0.4 | Be | $2 \times 10^{37}$ T | 9 |

## low-mass X-ray binaries

| name | position | $P(s)$ | $P_{orb}(h)$ | | | $L_x$ | ref |
|------|----------|--------|--------------|--|--|-------|-----|
| NGC6625 | $1820-30$ | | 0.19 | | | $8 \times 10^{37}$ | 10 |
| | $1627-67$ | 7.7 | 0.70 | | | $6 \times 10^{36}$ | 11 |
| | $1916-05$ | | 0.83 | | | $8 \times 10^{36}$ | 12 |
| transient | $0748-68$ | | 3.8 | | | $1 \times 10^{37}$ T | 13 |
| | $1755-34$ | | 4.4 | | | $6 \times 10^{36}$ | 14 |
| A0620-00 | $0620-00$ | | 7.3 | | | $2 \times 10^{38}$ T | 15 |
| Cen X-4 | $1455-31$ | | 15.1 | | | $1 \times 10^{38}$ T | 16 |
| Sco X-1 | $1617-16$ | | 19.2 | | | $3 \times 10^{37}$ | 17 |
| Cyg X-2 | $2142+38$ | | 235.2 | | | $1 \times 10^{38}$ | 18 |

## low-$B$ radio pulsars

| name | position | $P(ms)$ | $P_{orb}(d)$ | $e$ | $M_c(M_\odot)$ | $B(G)$ | ref |
|------|----------|---------|--------------|-----|----------------|--------|-----|
| | $1937+21$ | 1.6 | | | | $4 \times 10^8$ | 19 |
| | $1957+20$ | 1.6 | 0.38 | $< 0.001$ | 0.02 | | 20 |
| | $1855+09$ | 5.4 | 12.33 | 0.000021 | 0.2-0.4 | $3 \times 10^8$ | 21 |
| | $1953+29$ | 6.1 | 117.35 | 0.00033 | 0.2-0.4 | $4 \times 10^8$ | 22 |
| | $1831-00$ | 521.0 | 1.81 | $< 0.005$ | 0.06-0.13 | $< 8 \times 10^{10}$ | 23 |
| | $0820+02$ | 864.9 | 1232.40 | 0.0119 | 0.2-0.4 | $3 \times 10^{11}$ | 24 |
| | $1913+16$ | 59.0 | 0.32 | 0.6171 | 1.40 | $2 \times 10^{10}$ | 25 |
| | $0655+64$ | 195.6 | 1.03 | $< 0.00005$ | 0.7-1.3 | $1 \times 10^{10}$ | 26 |
| | $2303+46$ | 1066.4 | 12.34 | 0.6584 | 1.2-2.5 | $6 \times 10^{11}$ | 27 |
| M 28 | $1821-24$ | 3.1 | | | | $< 2 \times 10^{10}$ | 28 |
| 47 Tuc A | $0021-72$ | 4.5 | 0.02 | 0.33 | 0.25 | | 29 |
| 47 Tuc B | $0021-72$ | 6.1 | 7-95 | | | | 29 |
| M 4 | $1620-26$ | 11.1 | 191.4 | 0.025 | 0.35 | $3 \times 10^9$ | 30 |
| M 15 | $2127+12$ | 110.7 | | | | | 31 |

## Table 1

*This Table gives parameters for a number of massive and low-mass X-ray bina-ries, and for binary and/or low-magnetic field radio pulsars. From left to right are the name, position, pulse period, orbital period and eccentricity; then for the X-ray binaries the spectral type of the companion and the X-ray luminosity (at maximum for transient (T) sources), or for radio pulsars the probable companion mass and the surface magnetic field strength. For each source one reference is given. More references can be found in the review papers listed in the Introduction.* [1] Dennerl 1989 *PhD Thesis* München. [2] Cowley *et al.* 1983 *ApJ* **272**, 118. [3] Schreier *et al.* 1972 *ApJ* **172**, L79. [4] Primini *et al.* 1977 *ApJ* **217**, 543. [5] Webster & Murdin 1972 *Nature* **235**, 37. [6] Rappaport *et al.* 1976 *ApJ* **206**, L103. [7] Skinner *et al.* 1982 *Nature* **297**, 568. [8] Rappaport *et al.* 1978 *ApJ* **224**, L1. [9] Nagase *et al.* 1982 *ApJ* **263**, 814. [10] Stella *et al.* 1987 *ApJ* **312**, L17. [11] Middleditch *et al.* 1981 *ApJ* **244**, 1001. [12] White & Swank 1982 *ApJ* **253**, L61. [13] Parmar *et al.* 1986 *ApJ* **308**, 199. [14] Mason *et al.* 1985 *MNRaS* **216**, 1033. [15] McClintock & Remillard 1986 *ApJ* **308**, 110. [16] Chevalier *et al.* 1989 *A & A* in press. [17] Gottlieb *et al.* 1978 *ApJ* **195**, L33. [18] Cowley *et al.* 1979 *ApJ* **231**, 539. [19] Backer *et al.* 1982 *Nature* **300**, 615. [20] Fruchter *et al.* 1988 *Nature* **333**, 237. [21] Segelstein *et al.* 1986 *Nature* **322**, 714. [22] Boriakoff *et al.* 1986 *Nature* **304**, 417. [23] Dewey *et al.* 1986 *Nature* **322**, 712. [24] Manchester *et al.* 1980 *ApJ* **236**, L25. [25] Hulse & Taylor 1975 *ApJ* **195**, L51. [26] Damashek *et al.* 1982 *ApJ* **253**, L57. [27] Stokes *et al.* 1985 *ApJ* **294**, L21. [28] Lyne *et al.* 1987 *Nature* **328**, 399. [29] Ables *et al.* 1988 *IAU Circ.* # 4602. [30] Lyne *et al.* 1988 *Nature* **332**, 45. [31] Wolszcsan *et al.* 1988 *IAU Circ.* # 4552.

### 3.2 Potential in a binary frame: the Roche lobe.

The potential in a binary is determined by the gravitational attraction of the two stars, and by the motion of the two stars around one another. In the binary frame, one has

$$\Phi = -\frac{GM_1}{r_1} - \frac{GM_2}{r_2} - \frac{\omega^2 r_3^2}{2} \tag{4}$$

where $r_1$ and $r_2$ are the distances to the center of the stars with mass $M_1$ and $M_2$, respectively; $\omega = \sqrt{G(M_1 + M_2)/a^3}$ is the orbital angular velocity; and $r_3$ is the distance to the axis of rotation of the binary (see Figure 1). Writing Eq.(4) in dimensionless units (mass in units of the total mass, and distances in units of the semi-major axis $a$), one sees that the form of the surfaces of constant $\Phi$ depends only on the mass ratio $M_1/M_2$.

We can discriminate four types of surfaces of $-\Phi = C$, with $C$ a positive constant. For large $C$, the potential surface consists of two closed surfaces, one around each star. For a critical value of $C$, the two closed surfaces touch, in the

184

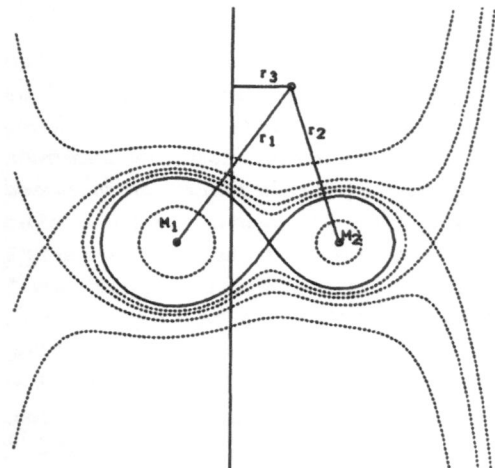

*Figure 1. Roche lobe geometry for a mass ratio $M_1/M_2 = 2$. Equipotential surfaces are shown for different values of $-\Phi = C$. For the largest value of $C$ the surface consist of two separate lobes, one around each star. The Roche lobe is the surface around both stars that passes through the inner Lagrangian point. Also shown are the surfaces containing the two outer Lagrangian points. The vertical line is the rotation axis.*

inner Lagrangian point. The surface at this value of $C$ is called the Roche lobe. For smaller values of $C$ we have a closed surface around both stars, and for very small values the surfaces become open.

The volume of the Roche lobe can be calculated numerically. A useful approximate formula for the average radius of the Roche lobe around the most massive star (with mass $M_1$) is:

$$\frac{R_L(M_1)}{a} \simeq 0.38 + 0.2\log\frac{M_1}{M_2} \tag{5a}$$

which is accurate to 2 % for mass ratios $0.2 < M_1/M_2 < 20$ (Paczyński 1966 , 1971). For the average radius of the Roche lobe around the less massive star, with mass $M_2$, one may use the approximate formula:

$$\frac{R_L(M_2)}{a} \simeq 0.46(\frac{M_2}{M_1 + M_2})^{1/3} \tag{5b}$$

which is accurate to 2 % for mass ratios $M_2/M_1 < 0.8$ (Paczyński 1967b, 1971).

A particle within the Roche lobe is attached to one star; a particle on the Roche lobe can move to the other star. Thus, if a star reaches the size of the Roche lobe, mass transfer may ensue. This can occur because the star expands in the course of its evolution, or because the binary shrinks. An evolving star in a binary can fill its Roche lobe for the first time as it expands on the main sequence (Case A), as it expands after hydrogen exhaustion (Case B), or as it expands after helium exhaustion (Case C). Which of the three cases applies, depends on the size of the Roche lobe, which in turn depends on the distance between the two stars and (to a lesser extent) on the mass ratio (see Eq.(5)).

## 3.3 Conservative mass transfer.

The angular momentum of a binary can be written:

$$J_{orb} = M_1 M_2 \sqrt{\frac{Ga}{M_1 + M_2}} \tag{6}$$

For the moment we consider *conservative* mass transfer, in which the total mass $M_1 + M_2$ of the binary is conserved, and in which $M_2$ is the mass donor. In this case, all the mass lost by star 2 is gained by star 1: $\dot{M}_1 = -\dot{M}_2$, and the time derivative of Eq.(6) can be written:

$$\frac{\dot{a}}{a} = 2\frac{\dot{J}_{orb}}{J_{orb}} - 2(1 - \frac{M_2}{M_1})\frac{\dot{M}_2}{M_2} \tag{7}$$

Consider first a binary whose angular momentum is conserved: $\dot{J}_{orb} = 0$. According to Eq.(7), transfer from the more massive star to the less massive star causes a decrease of the distance between the stars: if $M_2 > M_1$, $\dot{a} < 0$. Conversely, transfer from the less massive star causes $a$ to increase.

From conservation of angular momentum the ratio of final and initial semi-major axis follows from Eq.(6), and with Kepler's law

$$(\frac{2\pi}{P_{orb}})^2 = \frac{G(M_1 + M_2)}{a^3} \tag{8}$$

we find the ratio of final and initial periods:

$$\frac{a_f}{a_i} = (\frac{M_{1i}M_{2i}}{M_{1f}M_{2f}})^2 \qquad \frac{P_f}{P_i} = (\frac{M_{1i}M_{2i}}{M_{1f}M_{2f}})^3 \tag{9}$$

where $i$ and $f$ denote the initial and final parameters, respectively.

When mass is transferred, the radius of the mass losing star changes, and also the radius of its Roche lobe. When a star finds itself with a new mass, it first adjusts its structure on a dynamical time scale (Eq.(1)). During this short time the energy content of the star cannot change, i.e. the change is adiabatic. The star subsequently adjusts its thermal structure on the thermal time scale (Eq.(2)). The size of the Roche lobe changes due to the variations of $a$ and of $M_1/M_2$, according to Eq.(5). If the new stellar radius is larger than the new Roche-lobe radius, more matter is transferred, causing a further increase of the stellar radius with respect to the Roche-lobe radius, etc. The mass transfer in such a case is unstable. If the star is larger than the Roche lobe after adjusting on the dynamical time scale, the instability is dynamical; if it is larger after the thermal adjust-ment, the instability proceeds on a thermal time scale.

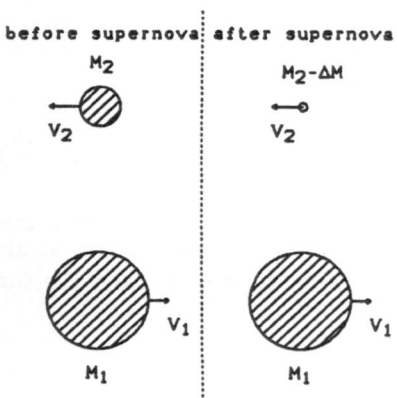

before supernova | after supernova

$M_2$        $M_2-\Delta M$

$V_2$        $V_2$

$V_1$    $V_1$

$M_1$       $M_1$

*Figure 2. Schematic representation of the assumption used to calculate the eccentricity after supernova explosion, in the absence of kick velocity: before and after the explosion, the positions and velocities of the stars are the same. Only the mass of the exploded star has changed.*

## 3.4 Consequences of sudden mass loss: the supernova explosion

During the supernova event, the neutron star progenitor loses mass, and in addition it may receive a kick velocity, if the explosion is asymmetric. For zero kick velocity, the orbit after explosion is found from the pre-explosion orbit with the assumption that the explosion is instantaneous. In that case, the position and velocity of the two stars is the same after and before the supernova event (see Figure 2). Consider an initially circular orbit, with radius $a_i$ which changes into an elliptical orbit with eccentricity $e$ and semi-major axis $a_f$, due to loss of $\Delta M$ in a supernova explosion. The periastron distance of the new orbit equals the radius of the initial orbit:

$$a_i = (1-e)a_f \tag{10}$$

and the periastron velocity of the new orbit equals the orbital velocity of the original orbit:

$$\frac{G(M_1+M_2)}{a_i} = \frac{G(M_1+M_2-\Delta M)}{a_f}\frac{1+e}{1-e} \tag{11}$$

Substituting Eq.(10) in Eq.(11) gives the eccentricity:

$$e = \frac{\Delta M}{M_1+M_2-\Delta M} \tag{12}$$

If more than half of the total mass of the binary is lost in the explosion, i.e. if $\Delta M > (M_1+M_2)/2$, the orbit becomes hyperbolic, $e > 1$, and the binary is disrupted. The system velocity $v_s$, i.e. the velocity of the center of mass, after explosion is given by:

$$v_s = \frac{M_2 v_2 - (M_1 - \Delta M)v_1}{M_1+M_2-\Delta M} = ev_1 = eM_2\sqrt{\frac{G}{a_i(M_1+M_2)}} \tag{13}$$

In all these equations, $M_1$ and $M_2$ are the stellar masses before the supernova event. $v_1$ and $v_2$ are the orbital stellar velocities with respect to the center of mass before the explosion (Figure 2).

Eqs.(12) and (13) show that a supernova event in a binary may lead to substantial system velocity and eccentricity, even if the kick velocity is zero. For a finite kick velocity, the resulting system velocity and eccentricity depend on the direction and magnitude of the kick.

If the eccentric orbit has a sufficiently short orbital period, tidal forces are able to circularize it. During this circularization, energy is dissipated but angular momentum is conserved. This enables us to calculate the semimajor axis $a_c$ of the circularized orbit, in units of                   the semimajor axis of the eccentric orbit $a_f$, or via Eq.(10) of radius $a_i$ of the pre-supernova orbit:

$$a_c = (1 - e^2)a_f = (1 + e)a_i \qquad (14)$$

3.5 The consequences of slow mass loss.

Binaries consisting of very massive stars may be affected by mass loss that such stars undergo because of their massive stellar winds. The effect of this mass loss depends on the amount of angular momentum that the winds carry away with them. This amount is not known. As a simple example, we consider a system in which the wind matter has the specific angular momentum of the orbit $j \equiv J_{orb}/(M_1 + M_2)$. After an amount of matter $\Delta M$ has been lost, the new orbital angular momentum is $J_{orb,f} = J_{orb} - j\Delta M$. Combining this with Eq.(6), one finds the new semi-major axis:

$$\frac{a_n}{a} = (\frac{M_1 M_2}{M_{1n} M_{2n}})^2 (\frac{M_{1n} + M_{2n}}{M_1 + M_2})^3 \qquad (15)$$

where $n$ indexes the parameters after mass loss.

If the specific angular momentum of the wind differs appreciably from the specific orbital angular momentum, Eq.(15) can not be used.

3.6 The spiral-in process

Mass transfer from the more massive to a less massive star causes a decrease in the distance between the stars (see Sect.3.3). It has been suggested that this can cause the less massive star to enter the atmospheric envelope of the mass loser, in particular in systems with an extreme mass ratio. The friction between the less massive star in its orbital motion and the donor envelope causes a transfer of angular momentum from the orbital motion to the envelope matter. The star plunges deeper into the envelope, the friction increases, the star plunges even deeper. This process can stop when the whole envelope is brought to corotation with the motion of the less massive star. Alternatively, the energy dumped into the envelope by the friction may become large enough to expel the envelope, leaving the naked core of the original donor in a binary with the less massive star.

This process of spiralling in is badly understood: we do not know how it begins, nor do we know how it ends. An excellent review of the topic and its uncertainties is given by De Kool (1987). In terms of energy we can describe the process as follows. The binding energy $E_{env}$ of the stellar envelope is of order

$$E_{env} \sim \frac{GM_c M_{env}}{R_{env}} \tag{16}$$

where $M_c$ is the mass of the stellar core and $M_{env}$ and $R_{env}$ are the mass and average radius of the envelope. The binding energy $E_b$ of a binary with semi-major axis $a$ is

$$E_b = \frac{GM_c M_2}{2a} \tag{17}$$

where $M_2$ is the mass of the star that spiralled in. In order to expel the envelope, the binding energy of the final binary must exceed the binding energy of the original envelope: $E_b > E_{env}$. One can define an efficiency factor $\alpha$ for the spiral-in process $\alpha = E_b / E_{env}$. Thus, for expulsion of the envelope, $\alpha > 1$.

Early calculations of the spiral-in process were one-dimensional, and assumed that the frictional energy released is deposited in a shell at the instantaneous radius of the stellar orbit within the envelope. These calculations indicated efficient expulsion of the envelope: $\alpha \gtrsim 1$. More recent calculations are two-dimensional, and assume that the energy is deposited in a ring, corresponding to the instantaneous orbit of the star that spirals in. In these calculations the efficiency of expulsion is much smaller: $\alpha \gg 1$.

Our understanding of the spiral-in process is very limited indeed. It is worth noting that the existence of a common envelope as such need not lead to a spiral-in process, as the stable existence of contact binaries (binaries in which both stars (over)fill their Roche lobes) shows.

## 3.7 Accretion, magnetosphere and spin-up

Accretion of matter onto a neutron star leads to energy release mainly at X-ray wavelengths. Therefore the X-ray luminosity $L_x$ is related to the accretion rate $\dot{M}$ via

$$L_x \simeq \frac{GM\dot{M}}{R} \tag{18}$$

where M and R are the neutron star mass and radius, respectively. The accretion rate has an upper bound, given by the Eddington limit: if the outward radiation pressure caused by $L_x$ is larger than the gravitational attraction, matter will be blown away rather than accreted. The limiting luminosity is thus given by the condition:

$$\frac{L_{Edd}}{4\pi r^2}\frac{\sigma}{c} = \frac{GM}{r^2} \Rightarrow L_{Edd} = \frac{4\pi cGM}{\sigma} \simeq 1.8 \times 10^{38}\frac{M}{1.4M_\odot}\text{erg s}^{-1} \tag{19}$$

where $\sigma$ is the radiation absorption coefficient, for which in the X-ray regime we can use the Thomson cross section. With Eq.(18) the luminosity limit can be translated into a limit to the accretion rate:

$$\dot{M}_{Edd} = \frac{4\pi cR}{\sigma} \simeq 1.5 \times 10^{-8}\frac{R}{10^6 \text{cm}}M_\odot \text{yr}^{-1} \tag{20}$$

An estimate of the surface magnetic field $B$ of a radio pulsar is usually made by equating the energy loss via radiation of the pulsar, assumed to be given by the magnetic dipole radiation formula, to the loss of rotation energy:

$$-\frac{2B^2R^6\omega^4}{3c^3} = \frac{\mathrm{d}}{\mathrm{d}t}(\frac{1}{2}I\omega^2) \Rightarrow P\dot{P} = (\frac{8\pi^2R^6}{3Ic^3})B^2 \tag{21}$$

where $I \simeq 0.4MR^2$ is the moment of inertia of the neutron star, and $P \equiv 2\pi/\omega$ its rotation period. The magnetic fields of the radio pulsars listed in Table 1 are derived from the measured $\dot{P}$-values with Eq.(21).

An accreting neutron star with a finite magnetic field is surrounded by a volume in which the motion of the accreting matter is dominated by the magnetic forces. Heuristically, a radius $r_m$ of this magnetosphere is estimated by equating the magnetic pressure $B(r_m)^2/8\pi$ of a dipole to a ram pressure $\rho v^2$. For spherical accretion, with $v$ equal to the free fall velocity, this leads to

$$\frac{B^2R^6}{8\pi r_m^6} = \frac{\dot{M}}{4\pi r_m^2}\sqrt{\frac{2GM}{r_m}} \Rightarrow \frac{r_m}{R} = (\frac{B^2R^{5/2}}{2\dot{M}\sqrt{2GM}})^{2/7} \tag{22}$$

For accretion via an accretion disk, this formula is thought to be a reasonable approximation as well. The interaction between the accretion disk and the neutron star drives the neutron star rotation towards an equilibrium period approximately given by the Keplerian rotation period at the magnetospheric radius. With Eq.(22) this gives:

$$P_{eq} = 2\pi(\frac{B^2R^6}{2\sqrt{2}\dot{M}})^{3/7}(\frac{1}{GM})^{5/7}$$

$$\simeq 6 \times 10^{-3}\mathrm{sec}(\frac{B}{10^9\mathrm{G}})^{6/7}(\frac{R}{10^6\mathrm{cm}})^{18/7}(\frac{1.4M_\odot}{M_{ns}})^{5/7}(\frac{10^{-9}M_\odot\mathrm{yr}^{-1}}{\dot{M}})^{3/7} \tag{23}$$

It is necessary to remember that Eqs.(21-23) are heuristic in nature. Our very limited understanding of the radiation mechanisms of radio pulsars, and of the physics of the accretion process does not enable us to derive rigorous formulae.

## 4. Formation and evolution of high-mass systems

### 4.1 Evolution of a massive binary.

In a binary consisting of two massive stars, the most massive star evolves first. During the supernova explosion it loses most of its mass, leaving a neutron star with a mass $M_{ns} \simeq 1.4M_\odot$. If the less massive star differs by more than 2 $M_{ns}$ from the massive star, more than half of the total mass of the binary is lost during the supernova explosion, and the binary is disrupted (see Eq.(12)). The disruption can be avoided when the initially more massive star transfers mass to the initially less massive star. A massive star stays on the main sequence as long as nuclear fusion of hydrogen into helium continues in its core. The main-sequence life time is given by Eq.(3). During this time the star expands a little; a 5 $M_\odot$ star at the end of its main-sequence life time has a radius about 50 % larger than at zero-age main-sequence. Upon exhaustion of the hydrogen in the core, the star adapts to the new situation by expanding on its thermal time scale, given by Eq.(2). This expansion continues until helium fusion starts. A 5 $M_\odot$ star has by then increased its radius to some 70 $R_\odot$. The star shrinks somewhat as the energy production is switched on again. Upon the exhaustion of the helium, a new rapid expansion follows.

For mass transfer during the main-sequence phase of the mass donor (Case A), the initial orbital period must be very short; for mass transfer during the expansion after helium exhaustion (Case C) the initial orbital period must be very long (Sect.3.2). Many massive X-ray binaries are thought to have undergone mass transfer in Case B, when the mass donor expands after hydrogen is exhausted in its core.

A mainly radiative star shrinks dynamically due to mass loss, but expands on the thermal time scale. When the mass donor is the more massive star in the binary, its Roche lobe shrinks (Sect.3.2). Thus mass transfer from a massive, mainly radiative star to a less massive star is unstable on the thermal time scale. Mass continues to be transferred until the mass receiving star has become more massive than the mass losing star: then the Roche lobe of the donor increases with further mass transfer. Once it increases faster than the donor radius, mass transfer can stabilize again. In practice, mass transfer stabilizes only after the mass donor has lost most of its envelope, and has reached a total mass close to its core mass. The evolution of the stellar and Roche-lobe radii during transfer from a massive star is shown schematically in Figure 3.

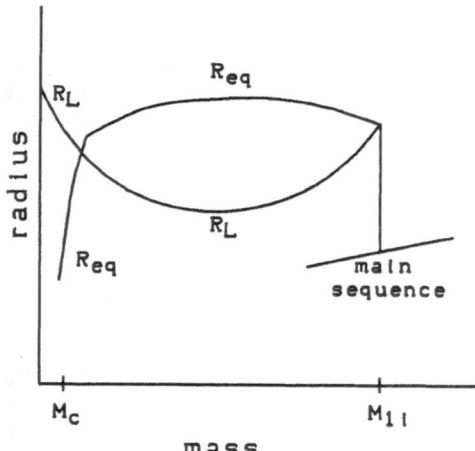

*Figure 3. Schematic representation of the change of the radius of the Roche lobe $R_L$ and of the equilibrium radius of a massive mass losing star $R_{eq}$. Mass transfer begins after the star has expanded from the main sequence to its Roche lobe. Mass transfer is unstable because the Roche lobe shrinks, whereas the equilibrium radius becomes larger. It stays unstable until $R_{eq}$ becomes equal to $R_L$. This occurs after reversal of the mass ratio, for donor mass close to the core mass $M_c$. (After Paczyński 1970).*

As an example of the evolution of a massive binary into an X-ray binary, we take a binary with orbital period $P_{orb} = 5$ days and masses $M_1 = 25M_\odot$ and $M_2 = 10M_\odot$ (see Figure 4). We choose a 25 $M_\odot$ donor in order to be able to compare our rough estimates with a calculation with a detailed stellar model by Kippenhahn (1969). The most massive star is the first one to exhaust hydrogen in its core, after a time estimated as about $3.2 \times 10^6$ yr, with Eq.(3) and the approximate mass-luminosity relation for massive stars (Table 2). (The detailed calculation gives $4.7 \times 10^6$ yr.) The expansion of the massive star causes it to fill its Roche lobe, and mass transfer ensues (Fig.4b). This mass transfer is unstable, and continues on a thermal time scale, of about $2.2 \times 10^4$ yr according to Eq.(2). (The full model gives 7200 yr.) When most of the envelope of star 1 is transferred, the equilibrium radius becomes equal to $R_L$. When helium fusion starts, star 1 shrinks to helium main-sequence radius, and mass transfer stops (Fig.4c). At this point, the mass of the helium star is about 8.5 $M_\odot$, and, in the conservative case, the mass of star 2 has increased to 26.5 $M_\odot$. The orbital period has increased to 6.84 days, according to Eq.(9).

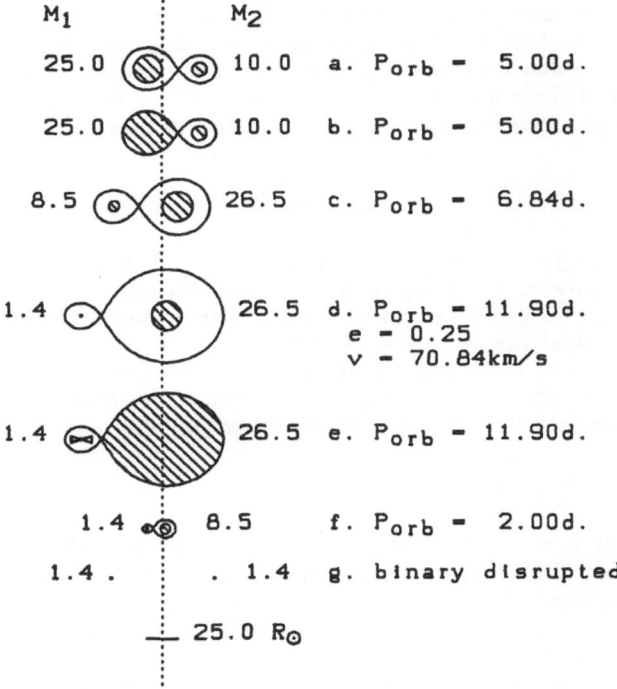

*Figure 4. Example of conservative evolution of a massive binary, leading to the formation of a massive X-ray binary. (After Van den Heuvel 1978.)*

The evolution from this point on was first suggested by Van den Heuvel & Heise

(1972). The helium star continues its evolution, and after some additional $5 \times 10^5$ yr explodes to leave a neutron star. For a zero kick velocity, and a neutron star of 1.4 $M_\odot$, the new orbital parameters are as follows. The system velocity (Eq.(13)) is $v_s = 71$ km/s, the eccentricity (Eq.(12)) is $e = 0.25$. The semi-major axis increases according to Eq.(10), from about $49R_\odot$ to $66R_\odot$, and the orbital period becomes 11.9 days (Fig.4d).

Star 2 exhausts its hydrogen in its turn, on the nuclear time scale corresponding to its new mass. Since its mass is just above 25 $M_\odot$, this time scale will be slightly shorter than the $4.7 \times 10^6$yr that it took the primary to evolve. After this time, star 2 begins to expand into a supergiant, and at the same time starts blowing a strong stellar wind. Accretion of a fraction of this wind matter by the neutron star causes the binary to light up as an X-ray source. The life time of this source is given by the expansion time of the 26.5 $M_\odot$ star, which equals its thermal time scale of $\sim 2 \times 10^4$ yr.

Star 2 continues to expand until it fills its Roche lobe, at which point mass transfer due to Roche-lobe overflow starts (Fig.4e). This mass transfer occurs also on the thermal time scale, and its rate is therefore about $M_2/\tau_{th}$, or $\sim 10^{-3} M_\odot$/yr. Roche lobe overflow does not start instantaneously: the mass donor does not have a sharp edge, but a density increasing at a finite pressure scale height $H_p$, which is a small fraction of the stellar radius. The time scale $\tau_{\dot{M}}$ for the mass transfer to rise is therefore given by this fraction of the expansion time scale:

$$\tau_{\dot{M}} = \tau_{th} \frac{H_p}{R} \tag{24}$$

It is only during this very short time that a massive donor can fuel an X-ray source by Roche-lobe overflow. After that the mass-transfer rate is higher than the maximum rate at which the neutron star can accrete, given by the Eddington limit, Eq.(20). It is thought that very high accretion rates smother the X-ray source, in the sense that much of the transferred matter forms a coccoon around the neutron star, which is optically thick to X-rays, and prevents the X-rays produced by the accretion of mass onto the neutron star from reaching us.

4.2 Comparison with observations. Problems.

Looking at Table 1, we see that the scenario discussed in Sect.4.1 nicely explains a system like Vela X-1, except for one detail: the observed eccentricity is smaller. This could mean two things: the mass loss in the supernova explosion may have been smaller, or some circularization can have occurred due to tidal interaction after the explosion. The very small eccentricities in the systems with the shortest orbital periods are certainly due to such tidal interaction, but in binaries with longer orbital periods such interactions are less efficient. The high eccentricity of the transient in the Large Magellanic Cloud may in part be the effect of a kick velocity given to the forming neutron star in an asymmetric supernova explosion. It appears then that the massive X-ray binaries with orbital periods in excess of about 5 days can be well explained via scenarios of conservative evolution.

In the example of Fig.4, the orbital period increased during the initial phase of mass transfer. As follows from Eq.(9), the orbital period decreases if the mass ratio after the mass transfer is more extreme than before. The first scenario discussed for the formation of a massive X-ray binary (Van den Heuvel and Heise 1972) , for Cen X-3, starts with a binary of a 16 $M_\odot$ and a 3 $M_\odot$ star, which after mass transfer have become a 4 $M_\odot$ helium star and a 15 $M_\odot$ B star. The orbital period has thereby decreased from the assumed initial period of 3.0 days to 1.5 days. The question then arises whether at this short period the newly massive B star still fits within its Roche lobe. Using the main-sequence mass-radius relation, we find that the star will in fact be slightly larger than its Roche lobe. This suggests that the evolution of systems with short orbital periods may involve loss of mass and angular momentum from the binary, in a phase during which the two stars share a common envelope.

This becomes even more likely if we consider the process of addition of matter to the mass-receiving star, in the first phase of mass transfer. As discussed in Sect.4.1 this mass transfer occurs on the thermal time scale of the mass donor. This time scale is shorter than the thermal time scale of the mass receiver, which has a smaller mass than the mass donor (see Eq.(2)). Thus, the mass-receiving star is not in thermal equilibrium as long as mass transfer proceeds, and is probably larger than an equilibrium main sequence star. This enhances the likelihood of a common envelope, and concomitant mass loss, in systems with short orbital periods.

Complications arise also in the calculation of the late stages in the evolution of massive stars. In these stages large differences in    chemical composition exist between the different layers of the star, and in this situation the proper treatment of convection is not well understood. In addition the evolution of massive stars is affected by mass loss. An interesting possibility is the formation of what one may call a 'suspended neutron star': a star which normally would evolve to a supernova and leave a neutron star, can be prevented from doing so by extended mass loss. It may then leave a Ne-O-Mg white dwarf, with a mass close to the Chandrasekhar limit. This white dwarf can be induced to collapse to a neutron star by accretion of matter from its companion (Nomoto 1987).

Yet another unknown parameter is the kick velocity which the neutron star gets at birth, due to possible asymmetries in the supernova explosion. Radio pulsars in the galactic disk have velocities up to several hundred km/s. If a supernova explosion occurs in a binary, the neutron star has an orbital velocity which may be sufficiently large to dissolve the binary. At the moment it is not settled whether the velocities of radio pulsars can be completely explained as remaining orbital velocities, or whether extra kick velocities are necessary to explain the observed velocity distribution. For discussions of this topic see Dewey & Cordes (1987) and Bailes (1989).

In an eccentric X-ray binary, the periastron moves because the mass donor is not perfectly spherical. The rate of periastron shift is small if the mass of the donor star is strongly concentrated to the core. In principle, therefore, the study of the motion of the periastron offers a clue to the stellar interior of the B star, and thereby to our theories of stellar structure (Joss & Rappaport 1983).

## 4.3 An example of non-conservative evolution

The evolution of very massive binaries can be affected by loss of matter via strong stellar winds. If only one of the two stars is very massive, the additional possibility arises of a spiral in during the first phase of mass transfer. It is very difficult to calculate non-conservative evolution, since we do not know how much matter is lost and how much angular momentum this mass loss entails. As an example, we discuss the scenario proposed by Van den Heuvel & Habets (1984) for the formation of LMC X-3 (see Figure 5).

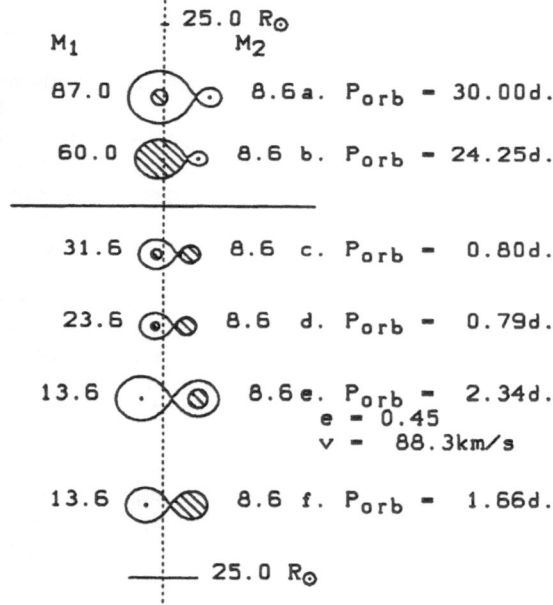

*Figure 5. Example of non-conservative evolution of a massive binary, leading to the formation of a massive X-ray binary. Notice the change in scale between b and c. (After Van den Heuvel & Habets 1984.)*

I will assume that the specific angular momentum lost with the stellar wind is that of the orbit, so that Eq.(15) can be applied. The original system consisted of a 87 $M_\odot$ and a 8.6 $M_\odot$ star. The 87 $M_\odot$ star has lost 27 $M_\odot$ in stellar wind before it leaves the main sequence and fills its Roche lobe, causing the orbital period to decrease, if Eq.(15) can be applied, to about 24 days (Fig.5b). Because of the extreme mass ratio, the mass transfer leads to a spiral-in of the B star in the envelope of the primary, from which the 31.6 $M_\odot$ core of the primary and the unchanged B star emerge, in a close binary. In order to lead to the current parameters of LMC X-3, the orbital period of this system must be about 0.8 days (Fig.5c). The Wolf-Rayet star loses another 8 $M_\odot$, causing a small reduction of the orbital period (Fig.5d), before it explodes as a supernova and leaves a 13.6 $M_\odot$ black hole. The eccentricity and system velocity follow from Eqs.(12,13): $e = 0.45$, $v_s = 88$km/s. The orbital period is 2.3 days (Fig.5e). Because the resulting system is very close, tidal forces circularize the system (see Eq.(14)). In this new orbit, the B star evolves until it fills its Roche lobe and transfers mass (Fig.5f). Notice that in order to fit within its Roche lobe in phase c, the B star must have a radius smaller than that given by the mass-radius relation of Table 2. In the Large Magellanic Cloud, B stars are indeed smaller, because of the lower metallicity.

## 4.4 The formation of radio pulsars from massive X-ray binaries

In a massive X-ray binary, the mass donor is often an order of magnitude more massive than the accreting neutron star. Such binaries are therefore thought to be prone to spiral-in of the neutron star. The end of the spiral-in process is a close binary consisting of the core of the mass donor and the neutron star. The final orbit depends on the mass and radius of the envelope and on the efficiency with which frictional energy release expells the envelope.

If the core is not sufficiently massive to implode, one is left with a binary of a massive white dwarf and a neutron star. The orbit after the spiral-in process is expected to be circular. If the core is sufficiently massive, it may continue to evolve and then implode to a neutron star. If the supernova event causes more than half of the binary mass to be lost, the binary is dissolved (Eq.(12)). However, if less mass is lost, a binary may remain that consists of two neutron stars. Such a binary is expected to be highly eccentric, in particular if the supernova explosion imparts a kick velocity to the new neutron star.

Looking at the observed eccentricities of the three binary radio pulsars with relatively massive companions (Table 1), we conclude that PSR1913 + 16 and PSR2303 + 46 must have neutron-star companions whereas PSR0655 + 64 must have a white dwarf companion.

Eqs.(10,12) enable us to calculate the parameters of the immediate progenitor of the newest neutron star, under the assumption that the supernova event occurred not long ago (as compared to the current timescale of orbital evolution). Both neutron stars have a mass of about 1.4 $M_\odot$, and the semi-major axis now is $a = 1.95 \times 10^{11}$cm (Taylor & Weisberg 1982). Eq.(12) gives the mass loss of the supernova event as 1.7 $M_\odot$, and Eq.(10) gives a pre-supernova semimajor axis $a_i \simeq 7.5 \times 10^{10}$cm. The orbital period of this system was just 1.5 hr! Eq.(13) gives the system velocity after the supernova explosion as $v_s \simeq 170$ km/s. These numbers have to be adapted if noticeable changes in $a$ and $e$ have occurred due to the evolution of the orbit via gravitational radiation upon formation of the second neutron star.

A nice touch in this evolutionary scenario is that the neutron star that we now see as the radio pulsar is in fact the one that formed first (Smarr & Blandford 1976). In the current view, every neutron star is formed with a magnetic field of order $\sim 10^{12}$G, which decays on a time scale of a few times $10^6$yr. The observed magnetic field of PSR1913+16 (as derived from its pulse period derivative) is about $2 \times 10^{10}$G. A pulsar whose field has decayed by so much must also have slowed down to a pulse period much longer than 59 ms. Thus the pulsar can not be the one that formed second. On the other hand, the pulsar that formed first will also have slowed down initially, during the time between its formation and the onset of mass transfer from its companion (phases d and e in Fig.4). During the mass transfer, it was spun up to a short period again. After the spin-up, the period derivative is lower, because of the lower magnetic field strength.

## 5. Formation and evolution of low-mass systems

### 5.1 Formation of low-mass X-ray binaries

The presence of a low-mass star close to a neutron star is very surprising. After all, a neutron star is formed in the most energetic event in stellar evolution: a supernova explosion. How can the low-mass star get closer to the neutron star than the main-sequence radius of the massive progenitor of the neutron star? How was disruption of the binary during the supernova event avoided? And why wasn't the low-mass star destroyed by the impact of supernova debris? Three scenarios have been proposed which avoid these problems.

1. In globular clusters, binaries can be formed in close encounters between single stars. Thus a neutron star can become member of a binary long after its violent birth. The discussion of this process is deferred to Section 6.

2. The evolution to the low-mass X-ray binary starts with a very wide binary, and involves large loss of matter and angular momentum in a spiral-in phase.

3. Possibly, a neutron star can be formed in a more quiet fashion if a white dwarf close to the Chandrasekhar limit is driven beyond this limit by accretion of matter from a companion. This process will be discussed in Sect.5.7.

Consider a wide binary, with semi-major axis $a \gtrsim 1000 R_{\odot}$, consisting of one massive and one low-mass star. Mass transfer starts once the massive star reaches a very large radius, i.e. case C (see Sect.3.2). At that point, the massive star has already lost much mass due to stellar wind, and its envelope is very extended and has a small binding energy (see Eq.(17)). This enables the energy released by the spiral-in of a low-mass secondary to expel the envelope. (In a closer binary, spiral-in would have occurred at much smaller radius, hence larger binding energy of the envelope, and would have led to complete merging of the two stars.) The binary formed after spiral-in consists of the highly evolved core of the massive star and the virtually unevolved low-mass star. If the core implodes to form a neutron star, a low-mass binary with a neutron star is formed, provided the binary survives. This demands that not too much mass is lost in the explosion, i.e. that the core is not too massive (see Eq.(12)), or alternatively that the kick velocity is in a direction opposite to the orbital velocity of the neutron star. Observationally, we have a probable example in the neutron-star companion to PSR1913+16 (see Sect.4.4).

The subsequent fate of the low-mass binary with a neutron star depends on the distance $a$ between the stars, after the spiral-in, and on the mass $M_2$ of the low-mass star.

If $a \lesssim 5R_\odot$ and $M_2 \lesssim M_\odot$, loss of angular momentum drives the two stars together until the low-mass star fills its Roche lobe and starts transferring mass. We then have a low-mass X-ray binary with a main-sequence mass donor.

If $a \gtrsim 10\,R_\odot$ and $M_2 \simeq M_\odot$, the low-mass star evolves into a giant before loss of angular momentum can affect $a$. We then get a low-mass X-ray binary with a (sub)giant mass donor. The low-mass star should not have a mass much smaller than $M_\odot$, because such a star does not evolve within the Hubble time (see Eq.(3)).

An interesting intermediate case can occur if $a \sim 7R_\odot$: mass transfer then starts due to a combination of loss of angular momentum and expansion of the donor star, but the mass loss halts the further nuclear evolution of the donor. The mass donor then becomes a main-sequence star with a helium-rich core (Pylyzer and Savonije 1988).

If $M_2 \gtrsim 1.5M_\odot$ the mass transfer may lead to a second spiral-in, from which a binary may emerge which consists of a neutron star and a white dwarf. For even more massive stars ($M_2 \gtrsim 5M_\odot$, say) spiral-in may lead to a close binary of a neutron star and a helium burning star.

Very wide binaries with extreme mass ratios are thought to be very rare compared to closer binaries with two massive stars. The formation rate of low-mass X-ray binaries according to the scenarios just discussed is therefore expected to be much smaller than the formation rate of massive X-ray binaries. This is as it should be. Massive X-ray binaries live only for a short time, up to about $10^5$yr, as discussed in Sect.4.2. Low-mass X-ray binaries live about $10^9$yr, as discussed below. However, the observed numbers of low-mass and massive X-ray binaries are similar. Thus the formation rate of low-mass X-ray binaries must be smaller than that of massive X-ray binaries by a factor of order $10^{-4}$ (Van den Heuvel 1983).

## 5.2 Orbital periods and the nature of the mass donor

According to the scenario discussed in Sect.5.1, low-mass X-ray binaries can be formed with a variety of donor stars: from small white dwarfs via the main-sequence stars to large giants. The orbital period of a low-mass X-ray binary gives a direct clue to the nature of the mass donor. To see this we combine Kepler's law, Eq.(8), with the approximate formula for the Roche lobe of the least massive star (Eq.5b), to find:

$$P_{orb} \simeq 9\mathrm{hr}(\frac{R_2}{R_\odot})^{3/2}(\frac{M_\odot}{M_2})^{1/2} \tag{25}$$

Here we use that the mass donor fills its Roche lobe, i.e. $R_2 = R_L(M_2)$. With mass-radius relations for stars of different type (Table 3), we can identify the nature of the mass donor in systems with known orbital periods. The observed period distribution of low-mass X-ray binaries is shown in Figure 6.

## Table 3

*This Table gives approximate mass-radius and mass-$P_{orb}$ relations for Roche-lobe-filling mass donor stars in thermal equilibrium. In the course of binary evolution, mass donors may deviate from thermal equilibrium: in those cases the following formulae become increasingly inaccurate.*

| | | |
|---|---|---|
| main-sequence | $R_2/R_\odot \simeq M_2/M_\odot$ | $P_{orb} \simeq 9\mathrm{hr}\, M_2/M_\odot$ |
| He main-sequence | $R_2/R_\odot \simeq 0.2 M_2/M_\odot$ | $P_{orb} \simeq 0.9\mathrm{hr}\, M_2/M_\odot$ |
| degenerate star | $R_2/R_\odot \simeq 0.013(1+X)^{5/3}(M_\odot/M_2)^{1/3}$ | $P_{orb} \simeq 48\mathrm{s}(1+X)^{5/2} M_\odot/M_2$ |

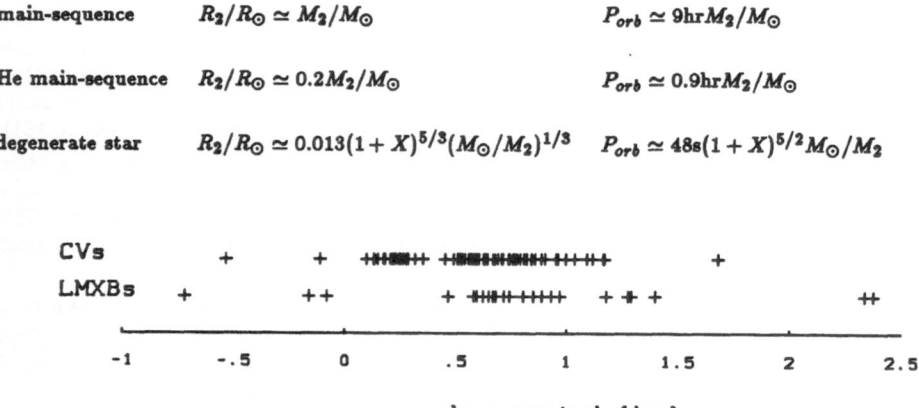

*Figure 6. Orbital period distributions for cataclysmic variables (CVs) and low-mass X-ray binaries (LMXBs). Data mainly from Ritter (1987). The distributions are roughly similar, but differ in detail.*

In systems with orbital period $P_{orb} \gtrsim 10\mathrm{hr}$, the long-period systems, the mass donor cannot be a main-sequence star with $M_2 < M_\odot$. Hence, it must be an evolved star. Mass transfer in such a binary occurs because of expansion of the (sub)giant donor. In systems with shorter orbital periods the mass donor can be a main-sequence star, or — in systems with $P_{orb} \lesssim 80\mathrm{min}$, the ultra-short period systems — a helium-rich star, i.e. a white dwarf, or a star on the He-burning main sequence. In these systems the evolution is driven by loss of angular momentum from the binary.

### 5.3 Evolution via loss of angular momentum

In low-mass X-ray binaries, the mass transfer is from the less massive star to the more massive star in the binary. The mass transfer causes the distance between the stars, and with it the size of the Roche lobe of the mass donor, to increase according to Eqs.(5b,7). For constant orbital angular momentum, $\dot{J}_{orb} = 0$, mass transfer would cease unless the radius of the donor star $R_2$ has increased more than $R_L$. In the short-period X-ray binaries, the mass transfer is kept going by loss of angular momentum from the system. During steady mass transfer, the mass donor fills its Roche lobe, $R_2 = R_L$, and keeps doing so, $\dot{R}_2 = \dot{R}_L$. The equation describing the binary evolution under influence of $\dot{J}_{orb}$ is found as follows. Combining Eq.(7) with the time derivative of Eq.(5b) we find:

$$\frac{\dot{R}_L}{R_L} = 2\frac{\dot{J}_{orb}}{J_{orb}} - 2(\frac{5}{6} - \frac{M_2}{M_1})\frac{\dot{M}_2}{M_2} \tag{26}$$

Write the mass-radius relation of the donor star as $R_2 \propto M_2^n$. The time derivative is:

$$\frac{\dot{R}_2}{R_2} = n\frac{\dot{M}_2}{M_2} \tag{27}$$

For stable mass transfer, Eqs.(26,27) can be combined into:

$$-\frac{\dot{J}_{orb}}{J_{orb}} = -\left[\frac{5}{6} + \frac{n}{2} - \frac{M_2}{M_1}\right]\frac{\dot{M}_2}{M_2} \tag{28}$$

The change in $M_2$ implies a change in orbital period via the relations given in Table 3. Eq.(28) also shows that mass transfer will be unstable for masses of the donor star $M_2/M_1 > 5/6 + n/2$.

The first mechanism to be proposed for the loss of angular momentum in a short-period low-mass binary is gravitational radiation (Kraft, Mathews & Greenstein 1962). According to general relativity, gravitational radiation of two masses $M_1$ and $M_2$ rotating around one another at a distance $a$ causes a loss of angular momentum given by (see Landau & Lifshitz 1962):

$$-\frac{\dot{J}_{gr}}{J_{orb}} = \frac{32G^3}{5c^5}\frac{M_1 M_2(M_1 + M_2)}{a^4} \tag{29}$$

By equating $\dot{J}_{orb}$ with $\dot{J}_{gr}$ we can determine the mass transfer rate in low-mass X-ray binaries, combining Eqs.(28,29). This can be done for main-sequence mass donors, which have $n = 1$ (Faulkner 1971), and for degenerate donors, with $n = 1/3$ (Paczyński 1967b, Vila 1971). The resulting mass-transfer rates are shown in Figure 7, as a function of the donor mass, and assuming a neutron star of 1.4 $M_\odot$.

For main-sequence mass donors, the mass-transfer rate $\dot{M} \equiv -\dot{M}_2$ due to $\dot{J}_{gr}$ is about $10^{-10} M_\odot/\text{yr}$ for donor masses between 0.2 and 1.0 $M_\odot$ (and virtually independent of $M_1$). Accretion of this rate onto a neutron star gives an X-ray luminosity according to Eq.(18) of about $10^{36}\text{erg s}^{-1}$ . The observed $L_x$ of many low-mass X-ray binaries is more than an order of magnitude higher. This suggests that additional loss of angular momentum from the binary occurs. A promising mechanism for such loss is magnetic braking of a stellar wind from the mass donor (Verbunt & Zwaan 1981). This idea is based on the observation that single stars of solar type are born rotating rapidly and slow down in the course of time. The reason for the slow-down is loss of angular momentum in the stellar wind. Although

200

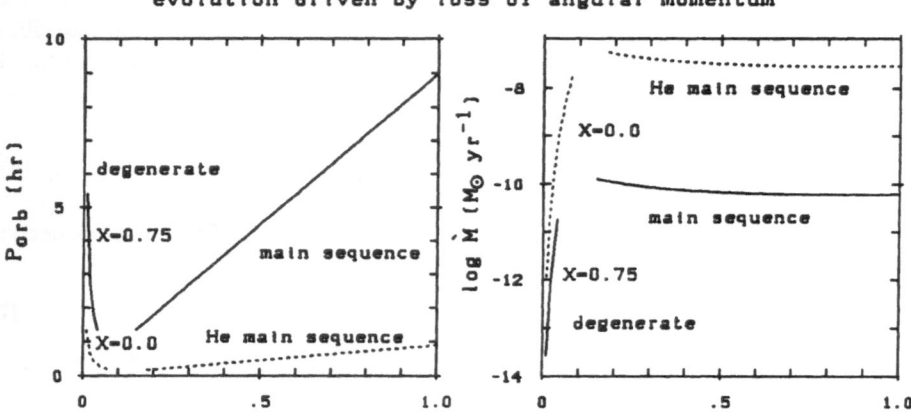

*Figure 7.* *The orbital period $P_{orb}$ and mass-transfer rate $\dot{M}$ in low-mass X-ray binaries that evolve via loss of angular momentum, as a function of the mass of the donor star, $M_2$. The relations are valid for mass donors in thermal equilibrium. Mass transfer is calculated for gravitational radiation only; if additional loss of angular momentum occurs, $\dot{M}$ will be higher. He-rich stars have shorter $P_{orb}$ and higher $\dot{M}$ than hydrogen-rich stars. (After Verbunt 1988a.)*

the amount of mass lost in the wind is very small ($\sim 10^{-14} M_\odot$/yr for the sun), the loss of angular momentum is appreciable because the magnetic field of the star forces the wind matter to corotate out to many stellar radii. According to observation, the braking is strong for rapidly rotating, young G stars, and less for old, slowly rotating stars. This is probably due to the correlation between rapid rotation and the buildup of strong magnetic fields via the dynamo mechanism in the convective surface layers of the star. A G star in a close binary cannot slow down its rotation since it is kept in corotation with the orbit by tidal forces. For such a G star, the braking is kept at a high level. Any loss of angular momentum is transferred by the tidal forces from the donor star to the orbit.

Our theoretical understanding of this process is very limited: there are no complete theories for the stellar wind and for the magnetic field generation in late-type stars. Several researchers have therefore taken semi-empirical braking laws, as derived from observations of contact binaries (Mochnacki 1981), single G-stars (Verbunt & Zwaan 1981), or cataclysmic variables (Patterson 1984). The generalization of the law for single G stars may serve as an example (Rappaport,

Verbunt & Joss 1983):

$$\frac{\dot{J}_{mb}}{J_{orb}} = \frac{-3.8 \times 10^{-30} R_{\odot}^{4-\gamma}(M_1 + M_2)R_2^{\gamma}\omega^2}{M_1 a^2} \tag{30}$$

Here $\omega$ is the angular velocity of the mass donor, equal to $2\pi/P_{orb}$ in corotation, and $\gamma$ is the parameter setting the dependence of the braking on the radius of the donor star. Combining this law with Eq.(28), one does find mass-transfer rates sufficiently high to explain X-ray luminosities up to $10^{38}$erg s$^{-1}$. However, it should be stressed that the real importance of magnetic braking for the evolution of low-mass X-ray binaries can not be established until a much better understanding is obtained of the physical processes involved, i.e. of the origin of the magnetic fields and of the stellar winds in late-type stars.

5.4 The minimum period. The period gap

As an example of the evolution of an X-ray binary, consider a binary of a 1.4 $M_{\odot}$ neutron star and a 1 $M_{\odot}$ main-sequence companion, at an orbital period of 10 hr. Loss of angular momentum drives the two stars together until the main-sequence star fills its Roche-lobe, and mass transfer sets in. The mass transfer lowers the mass of the donor, and the system evolves to shorter periods (see Table 3). The central entropy in thermal equilibrium is smaller in less massive stars. Thus a mass-losing star has to radiate energy to remain in thermal equilibrium. It can do this, as long as the time scale for the mass transfer, given by $M_2/\dot{M}$, is longer than the thermal time scale of the mass donor, given by Eq.(2). As the donor mass becomes lower, however, its thermal time scale increases, and at some point the donor starts deviating from the thermal equilibrium (Taam 1983a). The radius of the donor and the orbital period are then larger for given $M_2$ than indicated by the equations of Table 3. The effect of these deviations on the binary evolution are small in comparison to the uncertainty in $\dot{J}_{mb}$.

The temperature and pressure in the core of a very low-mass star is too small for hydrogen burning. At some point in the evolution of a low-mass X-ray binary, the mass of the donor will become so small that the hydrogen burning in its core drops significantly, causing the core to become degenerate. Further mass loss from the donor now leads to an increase in the donor radius and also to an increase in the orbital period (See Table 3). Thus, the system evolves through a minimum period (Faulkner 1971). The orbital period distribution of cataclysmic variables, shown in Figure 6, shows a lower cutoff at about 80 min that can be interpreted as this minimum period. This indicates that the mass transfer close to the minimum period, at $P_{orb} \lesssim 2hr$, say, is not higher than $\sim 10^{-10}M_{\odot}/\mathrm{yr}$: a higher $\dot{M}$ would cause the mass donor to be larger (at the same mass) and the minimum period to be longer. This indicates that the mass transfer at short orbital periods is driven mainly by gravitational radiation (Paczyński & Sienkewicz 1981, Rappaport, Joss & Webbink 1982). Once the system has passed the minimum period, the mass-transfer drops precipitously (see Figure 7), and the evolution of the binary almost comes to a standstill once $M_2/\dot{M}$ exceeds the Hubble time.

Low-mass X-ray binaries with normal main-sequence mass donors do not evolve to periods shorter than about 80 min. How can we explain the low-mass X-ray binaries with shorter orbital periods? One possible explanation is that the mass donor is a star on the helium main-sequence (Savonije, De Kool & Van den Heuvel 1986). These systems should be characterized by high $\dot{M}$ (see Figure 7b) and also by a high optical luminosity due to the helium-burning star. The minimum period which such systems can reach is about 10 min, beyond which period the mass donor becomes degenerate and mass transfer very low. Due to their different chemical composition and thermal history, the degenerate systems evolved from helium-burning stars lie on a different track in the $P_{orb} - M_2$ diagram as the degenerate systems that evolved from hydrogen-burning donor stars via the 80-min minimum period. Another explanation is that the mass donor is a main sequence star with a hydrogen-depleted core (Iben & Tutukov 1984, Nelson, Rappaport & Joss 1986). Such a main-sequence star can be formed when the start of mass transfer stops its evolution into a subgiant and causes it to revert to a main-sequence star (see Sect.5.1). The minimum period of such a system could be as low as $\sim 50$ min.

Comparison of the evolutionary tracks with the period distribution of cataclysmic variables leads to an additional detail in the description of the binary evolution. For a mass-transfer rate of $10^{-10} M_\odot \text{yr}^{-1}$, a system formed with a donor of mass $M_2 < 1 M_\odot$ evolves to a system with mass $M_2 = 0.25 M_\odot$ within the Hubble time. It would then be expected to be at an orbital period between 2 and 3 hr, according to Table 3. The observed period distribution of cataclysmic variables shows an absence of systems at these periods. The statistics for low-mass X-ray binaries are small, but compatible with a similar gap in the period distribution (Figure 6). A possible explanation for the existence of this gap is that magnetic braking (or any other mechanism for loss of angular momentum from the binary orbit) stops operating when the donor star becomes completely convective (Spruit & Ritter 1983, Rappaport, Verbunt & Joss 1983). This enables the star to relax somewhat closer to thermal equilibrium, i.e. to smaller radius. As the radius shrinks within the Roche lobe, however, mass transfer stops altogether, and the star shrinks all the way to thermal equilibrium, at main-sequence radius. Mass transfer will only resume when loss of angular momentum due to gravitational radiation has decreased $a$ sufficiently to bring the donor into contact with its Roche lobe again. A braking law according to Eq.(30) with $\gamma \simeq 2.5$ for stars with a radiative core, but with $\dot{J}_{mb}$ set to zero for fully convective stars, can explain a period gap between 2 and 3 hrs (Rappaport et al. 1983, Verbunt 1984).

## 5.5 Evolution driven by radius expansion

The low-mass X-ray binaries with $P_{orb} \gtrsim 12$hr must have evolved mass donors. In these systems the mass transfer can be driven by the radius expansion of the giant only. The radius and luminosity of a low-mass (sub)giant with a degenerate core and hydrogen-shell burning is determined mainly by the mass $M_c$ of the core and by the metallicity $Z$ of the envelope. The results of detailed model calculations can be represented with simple polynomial relations in $y \equiv \ln(M_c/0.25 M_\odot)$ (Webbink,

Rappaport & Savonije 1983):

$$\ln(R_2/R_\odot) = a_0 + a_1 y + a_2 y^2 + a_3 y^3 \tag{31}$$

$$\ln(L_2/L_\odot) = b_0 + b_1 y + b_2 y^2 + b_3 y^3 \tag{32}$$

The values for the fitting parameters $(a_i, b_i)$ are given in Table 4 for two values of $Z$.

### Table 4

*This Table gives the fitted constants for use in Eqs.(31-34), as calculated by Webbink, Rappaport & Savonije (1983).*

| | $a_0$ | $a_1$ | $a_2$ | $a_3$ | $b_0$ | $b_1$ | $b_2$ | $b_3$ | mass range |
|---|---|---|---|---|---|---|---|---|---|
| $Z = 0.02$ | 2.53 | 5.10 | -0.05 | -1.71 | 3.50 | 8.11 | -0.61 | -2.13 | $0.16 < M_c/M_\odot < 0.45$ |
| $Z = 0.0001$ | 2.02 | 2.94 | 2.39 | -3.89 | 3.27 | 5.15 | 4.03 | -7.06 | $0.20 < M_c/M_\odot < 0.37$ |

The luminosity is almost completely due to the hydrogen shell burning and therefore related to the growth of the core mass $\dot{M}_c$ by

$$\dot{M}_c = 1.37 \times 10^{-11}(\frac{L}{L_\odot})M_\odot \mathrm{yr}^{-1} \tag{33}$$

The change in radius is related to the change of core mass according to Eq.(31) as

$$\frac{\dot{R}_2}{R_2} = [a_1 + 2a_2 y + 3a_3 y^2]\frac{\dot{M}_c}{M_c} \tag{34}$$

The evolution of a binary with a low-mass giant donor is found by equating the change in size of the Roche lobe $\dot{R}_L/R_L$ as given by Eq.(26) to the change in the radius of the giant, which follows from Eq.(34). Usually, the orbital angular momentum is assumed constant: $\dot{J}_{orb} = 0$. The orbital period determines at which giant radius mass transfer starts (see Eq.(25)) and thus via Eq.(31) at which core mass mass transfer starts. The evolution of a giant accelerates as the core mass increases (see Eqs.(32,34)), and therefore systems with long orbital periods, which require the donor to evolve further before it fills the Roche lobe, have high mass-transfer rates. The evolution of low-mass X-ray binaries with different initial periods is shown in Figure 8.

After the onset of mass transfer, most of the envelope matter is transferred to the neutron star, but some of it is added to the core. In all cases shown in Figure

204

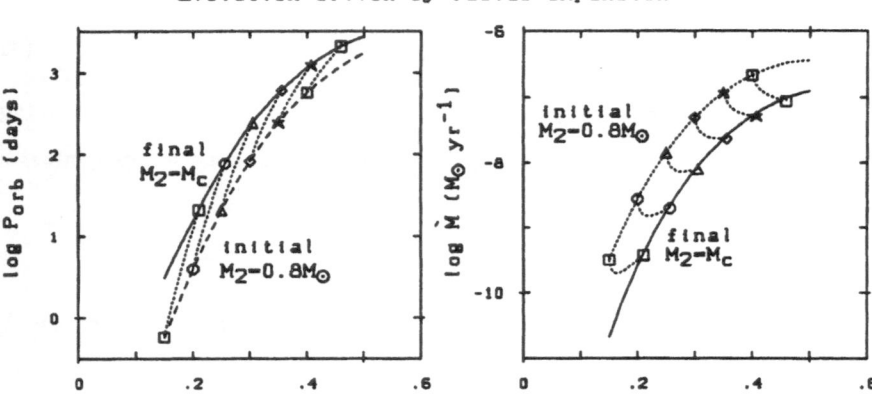

*Figure 8. The orbital period $P_{orb}$ and mass-transfer rate $\dot{M}$ for low-mass binaries driven by expansion of the mass donor, as a function of the mass of the core of the giant, $M_c$. The curves shown are for metallicity $Z = 0.02$. - - - onset of mass-transfer for total mass of giant $M_2 = 0.8M_\odot$. —— end of mass transfer at $M_2 = M_c$. These two curves assume a neutron star of 1.4 $M_\odot$. Start and end of evolutionary tracks .... are indicated with the same symbol for 6 initial core masses. These tracks are valid if all transferred mass is accreted by the neutron star. At the end of the track, the neutron star mass is ~ 2$M_\odot$, causing a small difference in $\dot{M}$ with respect to the 1.4 $M_\odot$ curve. (After Verbunt 1988a.)*

8, about 0.5 $M_\odot$ was transferred and about 0.05 $M_\odot$ added to the core during the evolution. At long orbital periods, the mass-transfer rate is higher than the Eddington limit, which presumably means that not all the transferred matter can be accepted by the neutron star. Eqs.(31,32) are only valid for giants at thermal equilibrium. Detailed calculations show that this is a good approximation until the envelope is almost exhausted (Taam 1983b).

After exhaustion of the envelope of the giant, the core forms a wide binary with the neutron star. It cools to become a white dwarf. The relation between the initial core mass of the donor giant and the orbital period at which mass transfer starts, given by Eqs.(25,31), leads to a relation between the final core mass and the final orbital period after mass transfer (Figure 8). The mass of a wide-orbit white-dwarf companion to a radio pulsar should therefore be correlated with the orbital period (Joss, Rappaport & Lewis 1987).

## 5.6 Evolution into radio pulsars. Spin-up via accretion

The evolution of a low-mass X-ray binary with a giant mass donor automatically leads to the formation of a wide binary of a white dwarf and a neutron star. Several of the known low-magnetic field radio pulsars are indeed in such binaries. In accordance with expectation in the evolutionary scenario, the eccentricities of these binaries are very small. The largest orbit has the largest eccentricity. This could be due to the high mass transfer rates at this period (see Figure 8), which may cause the orbital evolution to be proceeding more rapidly than the circularization by the tidal forces. The radio pulsar systems PSR1855 + 09, PSR1953 + 29 and PSR0820 + 02 can be satisfactorily explained via the conservative scenario. That mass transfer has occurred in these systems is not only shown by the rapid period to which the neutron star has been spun up, but also by the low mass of the white dwarf: a giant that doesn't lose its envelope leaves a white dwarf of mass $M \gtrsim 0.6 M_\odot$. PSR1831 − 00 has a rather short orbital period, indicating that its evolution was driven not only by expansion of the donor radius, but also by some loss of angular momentum (Pylyzer & Savonije 1988).

What about the single pulsar PSR1937 + 21? According to the evolutionary theories described so far, the neutron star in a low-mass X-ray binary is always left with a companion: a low-mass degenerate star if the original mass donor for evolution via loss of angular momentum and a white dwarf for evolution via radius expansion. A number of different mechanisms had been proposed in the literature, none of them very convincing (as discussed in Verbunt & Rappaport 1988 and Priedhorsky & Verbunt 1988). The discovery of PSR1957 + 20 suggests a possible solution: in this system, the mass of the companion is very low, which indicates that the pulsar is evaporating its companion. This rather spectacular possibility is further supported by the length of the eclipse of the radio pulsar, which indicates an eclipsing object larger than the Roche lobe of the companion (Fruchter et al. 1988). Evaporation of matter from the companion can lead to an increase in the orbital period (cf. Eq.(15)). If this is the case for PSR1957 + 20 the progenitor could have been a low-mass X-ray binary with a main-sequence donor.

The period of the fastest pulsar PSR1937 + 21 is about 1.6 msec. Interpreting this period as the equilibrium period during accretion, we can estimate the amount of accreted matter $\Delta M$, making the approximation that the angular momentum of the pulsar is completely due to accretion (a good approximation if the rotation period now is much shorter than at the onset of accretion). This gives

$$I\omega = \Delta M \sqrt{GMr_m} \qquad (35)$$

where $I$, $M$, and $\omega$ are the moment of inertia, mass and rotation frequency of the pulsar, and $r_m$ the radius of the magnetosphere during the accretion phase. With the magnetic field of the pulsar (see Table 1) Eq.(23) shows that $\dot{M} \gtrsim 4 \times 10^{-9} M_\odot \mathrm{yr}^{-1}$ at the time of the spin-up, and from Eq.(22) the radius of the magnetosphere follows as about $2 \times 10^6 \mathrm{cm}$. Substituting this in Eq.(35) we find a required amount of accreted matter $\Delta M \gtrsim 0.1 M_\odot$. This exercize tells us that the accretion rate during the spin-up time must have been relatively high, and also that the progenitor of the 1.5 msec pulsar is unlikely to have been a massive system since the total amount of matter accreted by a neutron star in a massive system must be less than the length of the mass-transfer phase multiplied with the Eddington limit, i.e. $\Delta M \lesssim 10^5 \times 1.5 \times 10^{-8} M_\odot \simeq 0.0015 M_\odot$.

### 5.7 Evidence for white dwarf collapse.

The life time of low-mass X-ray binaries as calculated in the preceding Sections is long compared to the time scale of $< 10^7$ yr on which the initially strong magnetic field of a neutron star decays. In agreement with this, the observations show no evidence for strong magnetic fields (in the form of X-ray pulse periods or hard spectra) in the majority of low-mass X-ray binaries. There are some systems, however, in which the neutron star is an X-ray pulsar (see Table 1). To explain this it has been suggested that young neutron stars can be formed in old binaries, when a massive white dwarf accretes enough matter to transgress the Chandrasekhar limit and implode. Theoretical calculations of this process, made most notably by Nomoto (for an overview see Nomoto 1987), indicate that the initial mass of the white dwarf must already be high, more than about $1.2 M_\odot$, for this process to be possible. The formation of a close binary with an accreting massive white dwarf may occur along similar lines as the formation of close binaries with accreting neutron stars, described in Sect.5.1. The difference lies in the somewhat smaller mass of the progenitor of a massive white dwarf. Because the binding energy of the envelope of such a progenitor is also smaller, the original binary may have been closer.

If we find an X-ray pulsar with a companion that has an aged companion, it is likely that the binary contained a massive white dwarf for most of its history. The pulsar was formed as the relatively short-lived mass-transfer stage started. An example of such a system is GX1 + 4: the mass donor in this system is an M6III giant. Another example is Her X-1. This system is exceptional in having a mass donor of about 2.35 $M_\odot$, intermediate in mass between those of massive and low-mass binaries. The donor therefore has a life time on the main sequence of about $10^9$ yr according to Eq.(3).

The values of the magnetic field strengths of the radio pulsars descended from low-mass X-ray binaries show that the decay of the magnetic field does not continue all the way to zero, but bottoms out at a level of order $10^{8-9}$ G. A radio pulsar which may have formed via white-dwarf collapse is PSR0820 + 02, which has a relatively strong field of $3 \times 10^{11}$ G (Table 1). According to the calculations by Nomoto (1987), high $\dot{M}$ is necessary to cause a massive white dwarf to implode and form a neutron star. The long orbital period of the binary with PSR0820 + 02, which corresponds to high $\dot{M}$ in the mass transfer phase (see Figure 8), further supports a collapse scenario for this pulsar.

Given the success of the white-dwarf collapse scenario in these three cases, one may wonder whether all low-mass X-ray binaries haven't formed that way. The observed absence of pulsars in most low-mass X-ray binaries then merely indicates that the neutron star is formed a short time after the onset of mass transfer and loses its high magnetic field during the long time that mass transfer continues.

There are several systems for which white–dwarf collapse is not possible, however. One of these is A0620−00, which contains a black hole with minimum mass of $5M_\odot$. Such a massive black hole cannot be formed via accretion onto a $< 1.4M_\odot$ white dwarf by a $< 1M_\odot$ donor! The direct formation scenario described in Sect.5.1 is more likely for this system. This shows that the direct scenario is viable and allows for the possibility that the formation of the majority of low-mass X-ray binaries does not involve the collapse of a white dwarf.

A second category of low-mass X-ray binaries for which formation cannot have proceeded via white-dwarf collapse is that of the globular cluster sources.

## 6. X-ray sources and pulsars in globular clusters

### 6.1 Formation of binaries via tidal capture

The reason to consider the X-ray sources in globular clusters in a separate chapter is given by their relative large number. The disk of our galaxy has a mass of about $10^{11}M_\odot$ and contains some 100 known low-mass X-ray binaries. All globular clusters together contain some $10^7 M_\odot$, and merely extrapolating from the numbers in the disk one would expect 0.01 X-ray binary to be observed in them, i.e. none at all. In fact, we know some ten X-ray sources in globular clusters in our galaxy. A similar statistic can be made for the globular clusters in M 31 in which 19 X-ray sources have been found (Van Speybroeck et al. 1979).

This large abundance of X-ray binaries in globular clusters indicates an efficient formation mechanism which only operates in globular clusters. Such a mechanism is the capture of a neutron star by a main-sequence or (sub)giant star, in a close encounter (Fabian, Pringle & Rees 1975, Sutantyo 1975). The basic principle can be understood with a simple calculation: consider a neutron star with mass $m$ and velocity $v$ at large distance ('infinity') relative to a target star with mass $M$ and radius $R$. The relative kinetic energy $E_k$ of the two stars is given by

$$E_k = \frac{1}{2}\frac{mM}{m+M}v^2 \tag{36}$$

As the neutron star closes in on the target star, it causes this star to deform. The height $h$ of the bulge and its mass $m_t$ can be estimated for distance $d$ with

$$h \simeq \frac{m}{M}\frac{R^4}{d^3}; \qquad m_t \simeq k\frac{h}{R}M \simeq k(\frac{R}{d})^3 m$$

where $k$ is the apsidal motion constant, which depends on the central condensation of the star, and indicates how easy it is to deform the star (see, e.g., Schwarzschild 1958). For a star with a deep convective envelope $k \simeq 0.14$ (Motz 1952). Thus the energy $E_t$ in the tidal deformation is of order

$$E_t \simeq m_t\frac{GM}{R^2}h \simeq k\frac{Gm^2}{R}(\frac{R}{d})^6 \tag{37}$$

If $E_t > E_k$, the two stars cannot escape from one another anymore, and a binary is formed. This condition can be written:

$$d \lesssim 3R\left(\frac{k}{0.14}\frac{m}{M}\frac{m+M}{2M_\odot}\frac{R_\odot}{R}\right)^{1/6}\left(\frac{10km\ s^{-1}}{v}\right)^{1/3} \tag{38}$$

Because of the strong dependence of $E_t$ on $d$, this rough estimate is in fact pretty accurate, as more detailed calculations confirm. These more detailed calculations take into account higher-order deformations for polytrope stellar models (Burke 1967, Giersz 1986, Lee & Ostriker 1986, McMillan, McDermott & Taam 1987) and for more realistic numerical stellar models (McMillan et al. 1987). Numerical calculations for stellar encounters (Gingold & Monaghan 1980, Benz & Hills 1987) show that mass loss in jets occurs during the encounter and removes angular momentum.

The initial binary orbit is highly eccentric, with $e \lesssim 1$, and with a velocity at periastron close to the escape velocity. Tidal forces are expected to circularize the orbit, during which process angular momentum is conserved. For an initial periastron velocity less than the escape velocity, the semi-major axis of the circularized orbit is given by

$$a_c \leq 2d \tag{39}$$

Thus the final orbital after circularization can have a semimajor axis of up to two times the capture distance.

The cross section $\sigma$ for closest passage within distance $d$ follows from conservation of energy and angular momentum in a Keplerian orbit:

$$\sigma = \pi d^2\left(1 + \frac{2G(m+M)}{v^2 d}\right) \simeq \pi d\,\frac{2G(m+M)}{v^2} \tag{40}$$

The second term within brackets gives the effects of gravitational focussing. This term dominates for the small relative velocities between stars in globular clusters, which justifies the subsequent approximation.

With number densities $n_c$ and $n$ for the neutron and target stars, respectively, the capture rate of neutron stars per unit volume can be written:

$$\Gamma = n_c n v \sigma \simeq 6 \times 10^{-11}\frac{n_c}{10^2 pc^{-3}}\frac{n}{10^4 pc^{-3}}\frac{m+M}{M_\odot}\frac{3R}{R_\odot}\frac{10km\ s^{-1}}{v}\ yr^{-1}pc^{-3} \tag{41}$$

To obtain the formation rate in a cluster, one must integrate Eq.(41) over the cluster volume. To give an idea of the characteristic numbers, a simple example may do. In a relatively dense core of a globular cluster, $n_c \sim 100pc^{-3}$ and $n \sim 10^5 pc^{-3}$. With a characteristic core volume of $\sim 1pc^3$ it follows that a close binary with a neutron star is formed every $10^9$yr. For an average life time of a bright source of $10^9$yr, we then expect to see of order 1 X-ray source in such a cluster, in accordance with observations.

The capture cross section for a main-sequence star or for a white dwarf is similar to that for a neutron star. The relative numbers of main-sequence stars, white dwarfs, and neutron stars captured in a cluster core are therefore roughly proportional to their respective number densities. In the galactic disk, on the other hand, binaries with white dwarfs and neutron stars rarely arise from binary evolution, as compared to the formation of single white dwarfs or neutron stars. Therefore, the fraction of main-sequence stars captured into a binary in globular clusters is small compared to the fraction of main-sequence stars in binaries in the galactic disk, but the fraction of white dwarfs and especially of neutron stars captured into binaries in globular clusters is very high compared to the fraction in binaries in the galactic disk.

## 6.2 Stars in a globular cluster

To calculate the number of neutron stars and of target stars involved in capture processes, one starts from the fact that all stars in a globular cluster were formed in a short time interval (compared to the age $\tau$ of the cluster) about $15 \times 10^9$yr ago. This follows from the absence of young stars in the globular clusters of our galaxy. Suppose that the Initial Mass Function, *i.e.* the number of stars $dN(m)$ formed in a mass interval $dm$ around mass $m$, is given by

$$dN(m) = C_o m^{-1-x} dm \qquad (42)$$

The most massive stars, with mass $m > m_a \simeq 8M_\odot$, evolved first, according to Eq.(3), and left neutron star remnants, or occasionally a black hole remnant. The number of these remnants follows from Eq.(42) as

$$N_{ns} = \frac{C_o}{x} m_a^{-x} \qquad (43)$$

Stars with intermediate mass $m_{to} < m < m_a$ evolved subsequently, leaving white dwarf remnants, to the number of

$$N_{wd} = \frac{C_o}{x}(m_{to}^{-x} - m_a^{-x}) \qquad (44)$$

Stars with masses less than the turnoff mass $m_{to} \simeq 0.8M_\odot$ are still on the main-sequence (see Eq.(3)). The stars that have just evolved away from the main sequence into (sub)giants have a mass close to the turnoff mass. Their number can be estimated from the dependence of the turnoff mass on the age of the cluster

$$log\frac{m_{to}}{M_\odot} = -0.28 \, log\frac{\tau}{10^9 yr} + 0.013 \, logZ - 0.75 \, logY + 0.453 \qquad (45)$$

$Z$ and $Y$ are the metal and helium contents of the cluster. (This approximate formula is a fit to the results of calculations with detailed computer models of low-mass stars, by Rood (1972), and as such differs from the dependence $logm_{to} \propto -0.5log\tau$ that one derives from Eq.(3) and Table 2.) Thus, a time interval $\Delta t$ corresponds to a mass interval $\Delta m_{to} \simeq 0.28 m_{to}\Delta t/\tau$, and the number of giants in evolutionary stage $i$, with life time $\Delta t_i$, is given by

$$N_{G,i} = 0.28 \, C_o \, \frac{\Delta t_i}{\tau} \, m_{to}^{-x} \qquad (46)$$

## Table 5

*Relative numbers of stars in globular clusters. For different assumed initial mass functions this table gives the number of giants, white dwarfs and neutron stars for each main-sequence star, for assumed mass ranges of their progenitors as indicated in the top line. f is the fraction of neutron stars remaining in the cluster after the supernova event.*

| stars | main seq. | giants | white dwarfs | neutron stars |
|---|---|---|---|---|
| mass-range | 0.3-0.8 $M_\odot$ | $\simeq 0.8 M_\odot$ | 0.8-8.0 $M_\odot$ | > 8.0 $M_\odot$ |
| $x=2$ | 1.0 | 0.012 | 0.16 | 0.0016 $f$ |
| $x=1$ | 1.0 | 0.023 | 0.54 | 0.0600 $f$ |

Table 5 gives the relative numbers of main-sequence stars, white dwarfs, giants and neutron stars, for different slopes $x$ of the initial mass function.

Inserting these numbers in Eq.(41),we can compare the number of captures by giants with that by main-sequence stars. For $x = 2$, the average mass of a main-sequence star is $\simeq 0.44 M_\odot$, and the average radius accordingly $R_{ms} \simeq 0.44 R_\odot$. The average radius of the giants is found by weighing the radius $R_i$ at each stage $i$ with the number of giants at that stage, according to Eq.(46).This gives the relative capture rate by giants to that by main-sequence stars as:

$$\frac{\Gamma_G}{\Gamma_{ms}} \simeq \frac{\Sigma(N_{G,i} R_i)}{N_{ms} R_{ms}} \simeq 0.07 \tag{47}$$

where we use radii and life times from the calculations by Rood (1972). The larger life times of subgiants makes up for their smaller radii, and about half of the captures by giants and subgiants occurs on the subgiant branch, half on the giant branch. The distribution of captures over the different giant stages is shown in Figure 9.

To find the actual numbers of stars of different type in a globular cluster, one has to determine the normalization constant $C_o$ and the slope $x$ of the initial mass function, for example by counting main-sequence stars of different masses. However, in deriving the number of neutron stars from Eq.(43), two serious problems arise.

In the first place, the extrapolation from measurements at $m \lesssim 0.8 M_\odot$ to progenitors of neutron stars at $m \gtrsim 8 M_\odot$ introduces a large uncertainty. There is no guarantee that the slope $x$ does not change between $m_{to}$ and $m_a$. A counter-example is provided by the cluster 47 Tuc. At the main-sequence, $x \simeq 0.02$. The number of white dwarfs and neutron stars found by extrapolating from this to higher masses implies a total mass for the cluster which is not compatible with measurements of velocities of cluster members. The actual number of white dwarfs and neutron stars must be much smaller (Meylan 1989).

Secondly, the escape velocity of a globular cluster is small, on the order of 20-40 km/s. If neutron stars in globular clusters are born with average velocities of several

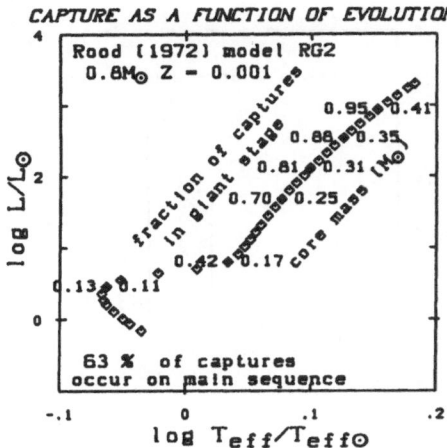

CAPTURE AS A FUNCTION OF EVOLUTION

Rood (1972) model RG2
0.8M⊙ Z - 0.001

fraction of captures
in giant stage

0.95 0.41
0.88 0.35
0.81 0.31 (M⊙)
0.70 0.25
0.42 0.17 core mass
0.13 0.11

63 % of captures
occur on main sequence

log L/L⊙

log T_eff/T_effO

*Figure 9. Capture as a function of evolutionary stage for a 0.8 M⊙ star. Parameters for the star are taken from Rood (1972). About 63% of all captures occur on the main sequence. Of the remaining captures, about 42 % occurs on the subgiant branch with core mass less than 0.17 M⊙; 58 % on the giant branch. For some stages, the cumulative fraction of capture after leaving the main sequence and the core mass are indicated. Thus 88 % of all captures on the (sub)giant branches occurs before the core mass of the (sub)giant has grown to 0.35 M⊙.*

hundred km/s, as their counterparts in the galactic disk, a large fraction will escape from the cluster. The fraction of neutron stars born with sufficiently small velocities to remain in the clusters is very uncertain. The best current estimate is that some 15 % of the neutron stars remain, but this estimate could well be completely wrong (Verbunt & Hut 1987).

## 6.3 Mass segregation

To obtain the formation rate of binaries in a cluster, we must integrate Eq.(41) over the cluster volume. Because the number density of stars drops very rapidly with the distance $r$ to the cluster centre, roughly as $n \propto r^{-4}$, virtually all captures occur at small radii $r \lesssim 2r_c$, where $r_c$ is the core radius of the cluster. In a globular cluster, the massive stars are more concentrated to the core than the low-mass stars. Due to this mass segregation, the relative numbers of stars of different type in the core therefore may differ markedly from that in the cluster as a whole. An example is shown in Figure 10.

The effect of mass segregation on the formation of binaries is twofold. First, due to the concentration of neutron stars in the core, the capture rate of neutron stars is higher than one would estimate when ignoring mass segregation. Second, the relative importance of capture by relatively massive stars, i.e. the (sub)giants and the stars with masses close to the turnoff mass, is enhanced by mass segregation. As shown in Figure 10, this can be a strong effect: in 47 Tuc the capture rate of neutron stars is enhanced by a factor 6 due to the concentration of neutron stars in the core; and the relative importance of capture by giants is enhanced by a factor 2-3 (Verbunt & Meylan 1988). 1 out of every 3 captures of a neutron star in 47 Tuc is by a giant or subgiant, a much higher fraction than estimated by ignoring mass segregation, with Eq.(47).

The importance of mass segregation is not so pronounced in every cluster, how-

212

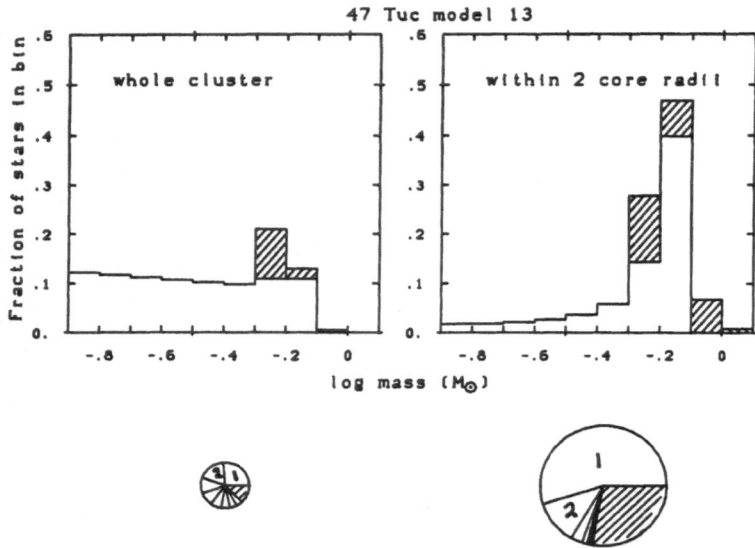

*Figure 10. The effect of mass segregation in 47 Tuc. The top two graphs give the mass functions for the cluster as a whole (left) and for the innermost region at $r < 2r_c$ (right). The data for the figure are from Verbunt & Meylan 1988. The hatched blocks indicate the compact stars: the white dwarfs are distributed in three groups with masses around 0.58, 0.74 and 1.1 $M_\odot$, the neutron stars all are 1.4 $M_\odot$. (The mass scale of the graph is not valid for the 1.1 $M_\odot$ white dwarfs or for the neutron stars.) The numbers of neutron stars shown in the Figure are based on the assumption that all neutron stars formed in the cluster still reside there. In reality, a large fraction of them probably was born with a velocity high enough to escape from the cluster. The relative importance of the more massive stars is enhanced in the core due to mass segregation.*

*The bottom graphs show the distribution of captures of neutron stars over the different types of main sequence stars and giants. Left: mass segregation ignored. Right: mass segregation taken into account. The size of the circle is proportional to the total number of captures: mass segregation enhances the total capture rate of neutron stars. The segments of the circles indicate the fraction of the captures by stars of different type: hatched the giants and subgiants, indicated with numbers the different bins of main-sequence stars shown in the top graphs, from 1 for the most massive to 8 for the least massive main-sequence stars. Mass segregation enhances the fraction of captures by massive stars.*

ever. Because of the large uncertainty in the absolute number of neutron stars, and because the effect of mass segregation is only known in detail for some 5 clusters, we cannot accurately calculate the expected formation rate of low-mass X-ray binaries by tidal capture in globular clusters. We can only state that the wide range of possible formation rates includes the formation rate required to explain the observed number of low-mass X-ray binaries in globular clusters.

## 6.4 Different types of low-mass X-ray binaries and their evolution

We will discuss the different possible capture processes involving neutron stars on the basis of Figure 11.

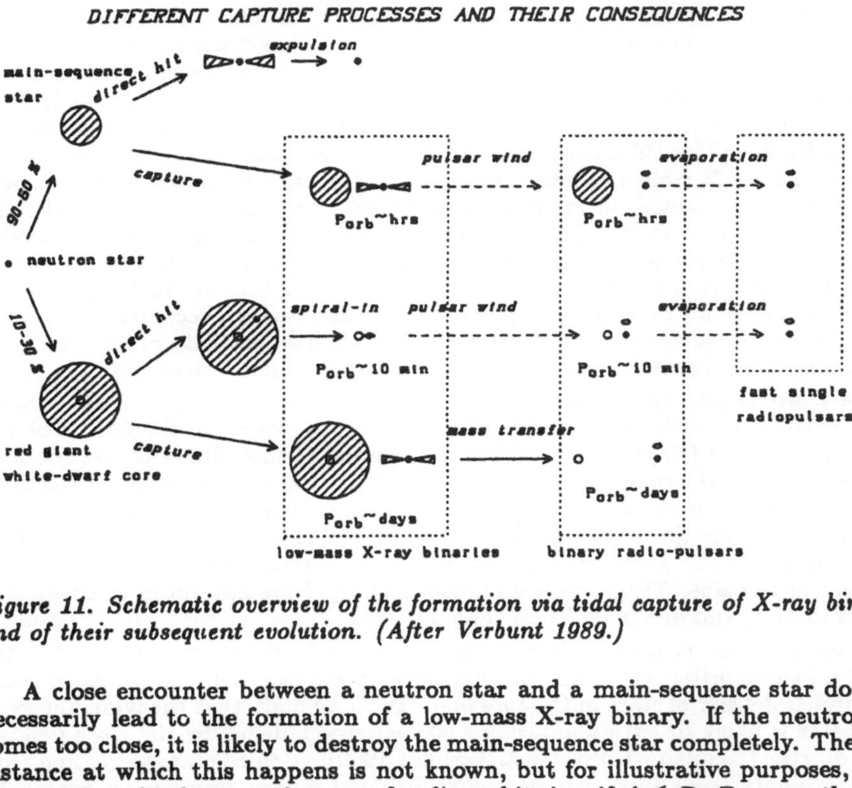

DIFFERENT CAPTURE PROCESSES AND THEIR CONSEQUENCES

*Figure 11. Schematic overview of the formation via tidal capture of X-ray binaries and of their subsequent evolution. (After Verbunt 1989.)*

A close encounter between a neutron star and a main-sequence star does not necessarily lead to the formation of a low-mass X-ray binary. If the neutron star comes too close, it is likely to destroy the main-sequence star completely. The exact distance at which this happens is not known, but for illustrative purposes, let us assume that this happens in case of a direct hit, i.e. if $d \lesssim R$. Because the cross section is proportional to $d$, see Eq.(40), about 1/3 of the captures of neutron stars by main-sequence giant stars will not lead to a binary but to a single neutron star again, possibly after an intermediate stage with some mass of the destroyed star still present around the main-sequence star in a massive disk (Krolik 1984). In the massive disk, $\dot{M}$ may be expected to be well above the Eddington limit of Eq.(20); if so, most of the disk matter will be expelled as soon as a small fraction accretes. The small amount of mass accreted cannot change the rotation of the neutron star (Verbunt et al. 1987).

Closest approaches with $R \lesssim d \lesssim 3R$ do lead to the formation of a low-mass X-ray binary with a main-sequence mass donor and hence orbital periods between 2 and 10 hr (Table 2). If for some reason the mass transfer in such a system stops, for example as the system enters the period gap (Sect.5.4), the neutron star becomes a radio pulsar. It rotates rapidly because it has been spun up by the accreted matter. If it rotates sufficiently rapidly, it may gradually evaporate its companion and eventually become a single radio pulsar. This may well be what happened to the single radio pulsar in M 28.

A direct hit of a neutron star on a (sub)giant has been proposed to lead to a spiral in. If this is indeed the case, a very close binary could emerge from the spiral in, consisting of the neutron star and the white-dwarf core of the giant, and with an orbital period in the order of minutes (Verbunt 1987). This scenario was proposed for the formation of the 11 min binary discovered in NGC6625 (see Table 1). The short period indicates a white-dwarf mass donor (Table 2), but the cross section of the white dwarf for tidal capture is small. A white dwarf in the core of a giant, however, can profit from the large cross section for capture of the giants as well as from the enhancement of capture by giants due to mass segregation. If mass transfer in the binary stops, the radio pulsar becomes visible, and it is conceivable that it will evaporate its companion, eventually leading to a single radio pulsar. The radio pulsar in a 1932 s orbit in 47 Tuc is a candidate for such a formation scenario, but it is not possible (so far?) to understand its high eccentricity because the spiral-in process leaves a circular orbit. (The small eccentricity of PSR1957+20 indicates that no eccentricity is induced by the evaporation process.)

If a neutron star approaches a (sub)giant to a distance $R \lesssim d \lesssim 3R$, a wide binary may be formed. An example of such a binary could be the 8.5 hr binary discovered in M 15 (Naylor et al. 1988). The period of this binary is just too long for a main-sequence donor with mass $\lesssim 0.8 M_\odot$, and more massive main-sequence stars do not exist in globular clusters! The donor is therefore probably a slightly evolved star. Once the whole envelope of the giant is transferred to the neutron star, or added to the giant core, a wide binary consisting of a white dwarf and a radio pulsar remains. Two such wide-binary radio pulsars have been discovered so far, one in M 4, and one in 47 Tuc (Table 1).

These wide-binary radio pulsars do not show any intrinsic evolution, but in the core of a dense cluster they may be affected by the passage of other cluster stars. The number density of stars in the core of 47 Tuc is so high that the wide binary in this cluster is likely to have a noticeable eccentricity, induced by the close passage of a third star (Rappaport, Putney & Verbunt 1989). A third star interacting with the wide binary sometimes can release an original binary member by taking its place, in a so-called exchange reaction. This is a good possibility for the formation of the single radio pulsar in M 15, a cluster with a very dense core (Rappaport et al. 1989). Alternatively, a rapidly moving star can have released the neutron star from the binary by disrupting it (Romani et al. 1987).

The general features of the observed X-ray binaries and radio pulsars in globular clusters are therefore well explained with the formation of X-ray binaries via tidal capture and subsequent evolution of the binary.

## 6.5 Other formation scenarios

Finally, we will briefly discuss two other mechanisms that have been proposed for the formation of low-mass X-ray binaries in globular clusters: exchange encounters and white-dwarf collapse.

If a single neutron star undergoes a close encounter with a binary consisting of two main-sequence stars, it may take the place of one of the original binary members, which itself is liberated. This is another way to form a close binary with a neutron star (Hills 1976). Compared with tidal capture by a single star, this process gains in having a bigger cross section (proportional to the binary semi-major axis $a$ rather than to a stellar radius $R$), but loses because the number of binaries is smaller than the number of single stars. The number of binaries in globular clusters is rather small, and exchange encounters are probably not an important mechanism for the formation of low-mass X-ray binaries (Verbunt & Hut 1987).

White dwarfs can be captured by giants or by main-sequence stars in the cores of globular clusters, leading to the formation of cataclysmic variables. A cataclysmic variable can be transformed into a low-mass X-ray binary if the white dwarf is pushed over the Chandrasekhar limit and implodes to form a neutron star (Sect.5.7). Capture of a white dwarf, followed by implosion to a neutron star, has been proposed for the formation of some X-ray sources in globular clusters (Van den Heuvel 1983). The problem with this scenario is that a study of the galactic disk shows that only a tiny fraction of cataclysmic variables ever undergoes white-dwarf collapse: there are $\sim 10^7$ cataclysmic variables in the disk, but only 100 low-mass X-ray binaries. Even assuming that all low-mass X-ray binaries in the disk are formed via white-dwarf collapse (which is certainly not the case, see Sect.5.7), we find that only 1 in $10^5$ cataclysmic variables ever forms a low-mass X-ray binary, if the life times of both types of binaries are similar. We can calculate the formation rate of cataclysmic variables in the same way as that of low-mass X-ray binaries. These calculations are in fact more accurate, as the number of white dwarfs can be more accurately estimated (see Sect.6.2): the required extrapolation from main-sequence stars to white-dwarf progenitors is small, white dwarfs have no kick velocities, and the effect of mass segregation on their distribution in the cluster is relatively small. The result is that the number of cataclysmic variables formed in all globular clusters together is of the order of $10^3$. As only 1 out of $10^5$ of these undergoes white-dwarf collapse, none of the low-mass X-ray binaries observed is likely to have formed from a cataclysmic variable (Verbunt & Hut 1987, Verbunt & Meylan 1988).

## Acknowledgment

It is my pleasure to thank Ed van den Heuvel for many discussions, over the years, that profoundly influenced my knowledge of and opinions on the topics described in this paper.

# References

Allen, C.W. 1976, *Stellar Quantities*, third ed., Athlone Press, London, p.209.

Bailes, M. 1989, *Astrophys. J.* in press.

Benz, W. & Hills, J.G. 1987, *Astrophys. J.* **323**, 614.

Bradt, H.V.D. & McClintock, J.E. 1983, *Ann. Rev. Astr. Astrophys.* **21**, 13.

Burke, J.A. 1967, *Mon. Not. R. astr. Soc.* **136**, 389.

de Kool, M. 1987, *PhD Thesis* Univ. Amsterdam, Chapter V.

Dewey, R.J. & Cordes, J.M. 1987, *Astrophys. J.* **321**, 780.

Fabian, A.C., Pringle, J.E. & Rees, M.J. 1975, *Mon. Not. R. astr. Soc.* **172**, 15P.

Faulkner, J. 1971, *Astrophys. J. (Letters)* **170**, L99.

Fruchter, A.S., Stinebring, D.R. & Taylor, J.H. 1988, *Nature* **333**, 237.

Giersz, M. 1986, *Acta Astr.* **36**, 181.

Gingold, R.A. & Monaghan, J.J. 1980, *Mon. Not. R. astr. Soc.* **191**, 897.

Hills, J.G. 1976, *Mon. Not. R. astr. Soc.* **175**, 1p.

Iben, I. & Tutukov, A.V. 1984, *Astrophys. J.* **284**, 719.

Joss,P.C. & Rappaport, S. 1983, in: *Accretion-driven stellar X-ray sources*, eds. W.H.G. Lewin and E.P.J. van den Heuvel (Cambridge; CUP), p.1.

Joss,P.C. & Rappaport, S. 1984, *Ann. Rev. Astr. Astrophys.* **22**, 537.

Joss,P.C., Rappaport, S. & Lewis, W. 1987, *Astrophys. J.* **319**, 180.

King, A.R. 1988, *Q.J.R.A.S* **29**, 1.

Kippenhahn, R. 1969, *Astron. Astrophys.* **3**, 83.

Kippenhahn, R. & Weigert, A. 1967, *Zeitschr. f. Astroph.* **65**, 251.

Kraft, R.P., Mathews, J. & Greenstein, J.L. 1962, *Astrophys. J.* **136**, 312.

Krolik, J.H. 1984, *Astrophys. J.* **282**, 452.

Kulkarni, S. 1988, in: *The physics of compact objects, theory vs. observation*, eds. N.E. White and L. Fillipov, Pergamon Press, Oxford, p.343.

Landau, L.D. & Lifshits, E.M. 1962, *The classical theory of fields* (2nd ed.; Oxford; Pergamon).

Lee, H.M. & Ostriker, J.P. 1986, *Astrophys. J.* **310**, 176.

Mason, K.O. 1986, in: *The physics of accretion onto compact objects*, eds. K.O. Mason, M.G. Watson and N.E. White (Berlin; Springer), p.29.

McMillan, S.L.W., Mc Dermott, P.N. & Taam, R.E. 1987, *Astrophys. J.* **318**, 261.

Meylan, G. 1989, *Astron. Astrophys.* in press.

Mochnacki, S.W. 1981, *Astrophys. J.* **245**, 650.

Morton, D.C. 1960, *Astrophys. J.* **132**, 146.

Mots, L. 1952, *Astrophys. J.* **115**, 562.

Naylor, T., Charles, P.A., Drew, J.E. & Hassall, B.J.M. 1988, *Mon. Not. R. astr. Soc.* **233**, 285.

Nelson, L.A., Rappaport, S.A. & Joss, P.C. 1986, *Astrophys. J.* **304**, 231.

Nomoto, K. 1987, in: *13th Texas Symp. on relativistic astrophysics*, ed. M.P. Ulmer (Singapore: World Scientific Press), p. 519.

Pacsyński, B. 1966, *Acta Astr.* **16**, 231.

Pacsyński, B. 1967a, *Acta Astr.* **17**, 193 & 355.

Pacsyński, B. 1967b, *Acta Astr.* **17**, 287.

Pacsyński, B. 1970, in: *Mass loss and evolution in close binaries*, eds. K. Gyldenkerne & R.M. West, Copenhagen University Observatory, p.139.

Pacsyński, B. 1971, *Ann. Rev. Astr. Astrophys.* **9**, 183.

Pacsyński, B. & Sienkewics, R. 1981, *Astrophys. J. (Letters)* **248**, L27.

Patterson, J. 1984, *Astrophys. J. Suppl. Ser.* **54**, 443.

Priedhorsky, W.C. & Verbunt, F. 1988, *Astrophys. J.* **333**, 895.

Pylyser, E. & Savonije, G.J. 1988, *Astron. Astrophys.* **191**, 57.

Rappaport, S., Joss, P.C. & Webbink, R.F. 1982, *Astrophys. J.* **254**, 616.

Rappaport, S., Putney, A. & Verbunt, F. 1988, *Astrophys. J.* submitted.

Rappaport, S., Verbunt, F. & Joss, P.C. 1983, *Astrophys. J.* **275**, 713.

Ritter, H. 1987, *Astron. Astrophys. Suppl. Ser.* **70**, 335.

Romani, R.W., Kulkarni, S.R. & Blandford, R.D. 1987, *Nature* **329**, 309.

Rood, R.T. 1972, *Astrophys. J.* **177**, 681.

Savonije, G.J. 1983, in: *Accretion-driven stellar X-ray sources*, eds. W.H.G. Lewin and E.P.J. van den Heuvel (Cambridge; CUP), p.343.

Savonije, G.J., de Kool, M. & van den Heuvel, E.P.J. 1986, *Astron. Astrophys.* **155**, 51.

Schwarsschild, M. 1958, *Structure and Evolution of the Stars*, Princeton University Press.

Smarr, L.L. & Blandford, R. 1976, *Astrophys. J.* **207**, 574.

Spruit, H.C. & Ritter, H. 1983, *Astron. Astrophys.* **124**, 267.

Sutantyo, W. 1975, *Astron. Astrophys.* **44**, 227.

Taam, R.E. 1983a, *Astrophys. J.* **268**, 361.

Taam, R.E. 1983b, *Astrophys. J.* **270**, 694.

Taylor, J.H. & Weisberg, J.M. 1982, *Astrophys. J.* **253**, 908.

van den Heuvel, E.P.J. 1978, in: *Physics and Astrophysics of Neutron Stars and Black Holes*, eds. R. Giacconi and R. Ruffini, North Holland Publishing Company, Amsterdam, p.828.

van den Heuvel, E.P.J. 1983, in: *Accretion-driven stellar X-ray sources*, eds. W.H.G. Lewin and E.P.J. van den Heuvel (Cambridge; CUP), p.303.

van den Heuvel, E.P.J. 1984, *J. Astroph. Astr.* 5,209.

van den Heuvel, E.P.J. & Habets, G.M.H.J. 1984, *Nature* 309, 598.

van den Heuvel, E.P.J. & Heise, J. 1972, *Nature* 239, 67.

van den Heuvel, E.P.J. & Rappaport, S. 1987, in *Physics of Be Stars*, eds. A. Slettebak & T.P. Snow, Cambridge University Press, p. 291.

van Paradijs, J. 1983, in: *Accretion-driven stellar X-ray sources*, eds. W.H.G. Lewin and E.P.J. van den Heuvel (Cambridge; CUP), p.189.

van Speybroeck, L., Epstein, E., Forman, W., Giacconi, R., Jones, C., Liller, W. & Smarr, L. 1979, *Astrophys. J. (Letters)* 234 45.

Verbunt, F. 1984, *Mon. Not. R. astr. Soc.* 209, 227.

Verbunt, F. 1987, *Astrophys. J. (Letters)* 312, L23.

Verbunt, F. 1988a, in: *The physics of neutron stars and black holes*, eds. Y. Tanaka, ISAS, Tokyo, p.159.

Verbunt, F. 1988b, in: *The physics of compact objects, theory vs. observation*, eds. N.E. White and L. Fillipov, Pergamon Press, Oxford, p.529.

Verbunt, F. 1989, in: *Timing neutron stars*, eds. H. Ōgelman & E.P.J. van den Heuvel, Kluwer, Dordrecht.

Verbunt, F. & Hut, P. 1987, in: *The origin and evolution of neutron stars*, IAU Symp. # 125 Nanjing P.R. China, eds. D.J. Helfand and J.H. Huang, Reidel, Dordrecht, p.187.

Verbunt, F. & Meylan, G. 1988, *Astron. Astrophys.* 203, 297.

Verbunt, F. & Rappaport, S.A. 1988, *Astrophys. J.* 332, 193.

Verbunt, F. & Zwaan 1981, *Astron. Astrophys.* 100, L7.

Verbunt,F., van den Heuvel, E.P.J., van Paradijs, J. & Rappaport, S.A. 1987, *Nature* 329, 312.

Vila, S.C. 1971 *Astrophys. J.* 168, 217.

Webbink, R.F., Rappaport, S. & Savonije, G.J. 1983, *Astrophys. J.* 270, 678.

# FIVE LECTURES ON ACCRETION-DRIVEN STELLAR X-RAY SOURCES

Jan van Paradijs* and Walter H. G. Lewin**


*    Astronomical Institute "Anton Pannekoek", University of Amsterdam,
     Roetersstraat 15, 1018WB Amsterdam, The Netherlands.

**   Massachusetts Institute of Technology, Physics Department and Center for
     Space Research, Room 37-627, Cambridge, MA 02139, USA.

## GENERAL INTRODUCTION

In this contribution we describe the five lectures that we have given in Erice. Since we have worked on these topics for many years in close collaboration, we decided to combine our contributions in one paper. Lecture #1 (by JvP) was on the galactic population of X-ray binaries, lecture #2 (by JvP) was a brief review on X-ray bursts, lecture #3 (by WHGL) was on the mass-radius relations of neutron stars, lecture #4 (by WHGL) was on the Rapid Burster, and lecture #5 (by JvP) was a brief review of quasi-periodic oscillations (QPO) in the X-ray flux of X-ray binaries.


## LECTURE #1 - GALACTIC POPULATIONS OF X-RAY SOURCES

The galaxy contains two major groups of accretion-driven stellar X-ray sources, the high-mass and the low-mass X-ray binaries. In this lecture the properties of these two types of sources are briefly described, with some emphasis on optical observations. The differences between their properties are likely related to a large difference in the magnetic fields of the neutron stars in these objects. A brief discussion is given of the evidence that magnetic fields of neutron stars decay.


### Introduction

The properties of the spectra of the first two optically identified X-ray sources, Sco X-1 (Sandage et al. 1966) and Cyg X-2 (Giacconi et al. 1967) led to early speculation that their X-ray luminosity is generated by accretion onto a compact star in a mass-transfer binary star. In the case of Sco X-1 the spectrum was found to be similar to those of old novae and U Gem type stars, which were known to be binary stars, in particular through the work of Crawford and Kraft in the 1950s and '60s (Crawford & Kraft 1956; Kraft 1962,1964). The optical spectrum of Cyg X-2 was found to be composite, showing the signatures of both a late type star and a component of much higher excitation; Cyg X-2 also showed significant radial-velocity variations. However, the single most important characteristic of a binary star, i.e., an orbital periodicity, was not found in either system (cf. Kraft & Miller 1969) until many years later, in spite of substantial effort (see e.g., Hiltner & Mook 1971; Kraft 1973, for discussions of early optical observations of X-ray sources, and references).

*W. Kundt (ed.), Neutron Stars and Their Birth Events, 219–252.*
*© 1990 Kluwer Academic Publishers.*

The idea that the bright galactic X-ray sources are mass exchanging binary stars was accepted, and became a paradigm of X-ray astronomy, with the discovery of the optical counterpart of Cygnus X-1 (Webster & Murdin 1972), and the regular eclipses of the pulsating X-ray sources Centaurus X-3 (Giacconi et al. 1971; Schreier et al. 1971). The optical identification of the rapidly variable X-ray source Cyg X-1 was established through an accurate position of a radio source (Braes & Miley 1971; Hjellming & Wade 1971), which underwent a large upward brightness transition, apparently correlated with a major change in the X-ray spectrum of Cyg X-1 (Tananbaum et al. 1972). The radio position (accurate to about one arcsecond) coincided with the bright star HD 226868. This early-type supergiant was soon found to be a single-lined spectroscopic binary, with an orbital period of 5.6 days (Webster & Murdin 1972). The observed mass function, in conjunction with an estimate of the mass of the optical star (from its position in the Hertzsprung-Russell diagram) led to the conclusion that the compact X-ray source in this system has a mass of at least 6 $M_\odot$. Thus, the first X-ray source for which the binary nature was reasonably well established, contained an accreting compact object that is likely a black hole (see McClintock 1988 for a recent review of the status of black-hole candidates in the stellar mass range).

The observation of variable delays of the pulse arrival times from Cen X-3, in phase with the periodic (2.1 days) eclipses of the X-ray sources, showed persuasively that in this system the X-rays are generated by accretion onto a strongly magnetized neutron star, rotating at the observed 4.8 s pulse period, in orbit around a companion star. The properties of the X-ray orbit showed that the mass-transferring companion star is very massive (> 10 $M_\odot$), a result which was later confirmed by the optical identification of this source with an O-type giant star (Krzeminski 1974). The discovery that Cen X-3 is an X-ray binary star was soon followed by more observations of eclipsing X-ray sources, some of them pulsating, and the identification of these X-ray sources with early-type stars (see e.g., Liller 1973; Penny et al. 1973; Vidal 1973). In addition, a general framework for the origin and evolution of a massive X-ray binary, as a rather normal episode in the life of a massive close binary star with successive stages of mass transfer between the two components, was readily accepted (Van den Heuvel & Heise 1973). Thus, within a few years the existence of a galactic population of high-mass X-ray binaries (HMXB) was well established.

The clustering of bright X-ray sources within ~30° of the direction of the galactic center without a strong background of unresolved sources showed already quite early that there is a group of sources located in the central regions of the Galaxy (Setti & Woltjer 1970; Ryter 1970). It was therefore suspected that apart from the above-described HMXB there is a class of low-mass X-ray binaries (see e.g., Salpeter 1973), but proof for this basic idea was hard to obtain. Apart from the difficulty of finding orbital periods, the apparent heterogeneity of the properties of low-mass X-ray binaries (LMXB) probably played a role. Compared to the HMXB the first handful of systems now classified as LMXB (Her X-1, Cyg X-3, Sco X-1, Cir X-1) show rather more diversity than similarity in their properties. As a result, the establishment of two broad groups of galactic X-ray sources has come slowly, and along various roads: only at the end of the 1970's it became clear that with respect to their sky distributions, X-ray spectral characteristics, optical properties, and types of X-ray variability, the LMXB are distinct from the HMXB as a group with "family traits" (see e.g., Lewin & Clark 1980). The LMXB comprise the globular cluster X-ray sources, X-ray bursters, soft X-ray transients, and the bright galactic bulge X-ray sources (most of which have recently been found to be QPO sources; see Lewin, Van Paradijs & Van der Klis 1988; Van der Klis 1989; Lamb 1989, for recent reviews).

In this paper we briefly describe the main properties of the HMXB and LMXB, with some emphasis on optical observations, and show how the differences between these two groups of X-ray sources may be linked to a difference in the strength of the magnetic fields of the neutron stars they harbor. Within the limits of these lectures I cannot strive for completeness. For background information on various topics related to X-ray binaries the interested reader is advised to consult individual chapters in books by Shapiro & Teukolsky (1983), Lewin & Van den Heuvel (1983), Eggleton & Pringle (1985), Frank, King & Raine (1985), Pringle & Wade (1985), Lewin, Trümper & Brinkmann (1986), Mason, Watson & White (1986), Pallavicini & White (1988), Filipov & White (1988). References on optically identified individual sources can be found in Bradt & McClintock (1983) and Van Paradijs (1983).

## Optical Counterparts

The optical counterparts of HMXB have normal early-type spectra, in the sense that they can be MK-classified without particular difficulty, on the basis of ratios of spectral line strengths. Some disturbance of the spectrum, indicative of anisotropic gas flow near the primary may show up as variable emission/absorption components, particularly in Hα, Hβ, He II λ4686, and the CIII-NIII λ4630-50 complex. However, when the latter two lines are strongly in emission (see e.g., Hensberge, Van den Heuvel & Paes de Barros 1973) this is likely due to a very early spectral type (Of characteristic) of the primary, and not to the presence of the X-ray source.

The reason that the X-ray source does not seem to affect the spectral properties of the primary much, is that the bolometric luminosity of the latter generally exceeds the X-ray luminosity, often by a large margin (see Van Paradijs 1983).

With respect to the spectral types of the optical counterparts the HMXB can be subdivided into two subgroups, as follows.
    (i) The spectral type is earlier than B2, and the luminosity class is I to III, i.e., the primary star has evolved off the main sequence. These stars are filling, or close to filling their Roche lobes, as is apparent from the amplitudes of their optical light curves (see below).
    (ii) The primary is a B-emission (Be) star, located in the Hertzsprung-Russell diagram rather close to the main sequence. The orbits of these Be/X-ray binaries are eccentric, and their periods tend to be long. The primaries underfill their Roche lobes.

As first suggested by Maraschi, Treves & Van den Heuvel (1975) the mass transfer in these two groups is driven by different mechanisms. In the first group mass is transferred via a strong stellar wind (in a few short-lived sources Roche-lobe overflow is important). In the Be/X-ray binaries the mass transfer is related to the anisotropic (often highly variable) shedding of mass as observed in all Be stars, which is believed to be the result of their rapid rotation (Slettebak 1987). This inferred difference in mass transfer mechanism is supported by the different relations between orbital period and X-ray pulse period, first pointed out in by Corbet (1984,1986) for these two groups of sources (see e.g., the discussion by Van den Heuvel & Rappaport 1987).

Most Be/X-ray binaries are highly variable, or transient. In some of them recurrent outbursts have been observed, which reflect the varying accretion rate onto the neutron star as it moves in its eccentric orbit through regions of varying density around the Be star. In addition, a more sudden turning on and off of the accretion can occur when the wind density becomes too low for the neutron star magnetosphere to be within the corotation radius, so that accretion becomes centrifugally inhibited (Stella, White & Rosner 1986). However, in many cases outbursts have been observed which are not related to the orbit of the Be/X-ray binary, but are due to a sudden enhancement of the mass loss of the Be star (for recent reviews of various aspects of Be stars see Slettebak & Snow 1987).

The optical counterparts of LMXB are rather faint stars. Their spectra show a few characteristic emission lines, particularly H$\alpha$, H$\beta$, HeII $\lambda$4686, and CIII/NIII $\lambda$4630-50, superposed on a flat (in frequency) continuum. These spectra, which definitely are not those of normal stars, are dominated by the emission from an accretion disk around a neutron star, which radiates mainly through reprocessing of incident X-rays into optical/UV photons. In very few cases the signature of a companion star can be discerned.

It appears that the optical properties of LMXB are rather uniform, and can in general be reasonably described by average values (Van Paradijs 1983). The colour indices B-V and U-B (reddening-corrected) have average values of $0.0 \pm 0.3$ and $-0.9 \pm 0.2$, respectively (errors are one sigma standard deviations), close to those expected for a flat continuum ($F_\nu$ = constant). The distribution of the ratio of X-ray to optical luminosity is rather sharply peaked. Expressed in terms of an "optical/X-ray colour index" $B_o + 2.5 \log F_X$ ($\mu$Jy), the peak occurs at 21.5, corresponding to a ratio of fluxes emitted in X-rays (2-11 keV) and in optical light (3000-7000 A) of ~350.

Absolute visual magnitudes $M_v$ have been estimated for optical counterparts of X-ray burst sources and some soft X-ray transients, for which a reasonable distance determination can be made. These absolute magnitudes (average value $M_v = 1.0$) scatter over a remarkably small interval of $\pm$ 1 mag (Van Paradijs 1983). This small scatter may be related to a small range in orbital periods of the systems used in this average. If the relative shape of the accretion disk (in particular its angular thickness as seen from the neutron star) is assumed to be independent of the orbital period then the X-ray irradiated accretion disk will (for the same X-ray luminosity) be hotter as the orbital period decreases since they are then smaller [$T^4$ (:) $a^{-2}$ (:) $P^{-4/3}$]. A relatively larger fraction of its emission will then be in the UV (i.e., the bolometric correction increases).

It is of some interest to compare the optical properties of LMXB with the closely related cataclysmic variables (CVs), which are different in that the accreting compact star is not a neutron star but a white dwarf. The spectra of CVs bear a general resemblance to those of LMXB, showing emission lines superposed on a continuum. However, in general the equivalent widths of these lines in LMXB spectra, in particular that of H$\beta$, are much smaller than those in CV spectra (Van Paradijs & Verbunt 1984). Since the absolute magnitudes of LMXB and CV differ by values ranging between ~3.5 mag (for nova-like variables, and dwarf novae in outburst) and at least 6 mag (for dwarf novae in quiescence) the luminosities in H$\beta$ are substantially higher in the LMXB than in CV. In general, the equivalent width of HeII $\lambda$4686 varies somewhat less between the LMXB and CV (except the AM Her systems); this suggests that in the LMXB the luminosity in this line is enhanced by X-ray reprocessing by a similar factor as the continuum flux.

The orbital-period distributions of CV and LMXB (Fig.1) are different. Compared to the CV there is a relatively larger fraction of LMXB with periods above about half a day; this may partly reflect the fact that CV systems with long orbital periods may be classified as symbiotic stars. Also, there are no LMXB in the period range between 1 and 2 hours (i.e., below the period gap), which is well populated by the CV. This is perhaps the result of evaporation of the companion stars by the large luminosity from the rapidly rotating neutron star (spun up by accretion torques), which becomes active as the mass transfer

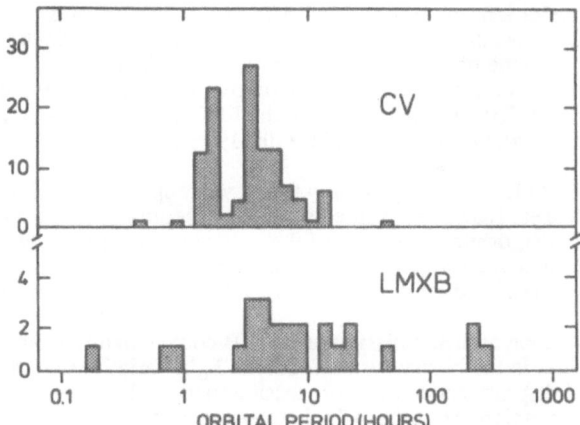

Fig. 1.    *Distributions of orbital periods for low-mass X-ray binaries and cataclysmic variables. Data have been taken from Mason (1986) and Ritter (1987), except that we have not included LMC X-2 (for which recent observations give no evidence for a 6.4 h period), and have included the 15.1 h period system Cen X-4 (Chevalier et al. 1988).*

stops when the system has reached the upper edge of the period gap (Ruderman *et al.* 1988; Van den Heuvel & Van Paradijs 1988).

## Optical Light Curves of HMXB and LMXB

Regular optical brightness variations at the orbital period have been observed for many HMXB and LMXB. In almost all HMXB an important contribution to the orbital optical light curve is due to a double-wave modulation with generally equal maxima and two somewhat different minima. This so-called ellipsoidal variation is the result of the tidal and rotational distortion of the companion star, and a non-uniform surface brightness distribution ("gravity darkening"), often described by Von Zeipel's theorem. The maxima in the light curve occur at quadratures of the system, the deepest minimum at inferior conjunction of the X-ray source. The amplitude of these ellipsoidal light curves is mainly determined by the mass ratio $q = M_{opt}/M_X$, the inclination angle i of the orbital plane, and a dimensionless potential parameter $\Omega$, which measures how far the companion star fills its Roche lobe.

Superposed on the regular orbital brightness variations significant irregularity is observed; consequently observations over many orbital cycles are required to obtain a fair estimate of the average light curve that hopefully contains information on the geometry of the primary.

In some systems (e.g., Cen X-3, SMC X-1) the presence of the X-ray source is noticeable through the heating by X-rays of the hemisphere of the companion star facing the X-ray source. This leads to a brightening of that side of the star and therefore a filling-in of the deepest minimum of the purely ellipsoidal light curve.

A further complication in the light curve arises when an accretion disk is present, which: (i) provides an additional source of light; (ii) can give rise to mutual eclipses with the companion; (iii) shields a fraction of the companion star from X-rays. Since those HMXB in which mass transfer occurs through Roche-lobe overflow (and which therefore have an accretion disk) are also the most luminous X-ray sources, the effects of X-ray heating and of the presence of the accretion disk tend to be present together. For detailed description of the light curves of HMXB, and discussions of the limitations and underlying assumptions, we refer to Tjemkes *et al.* (1986) and Zuiderwijk (1979).

Optical counterparts of LMXB are generally faint, with apparent visual magnitudes V > 17 in the majority of cases. Except for some notorious exceptions, e.g., Sco X-1 (Wright, Gottlieb & Liller 1973), orbital variations of the optical brightness are not too difficult to detect, particularly since the use of CCD photometers has become commonplace. As a result orbital light curves have now been determined for many optically identified LMXB.

As mentioned above, the optical emission of LMXB comes mainly from X-ray reprocessing in matter in the binary system. Orbital light variations are therefore due to deviations from axial symmetry of the reprocessing matter. This asymmetry is mainly provided by the companion star, whose "polar caps" are not shielded from X-rays by the disk, and therefore are heated (see below).

The orbital light curves of LMXB show "family characteristics"; their amplitude is correlated with the inclination, i, of the orbital plane (see Fig. 2), and also their shape seems to change with i in a characteristic way.

For low inclination angles (as apparent from the lack of X-ray eclipses and X-ray "dips", see Parmar & White 1988) the light curves are approximately sinusoidal, with an amplitude of a few tenths of a magnitude. For systems with somewhat higher

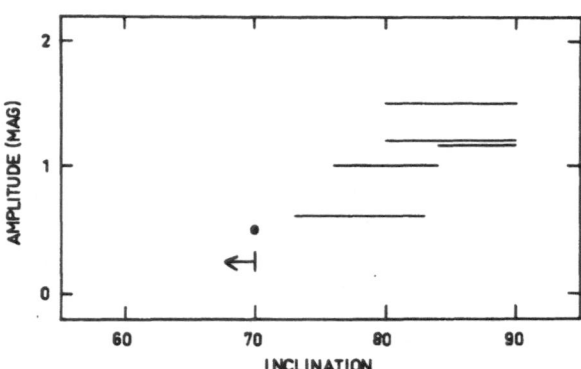

Fig. 2. *Relation between orbital inclination and amplitude of the optical light curve for low-mass X-ray binaries. Data have been taken from Mason (1986). Systems which show X-ray eclipses are represented by horizontal lines which indicate the allowed range of inclination. Three sources which show X-ray dips, but not eclipses, are assumed to have i = 70°. Sources which do not show eclipses or dips are likely to have smaller inclinations. Systems in which the companion star contributes significantly to the optical brightness have not been included in this figure.*

inclination angles (as evident from the presence of periodic dips in the X-ray intensity curve but absence of X-ray eclipses) the amplitude of this sinusoidal light curve increases to ~0.5 mag (e.g., X1755-338). For systems with the highest inclinations the amplitude reaches about 1.5 mag. For these systems the optical light curve can be decomposed into the sine wave that is also observed for systems at lower inclinations, and a rather sharp cusp superposed on the minimum of the sine wave. From a comparison of the phases of the optical and X-ray intensity curves it appears that the cusp (and therefore the minimum of the sine wave component) occurs at superior conjunction of the X-ray source. This indicates that the cusp in the optical light curve is due to the eclipse of the luminous accretion disk. The relative phasing in the non-eclipsing system X1755-338 of the sinusoidal optical light curve, and the X-ray dips (Mason, Parmar & White 1985), confirms that this picture is valid also for the lower-inclination systems. The correlation of the amplitude of the sinusoidal component with inclination angle is likely due to the fact that as the inclination angle decreases the average brightness of the accretion disk increases (larger projection factor, less self shielding), and the relative importance of the variable component (due to eclipses of the disk and the heating of the companion star) decreases.

## Galactic Distributions of Optically Identified HMXB and LMXB

The sky distributions of the optically identified HMXB and LMXB are shown in Fig.3. The HMXB are distributed along the galactic plane, with a narrow latitude distribution ($<b^{II}> = -0.5 \pm 3.9°$; if we leave out the very nearby high-latitude system X Per we find $0.2 \pm 1.9°$). The optical counterparts of LMXB have a much wider latitude distribution ($<b^{II}> = -1.6 \pm 10.7°$), and are also more concentrated in the general direction of the galactic center.

These distributions fit the idea that the HMXB and LMXB are parts of a very young galactic population of massive stars (population I), and of a much older population (population II, and old disk population), respectively.

A recent detailed analysis (Van Oyen 1988) of the kinematic properties of the optically identified HMXB indicates that these objects are runaway stars; this is perhaps the result of asymmetries of the supernova explosions in which the (now) accreting neutron stars were formed (Van Oyen 1988).

The radial velocities of the LMXB optical counterparts support their membership of an old galactic population (Cowley *et al.* 1987).

## X-ray Variability: Pulsations and Bursts

Almost all HMXB show X-ray pulsations, which indicates that the accreting compact stars in these systems are strongly magnetized neutron stars (for a review of various aspects of X-ray pulsars see e.g., Rappaport & Joss 1983; Joss & Rappaport 1984). Strong magnetic fields (a few $10^{12}$ G) have also been inferred from the presence of cyclotron lines in the hard X-ray spectra of some X-ray pulsars (see Kirk & Truemper 1983, for a review of this subject).

Observed pulse periods range over a factor $~10^4$, between 69 msec (for the LMC transient A0538-66) to 835 s in the Be/X-ray systems X Per. From a survey of X-ray pulsars with HEAO-1 White *et al.*(1984) found a correlation between the pulse profile and the X-ray

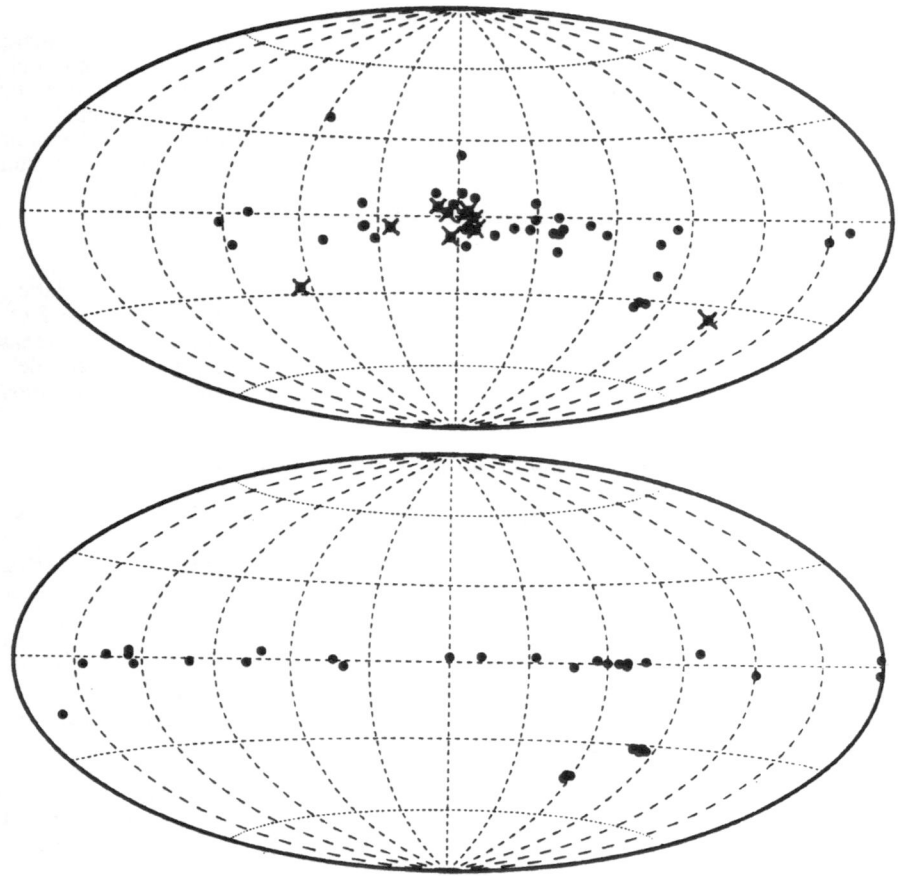

Fig. 3.   *Sky maps (in galactic coordinates) of the optically identified high-mass X-ray binaries (bottom panel) and low-mass X-ray binaries (top panel); the latter also contains the globular cluster sources (indicated by crosses).*

luminosity; their result are supported by recent EXOSAT observations of the transient sources EXO 2030+475 which cover a large range in luminosity (Parmar *et al.* 1988).

Pulse-arrival time measurements for pulsating HMXB, in combination with radial-velocity observations of their massive companions, have provided invaluable information on the masses of neutron stars. In Fig. 4 our present knowledge on neutron-star masses is summarized. These results are consistent with a standard neutron star mass between 1.2 and 1.6 $M_\odot$; however, mass differences of more than 0.5 $M_\odot$ cannot at present be excluded. The limiting factor in the accuracy of neutron star masses are the optical radial-velocity data (which so far have mainly been based on photographic and image-tube observations). Significant progress in this area in the form of much improved error ranges

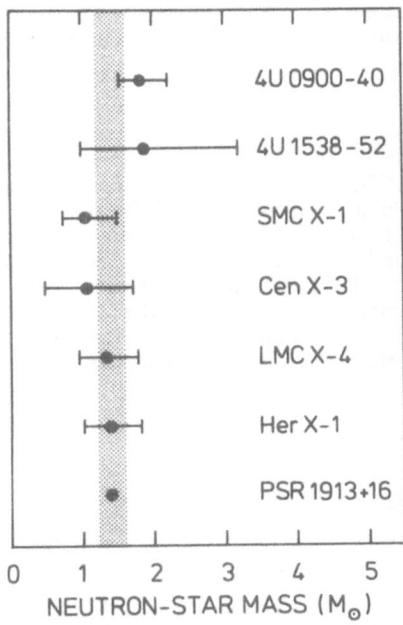

Figure 4.  *Summary of our present knowledge of neutron star masses from observations of binary X-ray pulsars and the binary radio pulsar PSR 1913+16. This figure is an update from Joss & Rappaport (1984) using more recent data from Pietsch et al. (1985).*

for these masses, and an extension of the sample of observed neutron stars, appears definitely possible with presently available CCD spectrographs.

X-ray pulsations occur only rarely in LMXB. This suggests that the magnetic fields of the neutron stars in these systems are generally much weaker than in HMXB. (The alternative that the magnetic and rotational axes of the neutron stars are aligned, is discussed below).

On the other hand, many LMXB emit X-ray bursts, which are the result of thermonuclear flashes in accreted matter on the surface of a neutron star (for reviews of X-ray bursts see Lewin & Joss 1983).

Not a single source is known that shows both pulsations and bursts. Apparently, the presence of a strong magnetic field suppresses the instability of the nuclear reactions that gives rise to bursts (as expected from models for thermonuclear flashes, see e.g., Joss & Li 1980). This mutual exclusion of bursts and pulsations indicates that it is a weaker magnetic field, and not only alignment of the field axis, which distinguishes the neutron stars in LMXB from those of HMXB.

The radiation observed during an X-ray burst originates directly from the surface of the neutron star (with possibly some modification due to e.g., Compton scattering in intervening hot plasma). Because of this, time-resolved spectral studies of bursts provide an observational method to study the properties of neutron stars, in particular their mass-radius relation. The main problem in this area appears to be the interpretation of X-ray burst spectra in terms of atmospheric model (see lecture 3 for references on this topic).

## X-Ray Spectra

The X-ray spectra of HMXB (most of which are pulsars) are generally much harder than those of LMXB ( Tananbaum 1973; Jones 1977; Ostriker 1977; White & Marshall 1984). This distinction is present for both the steady and transient X-ray sources (Cominsky *et al.* 1978). As is illustrated in Fig. 5 the difference in spectral hardness persists into the hard X-ray energy range, up to $\sim 10^2$ keV. From this figure it appears that the average difference in spectral hardness, as measured by the ratio of the count rates observed with the A4 experiment on HEAO-1 (Levine *et al.* 1984) and with the SSI on Ariel-5 (McHardy *et al.* 1981; Warwick *et al.* 1981), is about a factor 10.

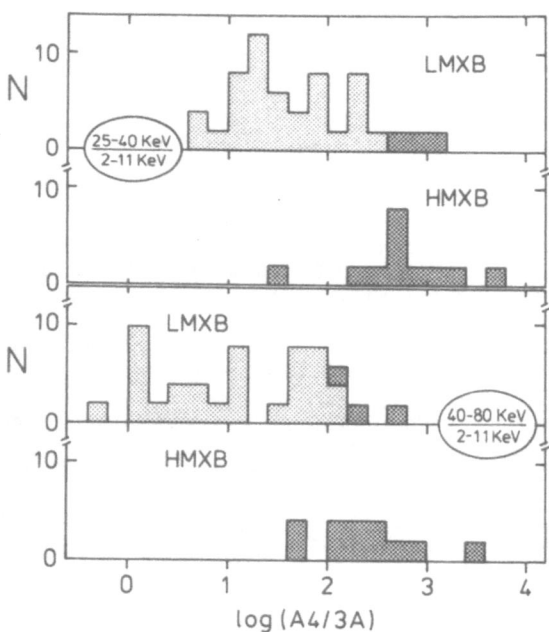

Figure 5. *Distributions of the ratios of count rates, as observed with the HEAO-1 A4 experiment (Levine et al. 1984) in the 25-40 keV and 40-80 keV bands, to that observed with Ariel 5 in the 2-11 keV band (McHardy et al. 1981; Warwick et al. 1981), shown separately for the high-mass and low-mass X-ray binaries. X-ray pulsars are indicated by a darker shade.*

It is remarkable that the few LMXB which show pulsations (GX 1+4, Her X-1, and 1627-673) have X-ray spectra which are as hard as those of HMXB (almost all of which are, likewise, pulsars). This result strongly suggests that the division into hard and soft

X-ray spectra is related to a difference in the geometry of the accretion flow. For neutron-star magnetic fields of the order of $10^{12}$ G, and sub-Eddington accretion rates, the accretion flow is dominated by the magnetic field within a relatively large distance (of the order of $10^3$ km) from the neutron star (magnetospheric radius, see e.g., Henrichs 1983); a large fraction of the inflowing matter reaches the neutron star via an accretion column on a relatively small area (near the polar caps). For magnetic fields $< 10^9$ G the magnetospheric radius becomes comparable to the radius of the neutron star; one then expects that the accreting material is distributed over a larger fraction of the neutron star surface.

It should be noted that alignment of the magnetic and rotational axis of the neutron star may also lead to the disappearance of pulsations, even for fields of $10^{12}$ G. However, since the magnetospheric radius will not be much affected by this alignment one does not expect the accretion flow within $\sim 10^3$ km of the neutron star to be much affected by the alignment; it is therefore unlikely that alignment alone can explain the systematic difference in the hardness of the X-ray spectra of pulsating and non-pulsating sources. Thus, the differences in these X-ray spectra support the inference, from the distribution of pulsars and bursters among the HMXB and LMXB, that the magnetic fields of the neutron stars in LMXB are systematically weaker than those of the neutron stars in HMXB.

In many LMXB a correlation has been observed between the X-ray intensity and the hardness of the X-ray spectrum (see e.g. Schulz, Hasinger & Truemper 1989). In addition, a number of LMXB have, at times, been observed in a different spectral state, in which the X-ray spectrum is relatively hard (but still much softer than the X-ray spectra of pulsars), and does not change much with source intensity. These different spectral states (which, after the location of the sources in an X-ray spectral hardness-intensity diagram, are called the normal branch, and the horizontal branch) are correlated with different fast-variability behavior, with high-frequency intensity-dependent QPO occurring in the HB state, and low-frequency QPO ($\sim 5$ Hz, independent of source intensity) occurring in the NB state (Hasinger 1987; Van der Klis et al. 1987). In some sources, a third spectral state (the "flaring branch" in the spectral hardness vs. intensity diagram) has been distinguished, which also is correlated with QPO behavior (Hasinger 1988). After the topology of these spectral branches in the hardness-intensity diagram, these sources are called "Z sources". Recently, Hasinger and Van der Klis (1989) have shown that with respect to their spectral behaviour another group of LMXB can be distinguished, the so-called "atoll sources", which also have different QPO properties.

For reviews of QPO in LMXB with emphasis on observations and data analysis we refer to Lewin, Van Paradijs & Van der Klis (1988), and Van der Klis (1989). The origin of the different spectral states, and the connection to the QPO properties are presently a subject of an intense study ; for recent reviews with emphasis on models for QPO we refer to Lamb (1989a, 1989b).

## Magnetic Field Decay?

The strong correlation between X-ray pulsars and HMXB on the one hand, and between X-ray bursters and LMXB on the other hand, and, in addition, the striking correlated difference in the persistent X-ray spectra of HMXB and LMXB (except for the LMXB pulsars), persuasively argues for a systematic difference in the magnetic field strengths of the neutron stars in these two groups of X-ray sources.

There are two possible ways to understand this difference. In the first place the magnetic fields of the (generally old) neutron stars in LMXB may be much weaker than those of the (young) neutron stars in HMXB, because they have always been very weak. This difference might be related to a difference in the formation mechanism of neutron stars in HMXB and LMXB, viz. via the normal evolution of a massive star, and via the accretion-induced collapse of a white dwarf, respectively (see e.g., Chanmugan & Brecher 1987). As discussed in more detail by Srinivasan et al. (1989) this idea is hard to reconcile with: (i) observed periods and period derivatives of new-born radio pulsars; (ii) the observed weak magnetic fields of the binary radio pulsars PSR 1913+16 and PSR 0655+64, which are late stages in the evolution of massive binary stars (see e.g., Van den Heuvel 1984); (iii) the statistics of CV's and LMXB in globular clusters (Verbunt & Hut 1987; Verbunt & Meylan 1988).

It is of interest to note that for two of the three (old) LMXB which show pulsations (Her X-1, GX 1+4) the accretion life time can be calculated from the system parameters; these lifetimes for both systems turn out to be very short (of the order of $10^7$ years) compared to a typical accretion lifetime of a LMXB (of the order of $10^9$ years). This coincidence strongly suggests that in these LMXB pulsars the neutron star magnetic field is high because the neutron star was formed recently through accretion-induced collapse of a white dwarf during the same stage of accretion in which we observe the system now as a source of X-rays.

It appears that the simplest description of the properties of LMXB vis a vis the HMXB is provided by the assumption that the magnetic fields of all neutron stars decay. This is in agreement with a simple interpretation of kinematic data of new-born radio pulsars (Lyne et al. 1982). We note that these kinematic data (and data on pulse profiles) have been interpreted as evidence for alignment of the rotational axis of the neutron star with the magnetic axis (Blair & Candy 1988). However, a recent study by Lyne & Manchester (1988) shows that although there is good evidence for alignment, it appears that, at least in some cases, the torque decay does not depend on the angle between the rotational and magnetic axes. It therefore appears that the interpretation of Blair & Candy (1988) does not exclude that magnetic field decay takes place.

It is possible that decay of a magnetic field takes place only in accreting neutron stars. As shown by Taam & Van den Heuvel (1986) one cannot distinguish between spontaneous field decay, and decay as a consequence of accretion; in particular the inferred B fields of neutron stars correlate well with the total amount of accreted material.

Observations of millisecond radio pulsars in binaries provide good evidence that the decay of the field of neutron stars does not continue indefinitely. From the colours of the optical counterpart of PSR 0655+64 (a white dwarf) Kulkarni (1986) estimated that the (cooling-) age of this system is ~2 $10^9$ years, yet the magnetic field of neutron star (as inferred from its period derivative) is a few $10^{10}$ Gauss. A similar conclusion follows from the observed number of millisecond pulsars in binaries which are thought to have descended from LMXB (Bhattacharya & Srinivasan 1986; Van den Heuvel, Van Paradijs & Taam 1986; see also Kulkarni & Narayan 1988). These results indicate that when the magnetic field has decayed to a value of the order of $10^{9-10}$ G (this value may be different for different sources) the decay time increases substantially (from ~$10^7$ to more than $10^9$ years).

## LECTURE #2 - X-RAY BURSTS (OF TYPE 1).

### Introduction

As mentioned in Lecture #1, X-ray bursts are a distinct type of X-ray intensity variations, which have, so far, been observed only in low-mass X-ray binaries. In this second lecture a brief description is given of the basic properties of type 1 bursts. This is followed by a discussion of recent results based, in particular, on observations with the European X-ray observatory EXOSAT. These observations allowed for the first time a detailed comparison of burst recurrence behaviour, as a function of the persistent X-ray luminosity, with expectations based on the thermonuclear flash model. For a comprehensive review of X-ray bursts, covering the entire field up to 1983, the reader is referred to Lewin & Joss (1981, 1983).

### X-ray burst properties

The characteristic which defines X-ray bursts is the sudden rise of the X-ray flux by typically at least an order of magnitude, usually within about a second (but rise times as large as ~10 s have been observed, see e.g. Murakami et al. 1980). This is followed by a subsequent decay, generally to the original flux level, in a time interval between ~10 s and about a minute. In rare cases bursts last longer than a minute.

Two types of X-ray bursts can be distinguished (Hoffman, Marshall & Lewin 1978), which are designated type 1 and type 2. The type 1 bursts are characterized by a distinct softening of the X-ray spectrum during the decay of the burst. Their recurrence times are generally of the order of hours and longer, but on occasion time intervals between bursts as short as five minutes have been observed. The spectral development in type 2 bursts is much less pronounced than that in the type 1 bursts.

The type 2 bursts have been observed only from the Rapid Burster (Lewin et al. 1976) [in a few other sources similar events may have been observed, see Lewin & Joss 1983 for references]. These bursts can vary enormously in duration, between a few seconds to ~12 minutes. The time intervals between the type 2 bursts can be as short as ~10 s, and as long as ~1h. They do not come at random, but in a characteristic pattern such that the total energy in a burst is approximately proportional to the time interval to the following burst (Lewin et al. 1976; Marshall et al. 1979); the Rapid Burster behaves like a relaxation oscillator. This strongly indicates that the type 2 bursts are the result of an accretion instability, which occurs when some critical level is reached in a "reservoir" that accumulates the inflowing matter, which somehow is temporarily prevented from accreting onto a compact star (Lewin et al. 1976; see also Lecture #4).

Many low-mass X-ray binaries, including the Rapid Burster (Hoffman, Marshall & Lewin 1978) are sources of type 1 bursts. Their sky distribution (Lewin et al. 1977) is strongly concentrated toward the center of the Galaxy (it is of interest to note that of the ten known low-mass X-ray binaries in globular clusters nine are X-ray bursters). One can infer from this that X-ray bursters are located at average distances of ~8 kpc. It follows that the total energy and maximum luminosity in type 1 bursts are of the order of $10^{39}$ erg, and ~$10^{38}$ erg s$^{-1}$, respectively.

Most X-ray burst sources also emit a persistent (but generally variable) X-ray flux, due to accretion of matter onto a neutron star. The ratio, $\alpha$, of the total energy emitted in the persistent flux, to that emitted in bursts, is typically of the order $10^2$ (see below). This observed fact has been an important argument (Hoffman, Marshall & Lewin 1978) leading to the acceptance that type 1 X-ray bursts are due to the thermonuclear-flash model of type 1 X-ray bursts. This model, that successfully explains the global features of type 1 X-ray bursts was put forward by Woosley & Taam (1976) and Maraschi & Cavaliere (1977), and has been worked out in substantial detail by Joss (1978), Fujimoto, Hanawa & Miyaji (1981), Taam (1982), Ayasli & Joss (1982), Woosley & Weaver (1984), and Fushiki & Lamb (1987). In this model, after each type 1 X-ray burst, a fresh layer of matter accretes onto the surface of a neutron star. When a sufficient amount of matter has accumulated on the neutron star surface, critical conditions may develop at the base of this envelope, causing unstable helium burning. The sudden release of nuclear energy gives rise to an X-ray burst. Typical values for the rise time, decay time, and recurrence time, and for the maximum luminosity and integrated energy for type 1 X-ray bursts, are well reproduced by this model.

The ratio, $\gamma$, of the persistent flux to the peak flux of bursts that show radius expansion (see lecture #4 for a more detailed discussionof these bursts) ranges between ~0.01 and ~0.3 (Van Paradijs *et al.* 1979; see below). This indicates that in X-ray burst sources the accretion rate is typically an order of magnitude below the critical Eddington rate. This agrees with the thermonuclear flash model , according to which X-ray bursts will not occur if the accretion rate, and therefore the persistent X-ray luminosity, is above a critical value, approximately a few tenths of the Eddington limit.

Swank *et al.* (1976) found that for a particular burst they observed the X-ray spectrum was best fit by a blackbody spectrum, with a temperature that decreased during the decay of the burst. The blackbody radius they found during burst decay (assuming a distance of 10 kpc) was ~15 km. Later work by Hoffman, Lewin & Doty (1977a,b) and Van Paradijs (1978) showed that in all cases the observed spectra were well described by a Planck function, and that the blackbody radii fitted the idea that the type 1 bursts originate from the surface of a neutron star. These results form the basis for attempts to study the mass-radius relation of neutron stars from observations of the spectral development of X-ray bursts (this subject is discussed extensively in lecture 4).

**Recurrence characteristics of X-ray bursts - comparison with models**

Current models of thermonuclear flashes on the surface of an accreting neutron star give a reasonable description of the global properties of X-ray bursts, such as typical values of burst energies and maximum luminosities. With respect to burst recurrence intervals, the simplest type of models would predict that the longer the waiting time to a burst is, the more energetic that burst will be, as more nuclear fuel has been accreted. The possible presence of such a correlation was first reported by Hoffman, Lewin and Doty (1977a,b) for MXB 1728-34, and later confirmed by Basinska *et al.* (1984) for the same source, and by Pedersen *et al.* (1982) for optical bursts from 4U/MXB 1636-53.

The study of burst recurrence behaviour made great progress through observations with the European X-ray observatory EXOSAT, which differed from all previous X-ray missions in that its orbit around the Earth had a period of ~4 days, as compared to orbital periods of ~100 minutes for satellites in near Earth orbits. This was particularly beneficial to the study of X-ray bursts, as it became possible, in much more detail than previously, to observe long, uninterrupted sequences of X-ray bursts.

In order to provide a framework for a discussion of the burst recurrence behavior, and its dependence on the mass accretion rate M, we will briefly summarize the salient features of models of thermonuclear flashes, based on the analysis by Fujimoto, Hanawa and Miyaji (1981: FHM), and Hanawa and Fujimoto (1984), for neutron stars accreting hydrogen-rich material. [For a recent, more general analysis of models of thermonuclear flashes, which addresses a wider range of the parameters such as accretion rate, envelope mass, and envelope temperature, see Fushiki and Lamb, 1987].

Both at very high and at very low accretion rates a helium flash is expected to occur in a proton-rich environment. The corresponding two critical values of M depend on several neutron star parameters, but are roughly $\sim 2\ 10^{-9}$, and $\sim 2\ 10^{-10}\ M_\odot$/yr.

At very high $\dot{M}$ the rate at which hydrogen burns steadily (through the CNO cycle), which is limited by the time scales of $\beta$-decay reactions, cannot keep pace with the rate at which new matter (assumed to be hydrogen rich) accumulates through accretion. As a consequence, the bottom of the hydrogen burning shell is pushed inward into the layer of unstable helium burning before all hydrogen is consumed in steady burning (FHM case 1).

For intermediate values of $\dot{M}$ the hydrogen burning shell settles into a steady state in which hydrogen is consumed as fast as it accumulates. As a consequence, the hydrogen burning shell has a constant mass and is located above the layers of unstable helium burning. A helium flash occurs in the latter when its mass has reached a (temperature-dependent) critical value. In these bursts the helium flash occurs in a proton-poor environment (FHM case 2); some hydrogen may be entrained in the flash through mixing in a convective layer which temporarily exists during the helium flash.

For very low values of $\dot{M}$ the stable burning of hydrogen is extinguished during the quiescent intervals between bursts. When the mass of the accreted layer reaches some critical value $\Delta M$ (which may depend on various parameters such as accretion rate), a hydrogen flash (unobservable) occurs at a relatively low temperature. If $\Delta M$ is sufficiently high this hydrogen flash develops continuously into a helium flash (low helium concentration) in a proton rich environment. For small values of $\Delta M$ a hydrogen flash is ignited, but afterwards the hydrogen shell burns steadily until a sufficient amount of helium has been accumulated to ignite; these helium flashes give rise to X-ray bursts.

Globally speaking, the involvement of hydrogen in a helium flash has two important observable effects. (i) The amount of energy liberated in a flash increases due to proton captures on seed nuclei produced in the helium flash. As a consequence, one expects that the ratio, $\alpha$, of the gravitational potential energy generated through accretion to the nuclear energy generated in bursts, decreases when the relative contribution of hydrogen increases. (ii) Proton captures on seed nuclei can give rise to prolonged energy generation after the exhaustion of helium in the flash. This gives rise to bursts of longer duration.

Results of long, uninterrupted, EXOSAT observations relevant to the study of burst recurrence, have been published so far for EXO 0748-676 (Gottwald et al. 1986; 1987), 4U/MXB 1636-53 (Sztajno et al. 1985; Lewin et al. 1987), and 4U/MXB 1735-44 (Van Paradijs et al. 1988). In all three sources there is a correlation between the burst fluence $E_b$ and the waiting time $\Delta t$ since the previous burst (and also with the integrated persistent flux $E_p$ during that interval), confirming unambiguously earlier results (see above). For small values of $\Delta t$ (a few hours) the relation between $E_b$ and $\Delta t$ is approximately linear (with a rather large scatter); for large $\Delta t$ (more than $\sim 6$ h) the relation "flattens", i. e., the increase of burst fluence with $\Delta t$ becomes smaller. Since most bursts after very long waiting times cause photospheric radius expansion, this deviation from linearity can at least partially be explained in terms of the energy spent in mass ejection. In addition, the flattening in the $E_b$-$\Delta t$ plot may be the result of a growing "leak" of nuclear fuel, due to steady burning during the intervals between bursts (Lewin et al. 1987). For the case of 1636-53 Fujimoto et al. (1987a) have argued that helium production due to stable CNO burning of hydrogen between bursts gives an adequate description of the dependence of both $E_b$ and the peak burst flux, $F_{max}$, on $\Delta t$. However, this mechanism cannot explain the enormous differences in waiting time (by a factor $\sim 50$), accompanied by only very modest changes in $E_b$, observed in 1735-44. It is likely that in this source most or all nuclear fuel is consumed by steady helium burning during considerable time intervals, probably related to the high mass accretion rate in this source (cf. Van Paradijs et al. 1979).

According to steady-state spherically symmetric models of thermonuclear flashes there exists a lower limit to the accumulated mass necessary to ignite flashes. This limit depends on M, and on the chemical composition of the flashing layer, and it is expected that there is a lower limit to the burst energy and the maximum luminosity of bursts (the latter limit is $\sim 0.6$ times the Eddington limit for hydrogen-rich material; see Hanawa and Fujimoto 1984). About half of the bursts from 1636-53 have peak luminosities below this lower limit. Such low peak luminosities can only be understood if weak flashes occur at relatively low pressure in the burning shell. Fujimoto et al. (1987a) have argued that this requires: (i) a reservoir of nuclear fuel which can survive the flashes, (ii) inward transportation and mixing of this "left-over" fuel, e. g., due to hydrodynamic instabilities related to the distribution of angular momentum of the accreted matter in the envelope (Fujimoto et al. 1987b).

In addition to providing a possible explanation for the so-called "dwarf bursts" in 1636-53 this mechanism may also account for the rather large deviations from a "smooth" $E_b$-$\Delta t$ relation, and possibly resolve the old problem of X-ray bursts with very short waiting times (as small as $\sim 5$-10 minutes) during which the accretion cannot account for the required fuel replenishment by a large margin. Fujimoto et al. (1987a) estimated the size of the fuel reservoir in 1636-53 at $\sim 9 \times 10^{38} f^{-1}$ erg (where $f$ denotes the fraction of the reservoir that contributes to the fluctuations in burst energy). The corresponding value for 0748-676 is $\sim 2 \times 10^{38} f^{-1}$ ergs (Gottwald et al. 1987).

The double- and triple-peaked bursts observed from 1636-53 (Sztajno et al. 1985; Van Paradijs et al. 1986) may be the result of two or three independent thermonuclear flashes, which occur at intervals of $\sim 10$ s. [Two bursts with an interval of 23 sec were recently observed with Ginga from the Galactic Center region; it cannot be excluded that these bursts originate from different sources (Tanaka 1987)]. As shown by Penninx, Van Paradijs and Lewin (1987) the observed correlation between $E_b$ and $\Delta t$ makes it very unlikely that the triple-peaked burst from 1636-53 is the result of scattering by a transient burst-induced accretion disk corona (Melia 1987).

The observations of 0748-676 are of particular interest, as bursts were observed over a large range in persistent flux $F_p$. The properties of these bursts showed a marked dependence on $F_p$: (i) The burst frequency was anti-correlated with $F_p$, varying between $\sim 0.75$ hr$^{-1}$ at $F_p \sim 3 \times 10^{-10}$ erg/cm$^2$s, and $\sim 0.12$ hr$^{-1}$ at $F_p \sim 10^{-9}$ erg/cm$^2$s. (ii) Over the above range in $F_p$ the average burst fluence, $E_b$, increased by a factor of $\sim 2$, the maximum burst flux, $F_{max}$, increased by a factor $\sim 10$, and the average burst duration $\tau$ [$= E_b / F_{max}$] decreased from $\sim 25$ s to $\sim 6$ s. (iii) The average ratio $\alpha$ of the total energy emitted in the persistent flux to that emitted in bursts increased strongly with $F_p$, from ~12 at $F_p$~ 3 $10^{-10}$ erg/cm$^2$s, to ~ 75 at $F_p$>$10^{-9}$ erg/cm$^2$s.

The positive correlation between $\alpha$ and $F_p$ for 0748-676 is in qualitative agreement with the above global picture, in which at very low accretion rates the helium flash occurs in a hydrogen rich environment, and hydrogen provides a major contribution to the burst energy. The anti-correlation between $\tau$ and $F_p$ for 0748-676 is consistent with this picture. It is unclear how the anti-correlation between burst frequency and the persistent flux fits into this picture; a similar anti-correlation has also been observed in some other sources (e.g., 1659-29: Lewin *et al.* 1978; 1820-30: Clark *et al.* 1977), but not in others (e.g., 1608-52: Murakami *et al.* 1980).

Following the above results of Gottwald *et al.* (1986), Van Paradijs, Penninx & Lewin (1988) attempted to attain an overall observational picture of the possible relation of burst properties ($\alpha$ values, and burst duration) with the mass accretion rate, through a search in the literature. The basic assumptions underlying their analysis are: (i) The maximum luminosity of X-ray bursts which show radius expansion is a standard candle. The justification for this assumption is the theoretical evidence that during the radius expansion the peak luminosity in a burst equals the Eddington limit (Kato 1983; Ebisuzaki, Hanawa and Sugimoto 1983; Paczynski 1983; Paczynski and Anderson 1986). Although this limit depends on the mass of the neutron star and on the composition of the envelope, it is likely that for most burst sources it has a similar value within a factor ~ 2 (in lecture #3 no such assumptions are made). (ii) They also made the assumption that the ratio of the amount of anisotropy during the burst and the persistent emission is the same for all sources (this ratio does not have to be 1.0, though that may be the case). Then the ratio $\gamma$ of the persistent flux, $F_p$, and the *net* peak flux, $F_{re}$, (not including the persistent emission during the burst) of bursts with radius expansion is proportional to the persistent luminosity (the latter should be a measure of the mass accretion rate).

The results of this study are summarized in Fig. 7, which displays the variation of log $\alpha$ (and log $\tau$) with log $\gamma$, averaged over time intervals of typically >1 day (not necessarily consecutive), during which the persistent flux is approximately constant. Clearly, both log $\alpha$ and log $\tau$ are strongly correlated with log $\gamma$, in the same sense as found by Gottwald *et al.* (1986) for 0748-676, but over a substantially larger range in $\gamma$.

The burst duration $\tau$ progressively decreases as the persistent luminosity increases (over the range corresponding to -2 < log $\gamma$ < -0.6), indicative of the decreasing importance of hydrogen in the energetics of the thermonuclear flashes. There is no evidence for a clear separation of bursts into two groups, one with long-, and another with short durations; it is therefore not too useful to make a distinction between different burst modes, such as "fast" and "slow" (cf. Murakami *et al.* 1980).

It is premature, however, to consider these results as good support for the global picture of X-ray bursts as presented by FHM. The reason is that the variation of log $\alpha$ with log $\gamma$ is very strong. A least-squares fit to the data in the range -1.7 < log $\gamma$ < -1.0 yields log $\alpha$ = 3.34 ($\pm$0.29) + 0.95 ($\pm$0.21) log $\gamma$. Outside the above log $\gamma$ interval the variation of log $\alpha$ with log $\gamma$ is even stronger. In other words, the ratio $\alpha$ / $\gamma$, which is proportional to the time-averaged burst luminosity, is approximately constant, independent of the persistent luminosity. In fact, the standard deviation of the distribution of values of log ($\alpha$ / $\gamma$), for -1.7 < log $\gamma$ < -1.0, equals only $\pm$0.19. Since the uncertainties in the values of log $\alpha$ and log $\gamma$ are not very different, it may be more appropriate to represent the average relation between log $\alpha$ and log $\gamma$ by the long axis of the dispersion ellipse of a bi-variate normal

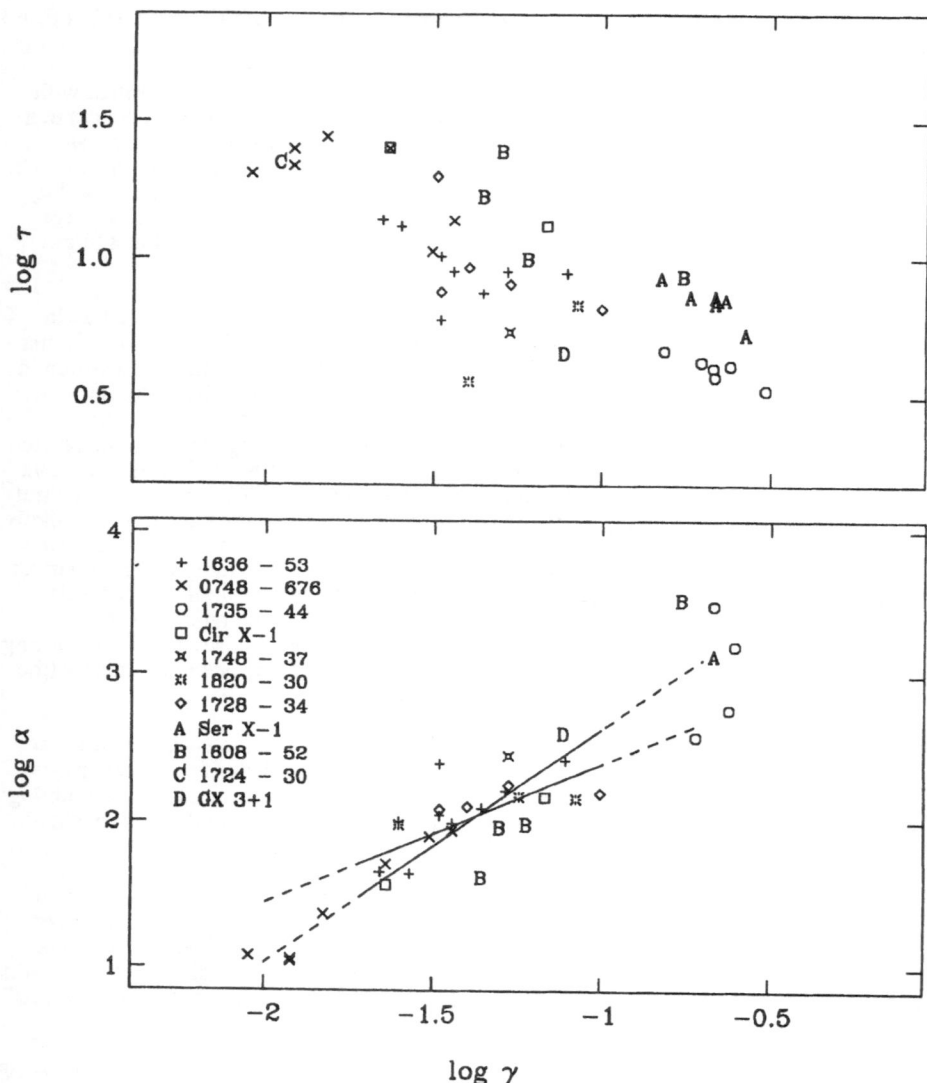

Figure 6.    *Relations between log τ and log γ (top panel) and between log α and log γ for a collection of burst sources which have produced X-ray bursts which caused photospheric radius expansion. Different symbols represent different sources (see inset). We show two straight lines in the bottom panel. One represents a least-squares fit to the data in the interval -1.7 < log γ < -1.0; the other (the steepest of the two) is the long axis of the dispersion ellipse of a bivariate (log α, log γ) distribution of the same data points.*

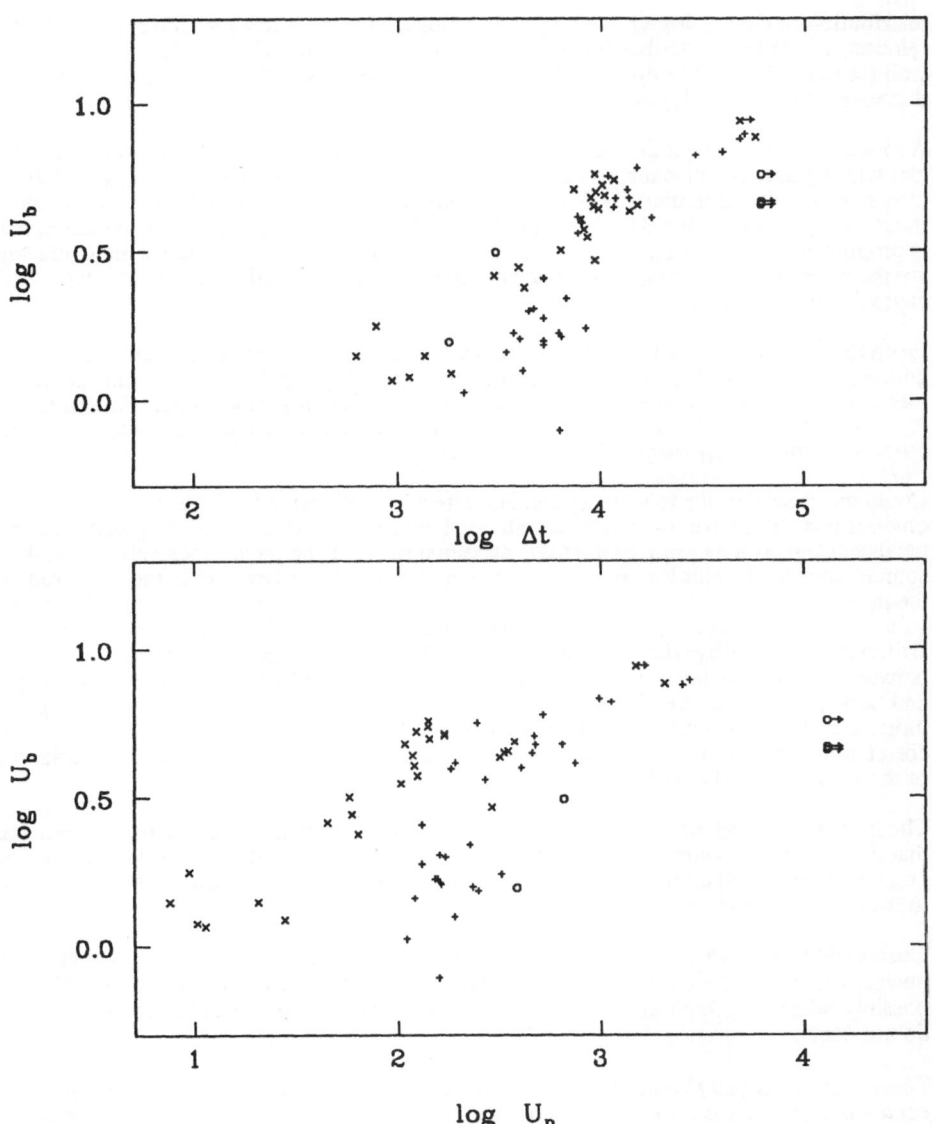

Figure 7.    *Dependence of the quantity, $U_b$, on the waiting time, $\Delta t$, since the previous burst (top panel) and between $U_b$, and $U_p$ (= γ $\Delta t$) (bottom panel), for 0748-676 (×), 1636-53 (+), and 1735-44 (○), with log γ values ranging between -2 and -0.6. Clearly, the ($U_b$,$\Delta t$) relation is much better defined than the ($U_b$,$U_p$) relation. This indicates that, independent of the persistent X-ray luminosity (which should be a measure of the mass-accretion rate), after a given waiting time, $\Delta t$, a burst source apparently produces an X-ray burst with always approximately the same energy.*

distribution of (log $\alpha$, log $\gamma$), which yields a slope d log $\alpha$ / d log $\gamma$ = 1.60 ± 0.35. This relation, which also describes the data points outside the interval -1.7 < log $\gamma$ < -1.0 quite well (see Fig. 7), would imply that the average luminosity which is emitted in bursts, decreases (quite strongly) as the persistent luminosity increases.

An instructive way to get this point across is to compare the relation between burst energy and waiting time $\Delta t$ since the previous burst, with that between the burst energy and the integrated persistent luminosity during the interval $\Delta t$. With the above assumptions, the burst energy is proportional to the quantity $U_b = E_b/F_{re}$; the integrated persistent luminosity is proportional to $U_p = \gamma \Delta t$ [the two constants of proportionality have the same value, and are the same for all sources; they are the reciprocal of an arbitrarily chosen "standard candle" luminosity].

From Fig. 8 it appears that over a large range in $\gamma$ the ($U_b$, $\Delta t$) relation is remarkably unique, whereas the ($U_b$, $U_p$) plot resembles a scatter diagram. If the persistent flux is a measure of the accretion rate, *this would lead to the rather surprising conclusion that, independent of the accretion rate, after a given waiting time $\Delta t$ a source produces an X-ray burst with always approximately the same energy.*

Could this result be due to some systematic effect, not related to the burst mechanism? An obvious possibility to explain a linear relation between $\alpha$ and $\gamma$ is a misinterpretation of the persistent flux as a measure of the mass accretion rate. It is perhaps conceivable that all sources accrete more or less at the same rate, and that large differences in the anisotropy of the persistent X-ray emission (but not of bursts!) cause large differences in the observed values of $F_p$. This would give rise to a linear relation between $\alpha$ and $\gamma$. This is rather artificial, but it could perhaps explain part (not all) of this strange result. The relation between the average burst duration and $\gamma$ (see Fig. 7), and the change in burst frequency and burst pattern observed in some sources, when the *observed* persistent flux changed, indicate that $\gamma$ is, at least partly, a measure of M. This also demonstrates that the correlations between $\alpha$ and $\gamma$, and between $\tau$ and $\gamma$, can not be due to systematic variations of the mass (and radius) of the neutron stars, depending on $\gamma$.

The large range over which $\alpha$ varies ( about two orders of magnitude; see Fig. 7) indicates that this variation cannot be ascribed to only a $\gamma$-dependent composition of the nuclear fuel, i.e., a hydrogen-rich environment vs. pure helium, as this would accommodate a variation in $\alpha$ of at most a factor of ~7.

Observations of 1735-44 provide clear evidence for the continuous burning of most or all nuclear fuel for extended time intervals (Lewin *et al.* 1980; Van Paradijs *et al.* 1988), possibly related to a high mass accretion rate, close to a critical value above which bursts do not occur.

*These results suggest that continuous burning of a sizeable fraction of the nuclear fuel occurs in less luminous sources as well; this fraction apparently is a gradually increasing function of the mass accretion rate.*

Finally, it may be noted that the correlations of $\alpha$, and $\tau$, with $\gamma$ may - in principle - be used for a statistical study of distances of X-ray burst sources from measurements of $\alpha$, $\tau$, and the persistent flux.

## LECTURE #3 - MASS-RADIUS RELATION OF NEUTRON STARS

### The Mass of Neutron Stars

There are several methods to derive the mass of neutron stars which, however, do not provide information on the star's radius (see, e.g., Joss and Rappaport 1984; Taylor & Weisberg 1982). To obtain information on the equation of state, one needs information on both the mass and the radius.

### Radii of Neutron Stars

#### *X-ray Bursts (part 1)*
It was first shown by Swank *et al.* (1977) that the spectrum of an X-ray burst (observed with OSO-8) was well described by that of a blackbody with a temperature (kT) varying between 0.9 and 2.3 keV. Hoffman, Lewin and Doty (1977a,b) confirmed this result for bursts observed with SAS-3 from MXB 1728-34 and MXB 1636-53. The blackbody interpretation of X-ray burst spectra was supported by the fact that during burst decay the observed flux F∞ varied approximately as the 4th power of the observed blackbody temperature $T_{c\infty}$ (the subscript *c* indicates "color" temperature; the subscript ∞ indicates that the observation is made by a distant observer). This implies that during a burst one observes an approximately constant burst emitting area. For an assumed homogeneous spherical blackbody emitter of radius $R_{\infty}$, at a distance d, we have:

$$L_{b\infty}=4\pi(R_{\infty})^2\sigma(T_{e\infty})^4 = 4\pi d^2 F_{b\infty} \qquad (1).$$

Here L stands for luminosity, and the subscripts *b* and *e* stand for "burst" and for "effective", respectively. If one knows an approximate source distance, the values of $R_{\infty}$ then follow from the flux measurements $F_{b\infty}$, and the associated observed blackbody temperatures (assuming that $T_{c\infty} \approx T_{e\infty}$; however, see below). For a distance of ~10 kpc, Swank (1977) and Hoffman, Lewin and Doty (1977a,b) found radii, $R_{\infty}$, roughly comparable to those expected for a neutron star.

The thermonuclear flash model for X-ray bursts was introduced by Woosley and Taam (1976) and Maraschi and Cavaliere (1977). However, the discovery of the Rapid Burster by Lewin *et al.* (1976) made this model unattractive as it was very clear from the Rapid Burster data that the abundant bursts were due to spasmodic accretion, and not due to thermonuclear energy.

The thermonuclear flash model was revived in the fall of 1977 when two different types of burst (type 1 and type 2) were observed from the Rapid Burster by Hoffman, Marshall and Lewin (1978). They proposed that the type 1 bursts, which occurred only once per few hours, were the result of thermonuclear flashes on the surface of a neutron star, and that the abundant type 2 bursts were due to spasmodic accretion. Subsequently, Joss (1978) demonstrated convincingly that thermonuclear flashes are responsible for the type 1 bursts.

Van Paradijs (1978) introduced the idea that the peak luminosity of an average X-ray burst might be a "standard candle", and he found that for an assumed average source distance of 10 kpc, and an assumed standard candle equal to the Eddington limit of a 1.4 $M_{\odot}$ object (with a hydrogen rich envelope), the radii of all sources were ~7 km. The standard candle

240

idea may turn out to be correct but if so, only for the brightest bursts that produce radius expansion of the photosphere of the neutron star (see e.g. Lewin 1984, and this lecture). It has also become clear that there are serious systematic errors in many radius determinations as the burst spectrum is not exactly that of a blackbody but only resembles one (Van Paradijs 1982; Czerny and Sztajno 1983; London, Taam and Howard 1984, 1986; Ebisuzaki and Nomoto 1986; Foster, Ross and Fabian 1986); as a result the observed color temperature does not equal the effective temperature as given in eq. (1). Van Paradijs (1979) showed that in principle the type 1 burst data provide information on the mass and radius of the neutron star. Since that time, the mass and radius determination of neutron stars, using the burst properties, has been at the focal point of many investigations. In this section we summarize our present knowledge on this issue, and we will emphasize the limitations (often overlooked) of the available methods.

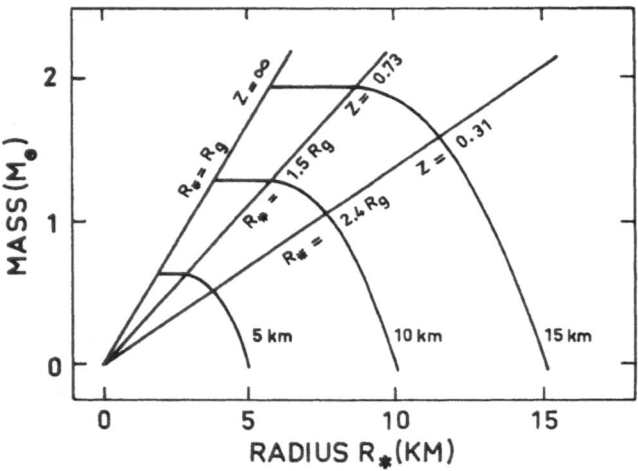

Figure 8.    Mass-radius relations for three imagined observed values of the blackbody radius, $R_\infty$ (5, 10, and 15 km). For clarity, we have not indicated error regions resulting from the uncertainties in the measurements. The straight lines indicate radii, $R_*$, equal to the Schwarzschild radius, $R_g$, 1.5 $R_g$, and 2.4 $R_g$, respectively. The latter could e.g. be the result of a burst with radius expansion (see Method I ), or of the determination of the gravitational redshift of an observed spectral feature (see text). For a given mass, the observed blackbody radius, $R_\infty$, has a minimum value of $1.5\sqrt{3}\, R_g$; conversely, for a given observed value of $R_\infty$ the mass cannot be larger than $R_\infty/(7.68\ km)\, M_\odot$.

## Mass-Radius Relation of Neutron Stars

### Gravitational Redshift - Emission/Absorption - Line/Edge

If a discrete feature is present in a spectrum during times that the radiation comes from the neutron star surface, and if this feature can be identified, one has a direct measurement of the gravitational redshift $1+z_*$ of the neutron star surface and thus of $M/R_*$ (see eq. 3). (The surface of the neutron star is indicated with a *).  This limits the allowed values in the M-$R_*$ diagram to a straight line through the origin (see Fig. 8).  Lines at 4.1±0.1 keV have been reported in bursts from 1636-53 (Waki *et al.* 1984), 1608-52 (Nakamura *et al.* 1988), and 1747-21 (Magnier *et al.* 1988). If the 4.1 keV line is due to helium-like iron, as suggested by Waki *et al.* (1984), $1+z_* \approx 1.6$; this seems high, but perhaps not impossible. Magnier *et al.* (1989) have shown that the observed equivalent line widths would require that the iron abundance is at least $\sim 10^2$ times higher than the cosmic abundance; this casts some doubt on the iron-line interpretation. An alternative explanation given by Fujimoto (1985) can now be dismissed as it requires a special geometry which could not by chance occur in three systems (Magnier *et al.* 1989).

Emission lines have also been observed during $\gamma$-ray bursts which have been interpreted as gravitationally redshifted 511 keV anihilation lines emitted from the surface of neutron stars (Mazets *et al.* 1981; for a review see Liang 1986). If these interpretations are correct, the redshift factors are $\sim 1.25 - 1.35$.

### *X-ray Bursts (part 2)*
Equation (1) could be used to reliably calculate the observed radius, $R_\infty$, if:
1. The effects of General Relativity could be omitted.
2. The emission were isotropic.
3. The burst spectrum were that of a blackbody (i.e. Planckian).
4. The source distance were known.
We will discuss each of these points separately.

Because of gravitational redshift (see e.g., Thorne 1977), the luminosity and blackbody temperature as observed locally at the surface of the neutron star (indicated by a subscript $*$) and those observed by a distant observer (subscript $\infty$ ) are related as follows:

$$L_{b\infty} = L_{b*}\,(1+z_*)^{-2} \tag{2a}$$

$$T_{e\infty} = T_{e*}\,(1+z_*)^{-1} \tag{2b}$$

with $1+z_* = (1 - 2GM/R_*c^2)^{-0.5}$ 　　　　　(3).

Here M is the gravitational mass of the neutron star, $R_*$ is the radius of the neutron star as observed by a local observer on the surface of the neutron star. For isotropic radiation, and a Planckian spectrum, this local observer could derive a burst luminosity $L_{b*}$, and an effective temperature $T_{e*}$ according to:

$$L_{b*} = 4\pi(\,R_*)^2\sigma(T_{e*})^4 \tag{4}.$$

Using the above relations, we find that

$$R_\infty = R_*(1+z_*) = R_*(1 - 2GM/R_*c^2)^{-0.5} = R_*(1 - R_g/R_*)^{-0.5} \tag{5},$$

$$R_g = 2GM/c^2 \approx 3(M/M_\odot)\ \text{km} \quad \text{(Schwarzschild radius)} \tag{6}.$$

These equations are strictly valid only for non-rotating neutron stars (Schwarzschild metric).

Equation (5) shows that an observation of the radius (by a distant observer) gives a relation between the mass M and the radius $R_*$ of the neutron star (Van Paradijs 1979).

It follows from the Schwarzschild metric (this is independent of the equation of state for neutron star matter) that $R_\infty$ has a minimum possible value when $R_* = 1.5\ R_g$. Then:

$$R_{min\infty} = (1.5\sqrt{3})R_g \approx 7.68\ (M/M_\odot)\ km \qquad\qquad (7).$$

For a 1.4 $M_\odot$ object, this minimum value is ~10.7 km (Van Paradijs 1979; Paczynski 1983).

For isotropic radiation (such as blackbody emission) eq. (5) does not hold when $R_* < 1.5$ $R_g$. This results from the fact that, when $R_* < 1.5\ R_g$, a fraction of the photons fall back to the neutron star surface; the number of photons observed by a distant observer is thus reduced (Misner, Thorne and Wheeler 1973; see also Van Paradijs 1979). The net result is that then

$$R_\infty = (1.5\sqrt{3})R_g \approx 7.68(M/M_\odot)\ km \qquad (R_g < R_* < 1.5\ R_g) \qquad\qquad (8).$$

When this is the case, eq. (2b) still holds, but (2a) does not. The radii $R_*$ of neutron stars are very likely larger than 1.5 $R_g$, we therefore do not have to "worry" too much about this possibility.

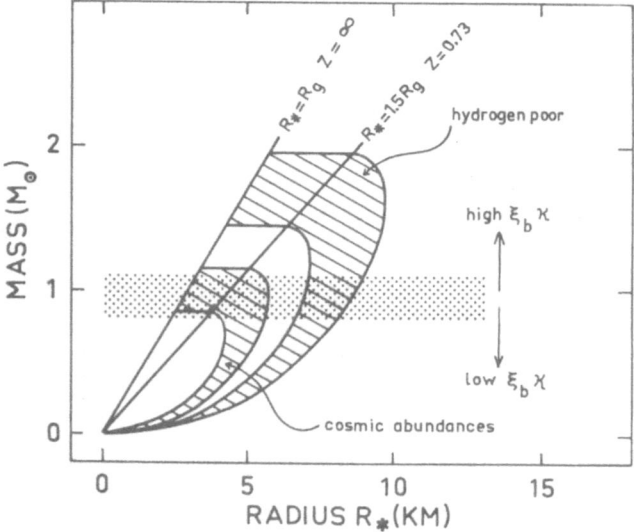

Figure 9.    *Mass-radius relation resulting from the observation of a burst with radius expansion (see Method II of the text). We have assumed, for the purpose of illustration only, that the quantity A of eq. (13) is in the range (0.85-1.15) $10^{-22}$ cm$^4$/g$^2$ (see text). The allowed regions in this diagram for assumed hydrogen-poor ($\kappa \approx 0.2$ cm$^2$/g), and cosmic compositions ($\kappa \approx 0.34$ cm$^2$/g), are indicated by the hatched areas. The horizontal band (shaded), indicates schematically the constraints on the mass obtained from a source with known distance which produces a burst with radius expansion (see text). The width of the band is due to the uncertainties. The position of the band depends both on $\kappa$, and on the anisotropy, $\xi_b$, as schematically indicated.*

In Fig. 8 we show M-R$_*$ diagrams for four values (5, 10, 15 and 20 km) of an observed value R$_\infty$ of the neutron star radius. For a given value of M, R$_\infty$ has a minimum (eq. 7). Conversely, for a given observed value of R$_\infty$, the mass of the neutron star can not be larger than (R$_\infty$/7.68) M$_\odot$.

In the above, it has been assumed, for simplicity, that the source distance $d$ (see eq. 1) is known, and that the radiation is emitted isotropically. In general, distances to burst sources are not accurately known; only the distances to the burst sources 1820-30 (in NGC 6624), and 1746-37 (in NGC 6441) are reasonably well known (Vacca, Lewin and Van Paradijs 1986; Haberl *et al.* 1986; Van Paradijs and Lewin, 1987; Sztajno *et al.* 1987). We will return to this issue.

*Spherical Symmetry - Anisotropy.*
Only if the burst radiation is isotropic can we make the conversion from flux to luminosity (eq. 1). Even if the burst emission were initially isotropic, it is quite possible that at large distances from the surface the emission is not isotropic e.g., due to the presence of an accretion disc or an accretion disc corona (see e.g., Lapidus and Sunyaev, 1985; Fujimoto 1988). The anisotropy depends on the properties of the corona and of the disc (e.g. its thickness) and on the inclination angle of the system; unfortunately, our knowledge on anisotropy in low-mass X-ray binaries is very poor. Following Sztajno *et al.* (1987) an anisotropy factor $\xi_b$ is introduced. Thus (see eq. 1),

$$L_{b\infty}=4\pi(R_\infty)^2\sigma(T_{e\infty})^4 = 4\pi d^2\,\xi_b\,\varepsilon_x\,F_{b\infty} \tag{9}.$$

We have also introduced here the bolometric correction factor $\varepsilon_x$. This takes into account that not all energy is observed in the X-ray energy band. The values of $\varepsilon_x$ are in general very close to 1 (but always >1); they vary throughout the burst as the spectral shape changes. Possible absorption of the burst radiation is also included in $\xi_b$. If the observed flux is lower than it would be in the absence of anisotropy (or absorption), $\xi_b$ >1. If the observed flux is higher, $\xi_b$ <1. In what follows we assume that $\xi_b$ is constant during the burst; this is not necessarily so, as the anisotropy could, in principle, change during the burst (Melia 1987; however, see also Penninx, Van Paradijs and Lewin 1987).

In order to find R$_\infty$ (eq. 9), we need to know $d^2\xi_b$. *As we will now discuss, $d^2\xi_b$ can be eliminated for bursts which cause a distinct radius expansion.*

*Method I.* During very strong bursts, the neutron star photosphere expands as a result of radiation pressure. According to model calculations, during the expansion and contraction phase of the photosphere, the luminosity remains to within a few percent of the Eddington luminosity (Kato 1983; Ebisuzaki, Hanawa and Sugimoto 1983; Paczynski 1983; Paczynski and Anderson 1986). During the expansion and contraction phase, at photospheric radius, R, the Eddington luminosity as observed by a distant observer is:

$$L_{Edd\infty}=(4\pi cGM/\kappa)(1- 2GM/Rc^2)^{+0.5} = 4\pi d^2\,\xi_b\,\varepsilon_x\,F_{Edd\infty} \tag{10}.$$

Here, $\kappa$ is the electron scattering opacity *during the expansion phase* (it is ~0.34 cm$^2$/g for matter with a cosmic abundance, and ~0.2 cm$^2$/g for hydrogen-poor matter). It is likely that the atmosphere *during expansion* is hydrogen-poor (Sugimoto, Ebisuzaki and Hanawa 1984). Notice that R is not R$_*$, but rather the radius of the photosphere which, at maximum expansion can be many times the stellar radius (Lewin, Vacca and Basinska, 1984; Tawara et al 1984; Vacca, Lewin and Van Paradijs 1986; Haberl *et al.* 1986). If the flux is measured during that part of the expansion phase where R$\gg$R$_*$, the gravitational redshift factor in eq. (10) is 1.0 to good approximation, and we find:

$$L_{Edd\infty} \approx 4\pi cGM/\kappa = 4\pi d^2 \, \xi_b \, \varepsilon_x \, F_{Edd\infty} \qquad (R \gg R_*) \qquad (11a).$$

We note that this simplification of negligible gravitational redshift (eq. 11a) is not strictly necessary (see e.g., Fujimoto and Taam 1986; Sztajno et al. 1987; Van Paradijs and Lewin 1987).

When the radius of the photosphere *just* at the end of the contraction phase is again $R_*$, the luminosity is still Eddington limited, and we have:

$$L_{Edd\infty} = (4\pi cGM/\kappa)(1 - 2GM/R_*c^2)^{+0.5} = 4\pi d^2 \, \xi_b \, \varepsilon_x \, F_{Edd\infty} \qquad (R=R_*) \qquad (11b).$$

One should note that $F_{Edd\infty}$ in eq. (11a) is the *observed* Eddington flux when the radius expansion is large, whereas $F_{Edd\infty}$ in eq. (11b) is the *observed* Eddington flux when the photospheric radius is $R_*$; a measurement of these two values of $F_{Edd\infty}$ leads immediately to the value for $M/R_*$ (by dividing the two equations, one eliminates $d^2\xi_b$; see also Paczynski and Anderson 1986). [We have assumed here that the anisotropy remains constant throughout the burst; this may not always be the case (see e.g. Melia 1987); the values of $\varepsilon_x$ are not the same during both measurements, however, they can be calculated from the observed spectrum]. A measurement of $M/R_*$ limits the allowed values in the M-$R_*$ diagram to a straight line through the origin.

*This method has, in principle, the great attraction that it is not plagued by our limited knowledge of how to convert color temperature to effective temperature (see below); this conversion plays no role here.*

Unfortunately, in practice this method does not give very accurate results (yet) as it is very difficult to derive an accurate value for the Eddington flux at the very moment that the photospheric radius has just contracted to $R_*$.

*Method II.* There is another way to get *different* information about the mass and radius by combining the observations during the expansion/contraction phase (eq. 11a) with those during the cooling phase (eq. 9); this way we can also eliminate $d^2\xi_b$. Again, this elimination is only possible if the anisotropy does not change during the burst. We then find with eq. (5):

$$(F_{b\infty}/F_{Edd\infty}) = R_*^2(1 - R_g/R_*)^{-1} \, \sigma(T_{e\infty})^4 \, (\kappa/cGM) \qquad (12).$$

"Observables" are: $F_{b\infty}$, $F_{Edd\infty}$, and $T_{c\infty}$ (we will discuss below the conversion from the observed color temperature to the effective temperature $T_{e\infty}$); the values for $\varepsilon_x$ in eqs. (9) and (10) in principle differ (different spectral shapes) and each varies throughout the burst, but the values can be reliably calculated from the spectral fittings; we have therefore deleted them from eq. (12).

*Eq. (12) is a relation between the mass M of the neutron star and its radius R\* independent of the source distance and the anisotropy of the burst emission.*

A measurement of the three observables (see above) then leads to a measurement of the quantity A:

$$A = R_*^2 \, (1 - 2GM/c^2R_*)^{-1} \, (\kappa/M) \qquad (R_* \geq 1.5 \, R_g) \qquad (13).$$

For $R_* < 1.5 R_g$ (see eq. 8), the quantity A becomes:

$$A \approx (7.68 M/M_\odot \; km)^2 \; (\kappa/M) \qquad\qquad (R_* < 1.5 R_g) \qquad\qquad (14).$$

For a given value of A, M is independent of $R_*$ (this was overlooked by Fujimoto and Taam 1986). In figure 9 we show in the M-$R_*$ diagram two curves for two different values of $\kappa$ (0.2 and 0.34 $cm^2$/g), for a typical observed value for A = $10^{-22}$ ($\pm$15%) $cm^4/g^2$.

Eq. (13) (in which the distance is absent), is essentially an expression of the unchanging angular size of the burst source (i.e., $R_\infty$/d). Thus, as one moves along one of the curves of Fig. 9, the angular size of the burst source remains constant, but $R_\infty$ and d change; this is not the case for the curves in Fig. 8.

During the decay phase of a burst which exhibits radius expansion (only one such burst per source is required), one can make several measurements of both the flux and the associated blackbody temperature. For blackbody radiation from a spherical object with a constant radius, $F_{b\infty}/(T_{e\infty})^4$ remains constant. Therefore, unless there are complications (see below), one derives the average value for several measurements during burst decay, and this value with its associated error is used to construct allowed regions in the M-$R_*$ diagram (Sztajno *et al.* 1987; Van Paradijs and Lewin 1987).

*Conversion from Color Temperature to Effective Temperature.*
The spectra emitted by hot neutron star atmospheres are not Planckian (Van Paradijs 1982; Czerny and Sztajno 1983; London, Taam and Howard 1984, 1986; Ebisuzaki and Nomoto 1986; Foster, Ross and Fabian 1986). Therefore, the observed color temperature $T_{c\infty}$ does not equal $T_{e\infty}$ as given in eq. (1). In spite of much theoretical work, the dependence between $T_c$ and $T_e$ is poorly understood.

There are cases where the observed quantity $F_{b\infty}/(T_{c\infty})^4$ remains nearly constant throughout the burst decay (Sztajno et al 1987; Van Paradijs and Lewin 1987). This clearly indicates that in those cases the percentage change in $T_{c\infty}$ as observed in the cooling phase during burst decay must be associated with an approximately equal percentage change in $T_{e\infty}$. The two could e.g. be linearly related; a somewhat more complicated relation is, of course, also possible as long as the percentage changes are approximately equal. Sztajno et al. (1987) have made an approximation using the numerical results of London, Taam and Howard (1986) for surface gravities of $10^{14}$ and $10^{15}$ $cm/sec^2$, in the range $0.05 < L/L_{Edd} < 0.85$. They find:

$$T_c/T_e \approx 1.5 \; (L/L_{Edd})^{0.04} \qquad\qquad (15).$$

Since this was derived from a fit to model calculations, the two luminosities have to be taken at the same radius (i.e., gravitational redshift). Since the observations combine fluxes during the cooling phase (R=$R_*$) and during the expansion phase (R$\gg R_*$), we have:

$$L/L_{Edd} = (F_{b\infty}/F_{Edd\infty}) \; (1-R_g/R_*)^{-0.5}, \text{ whence}$$

$$T_{c^*}/T_{e^*} \; (= T_{c\infty}/T_{e\infty}) \approx 1.5 \; (F_{b\infty}/F_{Edd\infty})^{0.04} \; (1- R_g/R_*)^{-0.02}.$$

This then leads to simple modifications of eqs. (4) and (9), and eq. (12) (which contains

$T_{e\infty}$), is replaced by eq. (16) *which contains the actual observable* $T_{c\infty}$:

$$(F_{b\infty}/ F_{Edd\infty})^{1.16} \approx (1.5)^{-4} R_*^{2}(1- R_g/R_*)^{-0.92} \sigma(T_{c\infty})^4 (\kappa/cGM) \qquad (16).$$

Thus the functional dependence of M and $R_*$, and the value of A in eq. (13) are now different as a result of the conversion from $T_{c\infty}$ to $T_{e\infty}$; this changes the curves of Fig. 9.

Using the approximation of eq. (15), for values of $F_{b\infty}/F_{Edd\infty}$ of 1.0, 0.5, and 0.25 respectively, the ratios $T_c/ T_e$ would be ~1.5, ~1.46, and ~1.42. The ratio is therefore approximately constant throughout the burst decay phase. This reflects our earlier remark that in at least two sources (Sztajno et al 1987; Van Paradijs and Lewin 1987) the two could be linearly related ($T_c \approx 1.5\ T_e$).

There exists evidence that the relation between $T_c$ and $T_e$ as expressed by eq. (15) does not hold in general, and this complicates matters enormously (Gottwald et al. 1986; Matsuoka 1986; Penninx, et al. 1989; Damen *et al.* 1989).

*If the observed color temperature, Tc, can be reliably converted into the effective temperature, Te, the method (II) described above is very powerful. However, for the time being, the largest uncertainty in the M-R\* determinations is probably due to the uncertainty in this conversion. More theoretical work is needed.*

### Reliable Distances
Reliable distances for burst sources (with radius expansion) are only available for 1820-30, in NGC 6624, and for 1746-37, in NGC 6441 (Vacca, Lewin and Van Paradijs 1985; Sztajno et al. 1987). In these cases one finds directly the value of $M/(\kappa\ \xi_b)$, as the Eddington limit (at large photospheres when the gravitational redshift is negligibly small) measured by a distant observer does not depend on the star's radius (see eqs. 10 and 11a), and this provides horizontal slices through the M-$R_*$ diagram as schematically shown in Fig. 9 (see Sztajno et al. 1987; Van Paradijs and Lewin 1987).

### Conclusions
In the absence of any knowledge of the distance (and anisotropy) of a burst source, a range of allowed masses and radii of the neutron star for a given source can be measured from the X-ray data of a single burst which causes radius expansion of the photosphere. If, in addition, the source distance is approximately known, and/or a gravitationally redshifted spectral feature is observed, the mass and radius can be determined separately. The largest uncertainty at present results from our incomplete knowledge of the conversion from the observed color temperature to the effective temperature. More theoretical work is needed in this area.

## LECTURE #4 - THE RAPID BURSTER (MXB 1730-335)

The Rapid Burster, discovered by Lewin *et al.* (1976), is unique among the galactic X-ray sources in producing two very different kinds of X-ray bursts (type 1 and 2; Hoffman, Marshall and Lewin 1978). The type 1 bursts, which are most likely due to thermonuclear flashes on the surface of an accreting neutron star (Woosley and Taam 1976; Maraschi and Cavaliere 1977; Hoffman, Marshall and Lewin 1978; Joss 1978), are characterized by a distinct spectral softening as they decay; they have been observed from another ~35 sources (for reviews see Lewin and Joss 1981,1983). The type 2 bursts do not show such a spectral evolution; they show a considerable range in duration (from a few seconds to ~12 minutes), and in recurrence interval (from ~10 seconds to ~1 hour). They behave like a relaxation oscillator: the time interval to the next burst is roughly proportional to the integrated flux in the previous burst (Lewin *et al.* 1976). Attempts have been made to describe this recurrence behaviour of the Rapid Burster in terms of deterministic chaos (Celnikier 1977).

The Rapid Burster is a recurrent transient with intervals of about half a year (Lewin 1977; Grindlay and Gursky 1977). During its active periods it has shown a variety of combinations of burst behaviour and persistent flux levels. At times only type 2 bursts are observed, with high or low level persistent emission between them, at other times both type 1 and type 2 bursts are observed, and there have also been periods when strong persistent emission and only type 1 bursts are observed (Lewin and Joss 1981,1983; Lewin 1985; Stella *et al.* 1988a). The Rapid Burster is located in the heavily reddened globular cluster Liller 1 (Liller 1977), at a distance of ~10 kpc (Kleinmann *et al.* 1976).

It is likely that the type 2 bursts are the result of instabilities in the accretion flow onto the neutron star. Well established models for the accretion instability can be divided into two classes, involving either a gating mechanism associated with the neutron star magnetosphere (Lamb *et al.* 1977; Baan 1977,1979; Michel 1977), or with viscous and thermal instabilities of the inner radiation-dominated region of the accretion disk (Taam and Lin 1984). Recently, Milgrom (1987) suggested that the instability occurs between the neutron star surface and the radius of the marginally stable orbits in a Schwarzschild metric (the possible relevance of this model for QPO has been discussed by Paczynski 1987).

During Hakucho observations of the Rapid Burster in 1979 pulsations with a frequency of ~2 Hz were detected in 2 out of 63 type 2 bursts (Tawara *et al.* 1982). Since the frequency measured during the two bursts differed by ~1%, Tawara *et al.* (1982) ruled out that the pulsations were the signature of the rotation of the neutron star.

During a 1983 *EXOSAT* observation (Barr *et al.* 1987) the Rapid Burster showed only persistent emission and type 1 bursts; no QPO were then detected, with upper limits to its rms variation of ~10%.

Two *EXOSAT* observations were made of the Rapid Burster in August 1985, lasting ~17 hours and ~13 hours, respectively. The time resolution of the fast timing data varied between ~1 msec and ~8 msec; energy resolved data were obtained at time resolutions of 156 msec and 312 msec (32 energy channels), 94 msec (8 energy channels), and 4 msec and 8 msec (4 energy channels). These observations revealed that the QPO behaviour of the Rapid Burster exhibits a complexity which matches that of its burst behaviour. For a detailed description of these results we refer the reader to Stella *et al.* (1988a,b).

There are some recent very interesting, unpublished results on the Rapid Burster. Some are from the Exosat databank (Tan *et al.*. 1989), and some came from observations with Ginga in August of 1988 (Dotani *et al.*. private communication). We refer the reader to the relevant publications which will probably appear by the end of 1989.

## LECTURE #5 - QUASI-PERIODIC OSCILLATIONS (QPO)

Quasi-periodic oscillations (QPO) have been observed during the past decade in the flux of many bright X-ray sources on time scales from $\sim 10^3$ sec to milliseconds. This area of research attracted lots of attention in 1984 when intensity-dependent QPO were found in GX 5-1 in the frequency range $\sim 20$-35 Hz (Van der Klis et al. 1985). Since then, a lot of observational and theoretical work has been done. Various forms of QPO have been identified, and a theoretical picture is beginning to evolve.

We refer the reader to the following reviews: Lewin, Van Paradijs & Van der Klis (1988); Van der Klis (1989); Lamb (1989).

### ACKNOWLEDGEMENTS

WHGLwas supported by the United States National Astronautical and Space Administration under grants NAS8-571, NAG8-674, and NSG 7643. He, and JvP, acknowledge support from NATO through grant RG0331/88. WHGL also acknowledges support from The Netherlands Foundation for the Advancement of Research, NWO.

### REFERENCES

Ayasli, S. & Joss, P.C. 1982, *Ap. J.*, **256**, 637,
Baan, W.A. 1977, *Ap. J.*, **214**, 245.
Baan, W.A. 1979, *Ap. J.*, **227**, 987.
Barr, P. et al. 1987, *Astr. Ap.*, **176**, 69.
Basinska, E. M., Lewin, W. H. G., Sztajno, M. *et al.* 1984, *Ap. J.*, **281**, 337.
Bhattacharya, D. & Srinivasan, G. 1986, *Current Sci.*, **55**, 327.
Blair, D. G. & Candy, B. N. 1988, preprint
Bradt, H.V. & McClintock, J.E. 1983, *Ann. Rev. Astr. Aph.*, **21**, 13.
Braes, L.L.E. & Miley, G.K. 1971, *Nature*, **232**, 246.
Celnikier, L.M. 1977, *Astr. Ap.*, **60**, 421.
Chanmugan, G. & Brecher, K. 1987, *Nature*, **329**, 696.
Chevalier, C. et al. 1988, *Astr. Ap.* (in press).
Clark, G. W., Li, F. K., Canizares, C.R. *et al.* 1977, *M. N. R. A. S.*, **179**, 651.
Cominsky, L. et al. 1978, *Ap. J.*, **224**, 46.
Corbet, R. H. D. 1984, *Astr. Ap.*, **141**, 91.
Corbet, R. H. D. 1986, in *Physics of Be Stars*, eds. A. Slettebak & T.P. Snow (Cambridge University Press), p. 63.
Cowley, A. P. *et al.* 1987, *Ap. J.*, **320**, 296.
Crawford, J. A. & Kraft, R. P. 1956, *Ap. J.*, **123**, 44.
Czerny, M. and Sztajno, M. 1983, *Acta Astron.*, **33**, 213.
Damen, E. *et al.* 1989, *M.N.R.A.S.* (in press).
Ebisuzaki, T., Hanawa, T. & Sugimoto, D. 1983, *Publ. Astr. Soc. Japan*, **35**, 17.
Ebisuzaki, T. & Nomoto, K. 1986, *Ap. J.*, **305**, L67.
Eggleton, P. P. & Pringle, J. E. 1985, *Interacting Binaries*, NATO ASI Series C. Vol. 150 (Reidel).
Filipov, L. & White, N. E. (eds) 1988, Proc. COSPAR/IAU Symposium *The Physics of Compact Objects: Theory versus Observations*, Sofia, July 1987, (Pergamon Press).

249

Foster, A. J., Ross, R. R. & Fabian, A. C. 1986, *M. N. R. A. S.*, **221**, 409.
Frank, J., King, A.R. & Raine, D. J. 1985, *Accretion Power in Astrophysics*, (Cambridge Univ. Press).
Fujimoto, M. Y. 1985, *Ap. J.*, **293**, L19.
Fujimoto, M. Y. 1988, *Ap. J.*, **324**, 995.
Fujimoto, M. Y., Hanawa, T. & Miyaji, S. 1981, *Ap. J.*, **247**, 267.
Fujimoto, M. Y. & Taam, R.E. 1986, *Ap. J.*, **305**, 246.
Fujimoto, M. Y. *et al.* 1987a, *Ap. J.*, **319**, 902.
Fujimoto, M. Y. *et al.* 1988, *Astr. Ap.*, **199**, L9.
Fushiki, I. & Lamb, D. Q. 1987, *Ap. J.*, **323**, L55.
Giacconi, R. et al. 1967, *Ap. J.*, **148**, L129.
Giacconi, R. et al. 1971, *Ap. J.*, **167**, L67.
Gottwald, M., Haberl, F., Parmar, A. N., & White, N. E. 1986, *Ap. J.*, **308**, 213.
Gottwald, M., Haberl, F., Parmar, A. N., & White, N. E. 1987, preprint.
Grindlay, J.E. & Gursky, H. 1977, *Ap. J. (Lett.)*, **218**, L117.
Haberl, F., Stellar, L., White, N. E. *et al.* 1987, *Ap. J.*, **314**, 266.
Hanawa, T. and Fujimoto, M. Y. 1984, *Publ. Astr. Soc. Japan*, **36**, 199.
Hasinger, G. 1988, *Astr. Ap.*, **186**, 153.
Hasinger, G. & Van der Klis, M. 1989, in preparation.
Henrichs, H. F. 1983, in *Accretion Driven Stellar X-ray Sources*, ed. W.H.G. Lewin & E.P.J. van den Heuvel (Cambridge University press), p. 393.
Hensberge, G., van den Heuvel, E. P. J. & Paes de Barros, M. 1973, *Astr. Ap.*, **29**, 69.
Hiltner W. A. & Mook D. E. 1970, *Ann. Rev. Astr. Ap.*, **8**, 139.
Hjellming, R.M. & Wade, C.M. 1971, *Ap. J.*, **168**, L21.
Hoffman, J. A., Lewin, W. H. G., & Doty, J. 1977a, *M. N. R. A. S.*, **179**, 57P.
Hoffman, J. A., Lewin, W. H. G., & Doty, J. 1977b, *Ap. J.*, **217**, L23.
Hoffman, J. A., Marshall, H., & Lewin, W. H.G. 1978, *Nature*, **271**, 630.
Jones, C. 1977, *Ap. J.*, **214**, 856.
Joss, P.C. 1978, *Ap. J.*, **225**, L123.
Joss, P. C. & Li, F. K. 1980, *Ap. J.*, **238**, 287.
Joss, P. C. & Rappaport, S. A. 1984, *Ann. Rev. Astron. Aph.*, **24**, 537.
Kato, M. 1983, *Publ. Astr. Soc. Japan*, **35**, 33.
Kirk, J. & Trümper, J. 1983, in *Accretion Driven Stellar X-ray Sources*, ed. W.H.G. Lewin & E.P.J. van den Heuvel (Cambridge University press), p. 261.
Kleinmann, D.E. *et al.* 1976, *Ap. J.*, **210**, L83.
Kraft, R. P. 1962, *Ap. J.*, **135**, 408.
Kraft, R. P. 1964, *Ap. J.*, **139**, 457.
Kraft, R. P. 1973, I.A.U. Symposium **55**, 36.
Kraft, R. P. & Miller, J.S. 1969, *Aph. J.* **155**, L159.
Krzeminski, W. 1974, *Ap. J.*, **192**, L135.
Kulkarni, S. 1986, *Ap. J.*, **306**, L85.
Kulkarni, S. & Narayan, R. 1988, *Ap. J.*, in press.
Lamb, F. K. 1989a, in *Timing Neutron Stars,* eds. H. Ögelman & E.P.J. van den Heuvel (Kluwer, Dordrecht).
Lamb, F.K. 1989b, *Ap. J.*, in press.
Lamb, F.K. et al. 1977, *Ap. J.*, **217**, 197.
Lapidus, I. I. & Sunyaev, R. A. 1985, *M. N. R. A. S.*, **217**, 291.
Levine A. M. *et al.* 1984, *Ap. J. Suppl.*, **54**, 581.
Lewin, W.H.G. 1977, *American Scientist* , **65**, 605.
Lewin, W. H.G. 1984, *AIP Proceedings* Vol. 115, 249.
Lewin, W. H. G., Doty, J., Clark, G. W. *et al.* 1976, *Ap. J.*, **207**, L95.
Lewin, W.H.G. *et al.* 1977, *Nature* , **267**, 28.

250

Lewin, W. H. G., Hoffman, J. A., Marshall, H. *et al.* 1978, *IAU Circular No.* 3190.
Lewin, W. H.G. & Clark, G. W. 1980, *Ann. New York Ac. Sci.*, **336**. 451.
Lewin, W. H. G. et al. 1980, *M.N.R.A.S.*, **193**,15.
Lewin, W.H.G. & Joss, P.C. 1981, *Space Sci. Rev.*, **28**, 3.
Lewin, W.H.G. & Van den Heuvel, E.P.J. (eds), 1983, *Accretion Driven Stellar X-ray Sources* (Cambridge University press).
Lewin, W. H. G. & Joss, P. C. 1983, in *Accretion-driven Stellar X-ray Sources*, ed. W. H. G. Lewin & E. P. J. van den Heuvel (Cambridge University Press), p. 41.
Lewin, W. H. G., Vacca, W. D., & Basinska, E. M. 1984, *Ap. J.*, **277**, L57.
Lewin, W. H. G., Penninx, W., Van Paradijs, J. *et al.* 1987, *Ap. J.*, **319**, 893.
Lewin, W. H. G., Trümper, J. & Brinkmann, W. (eds) 1986, *"The Evolution of Galactic X-ray Binaries"*, NATO ASI Series C, Vol. 167 (Reidel).
Lewin, W. H. G., Van Paradijs, J. & Van der Klis, M. 1988, *Space Sci. Rev.*,**46**, 273.
Liang, E.P., 1986, *Ap. J.*, **304**, 682.
Liller, W. 1973, *Ap. J.*, **184**, L23.
Liller, W. 1977, *Ap. J.*, **213**, L21.
London, R. A., Taam, R. E., & Howard, W. M. 1984, *Ap. J.*, **287**, L27.
London, R. A., Taam, R. E., & Howard, W. M. 1986, *Ap. J.*, **306**, 170.
Lyne, A. G. *et al.* 1982, *M. N. R. A. S.*, **201**, 503.
Lyne, A.G. & Manchester, R.N. 1988, *M.N.R.A.S.*, **234**, 477.
Magnier, E. et al. 1989, *M.N.R.A.S.*, in press.
Maraschi, L. & Cavaliere, A. 1977, *Highlights of Astronomy*, 4, part 1, p. 127.
Marshall, H.L. et al. 1979, *Ap. J.* **227**, 555.
Matsuoka, M. 1986, in: *The Evolution of Galactic X-Ray Binaries*, eds. J. Trümper, W.H.G. Lewin, & W. Brinkmann, Reidel, p 301.
Mazets, E.P. *et al.* 1981, *Nature*, **290**, 378.
Melia, F. 1987, *Ap. J.*, **315**, L43.
Maraschi, L., Treves A. & Van den Heuvel, E. P. J. 1975, *Nature*, **259**, 292.
Mason, K.O. 1986, in *The Physics of Accretion onto Compact Objects*, eds. K.O. Mason, M.G. Watson & N.E. White, (Springer), p. 29.
Mason, K. O., Parmar, A. N. & White, N. E. 1985, *M. N. R. A. S.*, **216**, 1033.
Mason, K. O. Watson, M. G. & White, N. E. (eds) 1986, *The Physics of Accretion Onto Compact Objects*, Lecture Notes in Physics, Vol. 266, (Springer).
McClintock, J.E. 1988, in *The Physics of Compact Objects: Theory versus Observations*, ed. L. Filipov & N.E. White (Pergamon Press).
McHardy, I.M., *et al.* 1981, *M. N. R. A. S.*, **193**, 893.
Michel, F.C. 1977, *Ap. J.*, **216**, 838.
Milgrom, M. 1987, *Astr. Ap.*, **172**, L1.
Misner, C., Thorne, K. S., & Wheeler, J. A. 1973, *Gravitation* (Freeman).
Murakami, T., Inoue, H., Koyama, K. *et al.* 1980, *Ap. J.*, **240**, L143.
Nakamura, N., Inoue, H. & Tanaka, Y. 1988, *Publ. Astr. Soc. Japan*, **40**, 209.
Ostriker, J. 1977, *Annals New York Acad. Sci.*, **302**, 229.
Paczynski, B. 1983, *Ap. J.*, **267**, 315.
Paczynski, B. 1987, *Nature*, **327**, 303.
Paczynski, B. & Anderson, N. 1986, *Ap. J.*, **302**, 1.
Pallavicini, R. & White, N.E. (eds) 1988, *X-ray Astronomy with EXOSAT,* (Memorie della Societa' Astronomica Italiana), in press.
Parmar, A. N. & White, N.E. 1988, in *X-ray Astronomy with EXOSAT*, ed. R. Pallavicini & N.E. White, in press.
Parmar, A. N. *et al.* 1988, *Ap. J.*, (in press).
Pedersen, H., Van Paradijs, J., Motch, C. *et al.* 1982, *Ap. J.*, **263**, 340.
Penninx, W., Van Paradijs, J. & Lewin, W.H.G. 1987, *Ap. J.*, **321**, L67.

Penninx, W. *et al.* 1989, *Astr. Ap.* in press.
Penny, A. J. *et al.* 1973, *M. N. R. A. S.*, **163**, 7p.
Pietsch, W. *et al.* 1984, *Space Sci. Rev.*, **40**, 371.
Pringle, J. E. & Wade, R. A. (eds) 1985: *Interacting Binary Stars*, (Cambridge University Press).
Rappaport, S. & Joss, P. C. 1983, in *Accretion Driven Stellar X-ray Sources*, ed. W.H.G. Lewin & E.P.J. van den Heuvel, (Cambridge University Press), p. 1.
Ritter, H. 1987, *Astr. Ap. Suppl.*, **70**, 335.
Ruderman, M. A. *et al.* 1988, *Ap. J.*, (in press).
Ryter, C. 1970, *Astr. Ap.*, **9**, 288.
Salpeter, E. E. 1973, *IAU Symposium*, **55**, 135.
Sandage A. *et al.* 1966, *Ap. J.* **146**, 316.
Schreier E. *et al.* 1972, *Ap. J.*, **172**, L112.
Schulz, N., Hasinger, G. & Trümper, J, 1989, *Astr. Ap.*, in press.
Setti, G. & Woltjer, L. 1970, *Ap. Space Sci.*, **9**, 185.
Shapiro, S. L. & Teukolsky, S. A. 1983: *Black Holes, White Dwarfs and Neutron Stars,* (John Wiley and Sons).
Slettebak, A. 1987, in *Physics of Be Stars*, ed. A. Slettebak & T.P. Snow, (Cambridge University Press), p. 24.
Slettebak, A. & Snow, T. P. (eds) 1987: *Physics of Be Stars*, (Cambridge Univ. Press).
Srinivasan, G. *et al.* 1988, in preparation.
Stella, L., White, N. E. & Rosner, R. 1986, *Ap. J.*, **308**, 669.
Stella, L. et al. 1988a, *Ap. J.*, **324**, 379.
Stella, L. et al. 1988b, *Ap. J.*, **327**, L13.
Sugimoto, D., Ebisuzaki, T., and Hanawa, T. 1984, *Publ. Astron. Soc. Japan*, **36**, 839.
Swank, J. H., Becker, R. H., Boldt, E. A. *et al.* 1977, *Ap J.*, **212**, L73.
Sztajno, M., Van Paradijs, J., Lewin, W. H.G. *et al.* 1985, *Ap. J.*, **299**, 487.
Sztajno, M., Fujimoto, M. Y., Van Paradijs, J. *et al.* 1987, *M. N. R. A. S.*, **226**, 39.
Taam, R.E. 1982, *Ap. J.* **258**, 761.
Taam, R.E. & Lin, D.N.C. 1984, *Ap. J.* **287**, 761.
Taam, R. E., & Van den Heuvel, E. P. J. 1986, *Ap. J.*, **305**, 235.
Tan, J. *et al.* 1989, in preparation.
Tananbaum, H. 1973, *IAU Symposium*, **55**, 9.
Tananbaum, H. *et al.* 1972, *Ap. J.*, **177**, L5.
Tawara, Y. *et al.* 1982, *Nature*, **299**, 38.
Tawara, Y., Kii, T., Hayakawa, S. *et al.* 1984, *Ap. J.*, **276**, L41.
Taylor, J.H. & Weisberg, J.M. 1982, *Ap. J.*, **253**, 908.
Thorne, K. S. 1977, *Ap. J.*, **212**, 825.
Tjemkes, S., Zuiderwijk, E. J. & van Paradijs, J. 1986, *Astr. Ap.*, **154**, 77.
Vacca, W. D., Lewin, W. H. G., and Van Paradijs, J.1966, *M. N. R. A. S.*, **220**, 339.
Van den Heuvel, E. P. J. 1984, *J. Ap. Astr.*, **5**, 209.
Van den Heuvel, E. P. J. & Heise, J. 1972, *Nature Phys. Sci.*, **239**, 67.
Van den Heuvel, E. P. J. & Rappaport, S. 1987, in *The Physics of Be Stars*, ed. A. Slettebak & T.P. Snow (Cambridge University Press), p. 291.
Van den Heuvel, E. P. J., Van Paradijs, J. & Taam, R. E. 1986, *Nature*, **322**, 153.
Van den Heuvel, E. P. J. & Van Paradijs, J. 1988, *Nature*, **334**, 227.
Van der Klis, M. 1989, *Annual Rev. Astr. Aph.*, (in press).
Van der Klis, M. et al. 1985, *Nature* **316**, 225.
Van der Klis, M. *et al.* 1987, *Ap. J.*, **313**, L19.
Van Oyen, J. 1988, *Astr. Ap.*, submitted.
Van Paradijs, J. 1978, *Nature*, **274**, 650.
Van Paradijs, J. 1979, *Ap. J.*, **234**, 609.

252

Van Paradijs, J. 1982, *Astr. Ap.*, **107**, 51.
Van Paradijs, J. 1983: in *Accretion Driven Stellar X-ray Sources*, ed. W.H.G. Lewin & E.P.J. van den Heuvel (Cambridge University Press), p. 189.
Van Paradijs, J.et al. 1979, *Nature*, **280**, 375.
Van Paradijs, J. & Verbunt, F. 1984, in *High Energy Transients in Astrophysics*, ed. S.E. Woosley, AIP Proc. Vol. **115**, p. 49.
Van Paradijs, J. *et al.* 1986, *M. N. R. A. S.*, **221**, 617.
Van Paradijs, J. & Lewin, W. H.G. 1987, *Astr. Ap.*, **172**, L20.
Van Paradijs, J. *et al.* 1988, *Astr. Ap.*, **192**, 147.
Van Paradijs, J., Penninx, W. & Lewin, W. H.G. 1988, *M.N.R.A.S.*, **233**, 437.
Verbunt, F. & Hut, P. 1987, in *The Origin and Evolution of Neutron Stars*, ed. D.J. Helfand & J.-H. Huang (Reidel), p. 187.
Verbunt, F. & Meylan, G. 1988, *Astr. Ap.*,**203**, 297.
Vidal, N. V. 1973, *Ap. J.*, **182**, L77.
Waki I., Inoue, H., Koyama, K., *et al.* 1984, *Public. Astron. Soc. Japan*, **36**, 819.
Warwick R. S. *et al.* 1981, *M. N. R. A. S.*, **193**, 865.
Webster, X. & Murdin, P. 1971, *Nature*, **237**, 35.
White, N. E. & Marshall, F. E. 1984, *Ap. J.*, **281**, 354.
White, N. E., Swank, J. H. & Holt, S. S. 1984, *Ap. J.*, **270**, 711.
Woosley, S. and Taam, R. E. 1976, *Nature*, **263**, 101.
Woosley, S.E. & Weaver, T.A.1984, in *High Energy Transients in Astrophysics*, ed. S.E. Woosley, AIP Proc. Vol. 115, 273.
Wright, E. L. Gottlieb, W. & Liller, W. 1975, *Ap. J.*, **200**, 171.
Zuiderwijk, E. J. 1979, Ph. D. Thesis, Univ. Amsterdam.

# SUPERNOVA REMNANTS I: HISTORICAL EVENTS

RICHARD G. STROM
*Netherlands Foundation for Research in Astronomy*
*Radiosterrenwacht*
*P.O. Box 2*
*7990 AA Dwingeloo*
*Netherlands*

ABSTRACT. Historical supernovae occupy a special place in the study of supernova remnants, being the only ones whose age can be definitely ascertained and for which anything can be said about the nature of the initiating event. After a brief introduction, this lecture considers the nine Galactic supernovae for which historical records exist, and summarizes the characteristics of their associated remnants. All have been observed at radio and X-ray wavelengths, and most also produce detectable optical and infrared emission. The evolution of the shock kinematics is then briefly reviewed, and the degree of deceleration summarized for the six remnants where this information is available. Finally, the statistics of SNR in the solar neighborhood is considered, and it is concluded that the historical record over the last two millenia is probably quite incomplete for supernovae beyond a few kpc.

## 1. Foreword

The three lectures on supernova remnants (SNRs) cover fairly broad and to some extent overlapping topics, although their separation will usually be clear. During the oral presentations in Erice this was perhaps less obvious, as they were not all of equal length and the beginning of the second and third simply took over where the previous one had finished. Anyone desiring a general introduction to the topics covered could do worse than peruse recent conference proceedings, including the Bad Honnef Workshop (Kundt, 1988a), IAU Colloquium 101 (Roger and Landecker, 1988), the Indian Academy Workshop (Srinivisan and Radhakrishnan, 1985) and (still useful though somewhat dated) IAU Symposium 101 (Danziger and Gorenstein, 1983). There have also been reviews (Weiler and Sramek, 1988; Raymond, 1984), and Green (1988a) maintains a catalogue of remnants. References more specific to each presentation will be cited in the appropriate Introduction.

## 2. Introduction

The appearance of SN1987A in the Large Magellanic Cloud was the first (recorded) naked-

*W. Kundt (ed.), Neutron Stars and Their Birth Events, 253–262.*
© *1990 Kluwer Academic Publishers.*

eye supernova (SN) in over 300 years. Such events are clearly rare. The historical SN/SNR considered here all lie in our Galaxy. The modern study of SN begins with that of 1885 in M31, and all those studied since (including SN1987A) are in external galaxies. They will not be considered further here, although we should not forget that with the improved sensitivity and resolution of modern instrumentation the study of SNR in external galaxies becomes increasingly feasible. Only there are we likely to be able to relate young SNRs to their initiating events in statistically significant numbers.

The last Galactic SN unambiguously recorded, SN1604 (usually associated with Kepler), appeared at the end of the pre-telescopic era. Consequently all information on the SN events considered here is derived from naked-eye observations. Most records before the Renaissance originate in Oriental sources, although for some events there are European and Arabic reports. As with all historical records, there are uncertainties regarding the objectivity of the reported event. In all the relevant societies one has to contend with cultural, religious and social attitudes, which may influence not only how a given phenomenon was recorded, but indeed *whether* it was. One need only consider the fact that astrology frequently played a significant role, so that the occurrence of any celestial event had potential predictive power. In addition, the role of history in Oriental societies (as, to be honest, all too frequently in our own) was partly to demonstrate the legitimacy of the dynasty in power. Finally, it is necessary to distinguish comets, meteors, novae and even meteorological phenomena from the SN which concern us here.

Despite the uncertainties with which they confront us, the information which we have been able to extract from ancient records has proven well worth the effort. The dates of specific SN provide us with the only incontrovertible SNR ages. In most cases we are able to estimate, or at least place limits upon, the peak apparent magnitude, and in two it has even been possible to construct tolerable light curves. And despite the small numbers, they actually do provide us with some statistical information.

## 2.1. GENERAL LITERATURE

The very readable monograph "The Historical Supernovae" (Clark and Stephenson, 1977) remains the best general reference, and is the source for much of the historical background presented here. Historically, one of the earliest investigations was Lundmark's (1921) search for candidate novae (before the nova/SN distinction was made). Pioneering work on SN1054 (progenitor of the Crab Nebula) by Duyvendak (1942) and Mayall and Oort (1942) deserves mention, as do Baade's investigations (1945, 1943) of SN1572 and SN1604. Ho Peng Yoke (1962, 1970) provides a compendium in translation of a great many Oriental records which include candidate SN.

## 3. Characteristics of the Historical SNR

About nine documented historical events can be considered candidate SN if we include Cas A. The evidence concerning them has been considered by Clark and Stephenson (1977). There have also been suggestions that there was a SN in 1408 (Li, 1979) and Strom *et al.* (1980) have proposed that it might be associated with the peculiar object CTB80, which includes a neutron star-driven component (Strom, 1988b; see also the third lecture in this series). However further investigation by Stephenson and Yau (1986) considerably weakens the case for a SN in 1408, and the matter will not be considered further here. We will now discuss each of the SN and its related remnant in turn. (Unless otherwise referenced, the historical information in all cases has been extracted from Clark and Stephenson, 1977.)

SN185, although the historical account is rather sketchy, appears to have been very bright, which may account for the fact that any record of it has survived to this day. It has left behind a somewhat distorted remnant, one of the most irregular radio shells among the young SNR (Kesteven and Caswell, 1987). Its shape agrees well with that seen in soft X-rays (Pisarski et al., 1984). The optical image is dominated by roughly circular, fairly bright filaments concentrated in the southwestern corner (Van den Bergh et al., 1973), where the radio and X-ray emission are most intense.

SN386 and SN393 were rather faint and presumably more distant events (if they were SN) about which consequently rather little is known. There have been recent suggestions that the former is associated with the Galactic nonthermal shell G11.2-0.3, which has been mapped at both radio and X-ray wavelengths (Downes, 1984). Green (1988b) presents higher resolution radio maps which show a clumpy shell with rather ragged edges, suggesting that it may be a somewhat evolved Cas A, as befits its suggested association with a SN some 1600 years ago. However, as Green points out, there are inconsistencies if one tries to reconcile this interpretation with its age and apparent size. Indeed, there would appear to be problems in trying to maintain that any essentially freely-expanding SNR is as much as 1600 years old: the clumpy, ragged appearance of G11.2-0.3 would seem to argue against an association with SN386. Kinematic studies, when they become feasible, may help resolve the issue. There is no convincing SNR candidate for SN393.

SN1006 probably had the brightest apparent peak magnitude of any recorded SN. The numerous accounts make for a convincing association with G327.6+14.5, a strikingly regular nonthermal shell about 30' arc in diameter. The radio structure (Reynolds and Gilmore, 1986) agrees well with the soft X-ray morphology (Pye et al., 1981). The optical emission, although weak, is in the form of narrow filaments along the northwestern edge, and it has been possible to determine their proper motion and hence the (local) expansion speed (Long et al., 1988). This is one of the remnants in which the Hα is found to consist of both broad and narrow components (Kirshner et al., 1987), enabling the shock speed to be estimated.

SN1054 produced the Crab Nebula, which still receives energy from the central neutron star, the pulsar PSR0531+21. As such it is the prototype neutron star-driven nebula and while the association is secure, we still have too little information on the light curve to make any statement about the SN type. Its properties are considered further in the third SNR lecture.

SN1181 produced the center-brightened (and hence morphologically similar to the Crab Nebula) radio source 3C58. Elongated, as is the Crab Nebula, a recent radio study shows the emission to mainly consist of filaments (Reynolds and Aller, 1988). The X-ray morphology also peaks near the center (Becker et al., 1982) where a probable point source may indicate the presence of a neutron star, although no pulsar has ever been detected. Optically, the remnant consists of rather faint filaments the spectra of which reveal motions as high as 900 km/s (Fesen, 1983).

SN1572 was the first SN to be extensively observed in Europe, by Tycho Brahe in particular. The position determination is so good that there is no doubt about its association with the shell remnant 3C10, which has been extensively studied at radio (Duin and Strom, 1975; Henbest, 1980), optical (Kamper and Van den Bergh, 1978; Chevalier *et al.*, 1980), infrared (Braun, 1987) and X-ray (Seward *et al.*, 1983) wavelengths. Its radio structure is shown in Figure 1. The expansion has been determined at both optical (Kamper and Van den Bergh, 1978) and radio (Strom *et al.*, 1982) frequencies.

SN1604 has, like that of 1572, a well-determined light curve. Both have long been classified as of Type I (for a discussion of SN types, and indeed of what happens during the supernova, the reader should refer to the lectures given by David Branch during this school). The remnant is a strong radio and X-ray source, in the form of a nearly circular shell. The X-ray emission has been studied by White and Long (1983), while Matsui *et al.* (1984) carried out a detailed radio study including a comparison with the X-ray properties. Recently, Dickel *et al.* (1988) have measured the rate of expansion of the radio shell and find it to be much greater than motions previously detected in the brightest optical nebulosity (Van den Bergh and Kamper, 1977), although more recent optical images (D'Odorico *et al.*, 1986) reveal faint emission tracing out more of the shell seen in radio and X-rays.

Finally we come to Cas A, which we know from kinematic arguments to be the youngest of the group, although it is uncertain that the event which produced it was ever recorded. From the fastest moving optical filaments, which show no evidence of having sufferred significant deceleration, its origin has been dated to 1658±3 years (Van den Bergh and Kamper, 1983). There have been claims that the SN was observed by Flamsteed in 1680 (Ashworth, 1980), although the evidence is sketchy. It seems likely that the event was in any case faint. Radio emission from Cas A is very strong, and maps show an irregular circular shell, with a weak plateau of emission outside the main ring (Figure 2). The compact radio knots show an overall pattern of expansion (Tuffs, 1986) with much irregular motion which Braun *et al.* (1987) show to be the result of the movement of bow shock-like features presumably formed by particle acceleration in the wakes of fast moving clumps. The X-ray emission from Cas A is also ring-shaped (Fabian *et al.*, 1980), as is the infrared (Braun, 1987).

Let us briefly contrast the observed properties of the eight SNR which arose from supernovae seen in the last two millenia. All have been observed in the radio and X-ray bands, all but one (SN386) have detected optical emission, and all but three (SN386, SN1006 and SN1181) have been detected in the infrared. The gross radio and X-ray morphologies are similar in each of the eight, with six displaying a shell structure and the other two being elongated, amorphous and brightest near the center. The center-brightened SNR, SN1054 (the Crab Nebula) and SN1181 (3C58), belong to the category of neutron star-driven nebulae (although no pulsar has yet been found in 3C58). There is every reason to believe that their X-ray emission is mainly produced by the synchrotron effect, as is much of the optical emission seen in the Crab Nebula. As they are considered further in my last lecture, the rest of this discussion will concentrate on the historical shell SNR.

The X-ray emission seen in the six historical shell remnants is thermal in nature, originating in gas heated by the expanding blast wave and by the interaction between fast-moving ejecta and the ambient medium. The infrared is most likely produced by dust particles heated in the same interaction (*e.g.*, Draine, 1981). The X-ray and infrared data can be used to derive gas densities, temperatures, shock speeds and other fundamental parameters, and this has been done for Tycho, Kepler and Cas A by Braun (1987). In these three and SN1006, motion has been observed which in Tycho, Kepler and SN1006

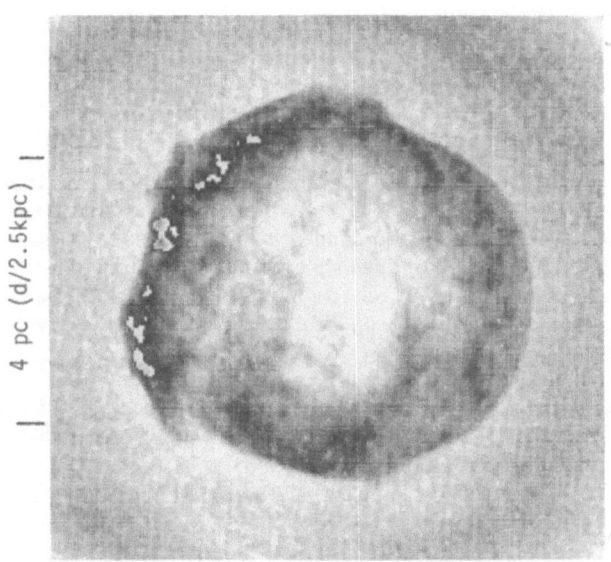

**Figure 1.** A λ6 cm map of 3C10, the remnant of SN1572, made with the Westerbork Telescope. Note in particular the near-circularity of the western edge.

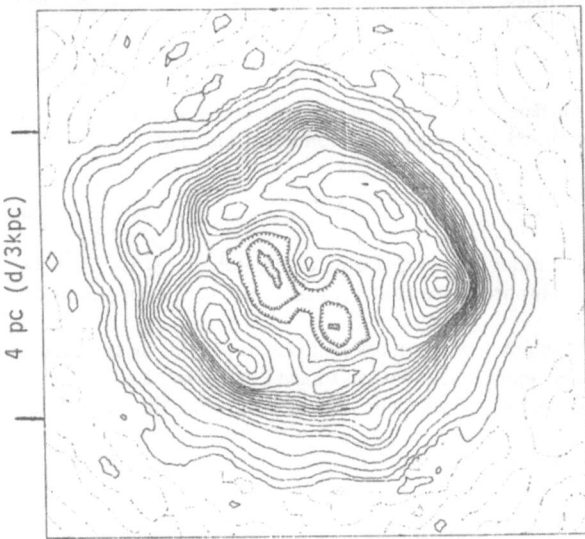

**Figure 2.** A λ49 cm map of Cas A, the remnant of a SN which exploded in the seventeenth century. This map was made with the Westerbork Telescope, and the lowest contour is below 0.1% of the peak brightness.

indicates substantial deceleration, while Cas A exhibits a range from undecelerated to virtually stationary features.


## 4. Discussion

SNR kinematics can be best discussed in the context of models in which those based upon the gas dynamics of spherical shock fronts have generally found the greatest favor. An alternative viewpoint vigorously expounded by Kundt (1988b) argues that SNRs are dominated by fragments of ejecta, an idea which would seem to be most applicable to the knotty, undecelerated components of objects like Cas A, the Crab Nebula and perhaps Puppis A. In the case of the diffuse spherical shock fronts, it was Woltjer (1972) who summarized the situation in terms of four evolutionary stages:

Phase I. Free expansion of ejecta. This phase persists as long as the mass of swept-up ambient material is much less than that of the ejecta. The matter expelled in the explosion essentially coasts, undecelerated, at its original velocity, so

$$v = \frac{r}{t}$$

Phase II. Adiabatic expansion. The ejecta are now overwhelmed by the mass of ambient material which has been encountered and have essentially lost their identity, the shock being strongly decelerated. Energy is conserved (radiation losses are negligible) and there is a similarity solution which yields,

$$v = \frac{2}{5}\frac{r}{t}$$

Phase III. Momentum conserving expansion. Once radiation losses become dominant, the expansion is governed by momentum conservation. The velocity then depends upon,

$$v = \frac{1}{4}\frac{r}{t}$$

Phase IV. The remnant disperses and completely loses its identity. This will occur when its velocity is comparable to that typical of the clouds in the interstellar medium.

The majority of SNR which have been studied can be placed, on the basis of both kinematics and the nature of their emission, in either Phase I or II (most frequently the latter). There are essentially no remnants of which it can be said that they are in one of the last two phases.

It will be clear that for Phases I-III, we can write for the shock velocity,

$$v = \eta\frac{r}{t}$$

where $\eta$ takes on the values 1, 2/5, and 1/4 as the SNR evolves from I to III. For all of the SNRs we have been considering we know the age and size, hence we can determine an average (angular) expansion velocity, $r/t$. In six cases we have information about the present rate of expansion, $v$ (in the case of 3C58 this is based upon spectroscopic data, so we also require the distance, but for the other five we have proper motion determinations so the distance cancels out). Using this information, we are able to determine $\eta$ for each remnant:

SN1006, $\eta = 0.32$
SN1054, $\eta = 1.08$
SN1181, $\eta \sim 0.3$
SN1572, $\eta = 0.47$
SN1604, $\eta = 0.5$
SN1680, $\eta = 0.95 - < 0.1$

For Cas A (assuming that it actually was observed in 1680) a range of values is given, corresponding to the fast moving filaments at one extreme, and nearly stationary material at the other. The Crab Nebula shows the effects of continuing acceleration produced by the neutron star, while the other center-filled remnant 3C58 (SN1181) has clearly undergone much deceleration and is consequently past the stage where its neutron star has much influence. All of the shells (excepting Cas A) exhibit substantial deceleration, falling roughly in Phase II, although none of them can be said to conform precisely to the expected adiabatic behavior. In this context, however, it should be noted that both Tycho/SN1572 and Kepler/SN1604 exhibit considerable variability in the values of $\eta$ determined at different positions around the shell (Strom *et al.*, 1982; Dickel *et al.*, 1988). These differences have been attributed to changes in the ambient density, and it is clear that more study will be required before firm conclusions can be drawn.

Before one can be quantitative about the physical properties of these SNRs, we obviously require information on their distances. The most reliable estimates are either based on the kinematics of the objects themselves, or are derived from the kinematic location of the SNR in our Galaxy using, for example, line absorption by intervening gas. Green (1988b) gives distance estimates for many SNR in his catalogue. For four of the shell remnants considered here, Strom (1988a) has estimated their relative distances using the assumption that they are in the adiabatic phase. It is, in any event, fairly certain that the eight with associated SNRs must lie within 4-5 kpc of the sun. The best estimates are:

SN 185, d=1.2 kpc
SN 386, d=5? kpc
SN1006, d=1.4 kpc
SN1054, d=2 kpc
SN1181, d=2.6 kpc
SN1572, d=2.5 kpc
SN1604, d=4.2 kpc
SN1680, d=2.9 kpc

It has been said that if an astronomer discovers an object it will be used to define a class, and should a second one be found they constitute a set which can be subjected to statistical analysis. With eight SN in the last 2000 years within about 5 kpc of the sun, we can make a crude estimate of the Galactic rate (assuming the solar neighborhood is not atypical). Taking the Galaxy to be a disk of radius 15 kpc and assuming the SN in our sample all belong to the disk population, we have

$$f_{SNR} = \frac{n}{\Delta t} \frac{V_{Gal}}{V_{sample}} = \frac{8}{2000} \frac{15^2}{5^2} = \frac{1}{28} yr^{-1}$$

The statistical uncertainty in $n$ simply on the basis of small numbers is $\pm 3$, so the time interval between SN could be anywhere between 20 yr and 44 yr, giving us a Galactic SN rate of $3^{+2}_{-1}$ per century. This, however, ignores the possible incompleteness of the historical record.

Considering the distribution in time, there is no obvious sign that the earlier observations are more or less complete than the more recent ones (the clustering of three SN between 1000 and 1200 and again in the period 1570-1680 is presumably insignificant). Figure 3 shows the distribution with distance. If the SN are randomly distributed in the Galactic disk and our sample is complete, we would expect their numbers to grow as $d^2$. This clearly does not happen. Rather we see in Figure 3 that the numbers *decrease* beyond 2-3 kpc, suggesting that the historical record may be incomplete beyond this distance. This, however, presents us with something of a dilemma. Van den Bergh *et al.* (1987) argue that the Galactic SN rate based upon observations of other galaxies should be about 2 per century. While this does not seriously conflict with the rate found above, the incompleteness of the record means that the true discrepancy is, if anything, larger. The estimate of $f_{SNR}$ is of course strongly dependent upon the distance determination, so we can probably tentatively conclude that given the uncertainties, the two rates may well be consistent.

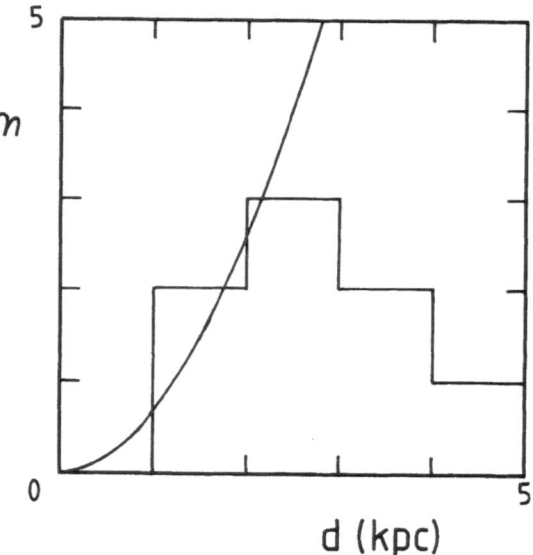

**Figure 3.** A histogram showing the distribution of historical SNR with distance from the sun. The curve shows the expected $d^2$ increase for a complete sample in a disk.

## 5. Acknowledgements

The Westerbork Telescope is operated by the Foundation for Research in Astronomy, which is financially supported by the Netherlands Organisation for Scientific Research (NWO).

## 6. References

Ashworth, W.B. (1980) *J. Hist. Astron.* 11, 1.
Baade, W. (1943) *Astrophys. J.* **97**, 119.
Baade, W. (1945) *Astrophys. J.* **102**, 309.
Becker, R.H., Helfand, D.J. and Szymkowiak, A.E. (1982) *Astrophys. J.* **255**, 557.
Braun, R. (1987) *Astr. Astrophys.* **171**, 233.
Braun, R., Gull, S.F. and Perley, R.A. (1987) *Nature* **327**, 395.
Chevalier, R.A., Kirshner, R.P. and Raymond, J.C. (1980) *Astrophys. J.* **235**, 186.
Clark, D.H. and Stephenson, F.R. (1977) *The Historical Supernovae*, Pergamon, Oxford.
Danziger, J. and Gorenstein, P. (eds.) (1983) *Supernova Remnants and Their X-ray Emission*, Reidel, Dordrecht.
Dickel, J.R., Sault, R., Arendt, R.G., Matsui, Y. and Korista, K.T. (1988) *Astrophys. J.* **330**, 254.
D'Odorico, S., Bandiera, R., Danziger, J. and Focardi, P. (1986) *Astr. J.* **91**, 1382.
Downes, A.J.B. (1984) *Mon. Not. Roy. Astr. Soc.* **210**, 845.
Draine, B.T. (1981) *Astrophys. J.* **245**, 880.
Duin, R.M. and Strom, R.G. (1975) *Astr. Astrophys.* **39**, 33.
Duyvendak, J.J.L. (1942) *P.A.S.P.* **54**, 91.
Fabian, A.C., Willingale, R., Pye, J.P., Murray, S.S. and Fabbiano, G. (1980) *Mon. Not. Roy. Astr. Soc.* **193**, 175.
Fesen, R.A. (1983) *Astrophys. J.* **270**, L53.
Green, D.A. (1988a) *Astrophys. Sp. Sci.* **148**, 3.
Green, D.A. (1988b) in Kundt, W. (ed.), *Supernova Shells and Their Birth Events*, Springer-Verlag, Berlin, p. 120.
Henbest, S.N. (1980) *Mon. Not. Roy. Astr. Soc.* **190**, 833.
Ho Peng Yoke (1962) *Vistas in Astron.* **5**, 127.
Ho Peng Yoke (1970) *Oriens Extremus* 17, 63.
Kamper, K.W. and Van den Bergh, S. (1978) *Astrophys. J.* **224**, 851.
Kesteven, M.J. and Caswell, J.L. (1987) *Astr. Astrophys.* **183**, 118.
Kirshner, R.P., Winkler, F.P. and Chevalier, R.A. (1987) *Astrophys. J.* **315**, L135.
Kundt, W. (ed.) (1988a) *Supernova Shells and Their Birth Events*, Springer-Verlag, Berlin.
Kundt, W. (1988b) in Kundt, W. (ed.), *Supernova Shells and Their Birth Events*, Springer-Verlag, Berlin, p. 1.
Li Qi-bin (1979) *Chin. Astr.* **3**, 315.
Long, K.S., Blair, W.P. and Van den Bergh, S. (1988) *Astrophys. J.* **333**, 749.
Lundmark, K. (1921) *P.A.S.P* **33**, 225.
Matsui, Y., Long, K.S., Dickel, J.R. and Greisen, E.W. (1984) *Astrophys. J.* **287**, 295.
Mayall, N.U. and Oort, J.H. (1942) *P.A.S.P* **54**, 95.
Pisarski, R.L., Helfand, D.J. and Kahn, S.M. (1984) *Astrophys. J.* **277**, 710.

Pye, J.P., Pounds, K.A., Rolf, D.P., Seward, F.D., Smith, A. and Willingale, R. (1981) *Mon. Not. Roy. Astr. Soc.* **194**, 569.

Raymond, J.C. (1984) *Ann. Rev. Astr. Astrophys.* **22**, 75.

Reynolds, S.P. and Aller, H.D. (1988) *Astrophys. J.* **327**, 845.

Reynolds, S.P. and Gilmore, D.M. (1986) *Astr. J.* **92**, 1138.

Roger, R.S. and Landecker, T.L. (eds.) (1988) *Supernova Remnants and the Interstellar Medium*, Cambridge University Press, Cambridge.

Seward, F., Gorenstein, P. and Tucker, W. (1983) *Astrophys. J.* **266**, 287.

Srinivasan, G. and Radhakrishnan, V. (1985) *Supernovae, their Progenitors and Remnants*, Indian Academy of Sciences, Bangalore.

Stephenson, F.R. and Yau, K.K.C. (1986) *Quart. J. Roy. Astr. Soc.* **27**, 559.

Strom, R.G. (1988a) *Mon. Not. Roy. Astr. Soc.* **230**, 331.

Strom, R.G. (1988b) in Kundt, W. (ed.) *Supernova Shells and Their Birth Events*, Springer-Verlag, Berlin, p. 91.

Strom, R.G., Angerhofer, P.E. and Velusamy, T. (1980) *Nature* **284**, 38.

Strom, R.G., Goss, W.M. and Shaver, P.A. (1982) *Mon. Not. Roy. Astr. Soc.* **200**, 473.

Tuffs, R.J. (1986) *Mon. Not. Roy. Astr. Soc.* **219**, 13.

Van den Bergh, S. and Kamper, K.W. (1977) *Astrophys. J.* **218**, 617.

Van den Bergh, S. and Kamper, K.W. (1983) *Astrophys. J.* **268**, 129.

Van den Bergh, S., Marscher, A.P. and Terzian, Y. (1973) *Astrophys. J. Suppl.* **26**, 19.

Van den bergh, S., McClure, R.D. and Evans, R. (1987) *Astrophys. J.* **323**, 44.

Weiler, K.W. and Sramek, R.A. (1988) *Ann. Rev. Astr. Astrophys.* **26**, 295.

White, R.L. and Long, K.S. (1983) *Astrophys. J.* **264**, 196.

Woltjer, L. (1972) *Ann. Rev. Astr. Astrophys.* **10**, 129.

# SUPERNOVA REMNANTS II: SHELLS

RICHARD G. STROM
*Netherlands Foundation for Research in Astronomy*
*Radiosterrenwacht*
*P.O. Box 2*
*7990 AA Dwingeloo*
*Netherlands*

ABSTRACT. Shells are the dominant morphology among the known supernova remnants. After presenting some statistics on galactic remnants, this lecture considers the properties of shells and in particular several well-studied examples. In the Cygnus Loop and IC443 there is good evidence that the brightest emission emanates from a pre-existing shell-like structure which surrounded the progenitor. This conclusion is especially backed up by the existence of several interconnecting shells which comprise the IC443 complex, and by the existence of morphologically similar structures associated with other remnants like VRO 42.05.01. Some consequences of this possibility are discussed.

## 1. Introduction

As we saw in the first of these lectures on supernova remnants (SNRs), those produced by historical events, the majority (in that case, six out of eight) display a ring or shell-like morphology (see for example Figures 1 and 2 from that lecture). Here we will be exclusively concerned with the ubiquitous shell remnants whose idealized three-dimensional structure is that of the surface of a sphere. Why should this be so? Clearly, if most of the ejecta from a supernova (SN) is expelled with only a small spread in speed, then they will always be roughly equidistant from the explosion center. But what about deceleration and the effects of an irregular ambient medium? As we shall see, once the shock has become adiabatic its propagation is relatively insensitive to the density of the surrounding gas. So common is the shell morphology that it is fair to say that any galactic radio source with a ring-like structure and nonthermal spectrum is bound to be a SNR (*e.g.*, Woltjer, 1972).

The vast majority of remnants, whatever their morphology, have been discovered because of, and are easiest to detect through, their radio emission. Although SNR usually are strong radio emitters, this is not the entire explanation, for the X-ray luminosity often exceeds that of the radio emission. However, it is clear that strong shocks are good at both heating the gaseous medium and accelerating charged particles, and they may, moreover, amplify magnetic fields. What is crucial is that intervening dust particles (which block the – often feeble – optical emission) and gas, whether neutral or ionized, for the range of

*W. Kundt (ed.), Neutron Stars and Their Birth Events, 263–273.*

densities encountered in the interstellar medium, are essentially transparent to radio waves. Soft X-rays (up to a few keV, the energy range for imaging X-ray telescopes) on the other hand are strongly attenuated after travelling relatively short distances in the Galactic disk. Infrared radiation will also penetrate to large distances, and SNR are often strong infrared emitters (Dwek, 1988), but in the Galactic disk one of the problems is the confusion caused by other (generally thermal) objects. In this respect the radio has an additional advantage, for its nonthermal nature almost always means that SNRs will dominate over thermal sources at long wavelengths.

Of the 155 known Galactic SNR (Green, 1988), less than a third have been detected at optical and X-ray wavelengths. In the observing bands the numbers ($n$) which have been detected are:

> Radio: $n = 155$
> Optical: $n = 42$
> X-ray: $n = 43$

A morphological classification of the 155 SNR (Green, 1988) into shell, center-filled and combination (displaying both types of structure) yields the following distribution:

> Shell: $n = 119$
> Filled: $n = 11$
> Combination: $n = 11$
> Uncertain: $n = 14$

where the last category consists of objects insufficiently resolved or too complex to reliably classify (there are also some questionable cases among the other three categories).

## 1.1. GENERAL LITERATURE

For general background, the references cited at the beginning of the first lecture (Kundt, 1988; Roger and Landecker, 1988; Green, 1988; Srinivisan and Radhakrishnan, 1985; Raymond, 1984; and Danziger and Gorenstein, 1983) can be relied upon. Green (1988) gives extensive references for each object which can be used to find information about a particular aspect (radio map, optical spectrum, etc.). Optical morphologies have been catalogued by Van den Bergh et al. (1973) and Van den Bergh (1978). An up to date general review is that of Weiler and Sramek (1988). Much work has also been done on SNRs in the Magellanic Clouds at optical, radio and X-ray wavelengths (Mathewson et al., 1984; Mathewson et al., 1983; Mills et al., 1982).

## 2. Supernova Shells and their Properties

There is considerable diversity among remnants whose morphology is undeniably shell-like. Few display the near-circularity seen in young shells like Tycho (see Figure 1 of my first lecture), although it is not uncommon to see a shell which is nearly circular for about half of its circumference. A good example is the SNR VRO42.05.01 (Figure 1) which

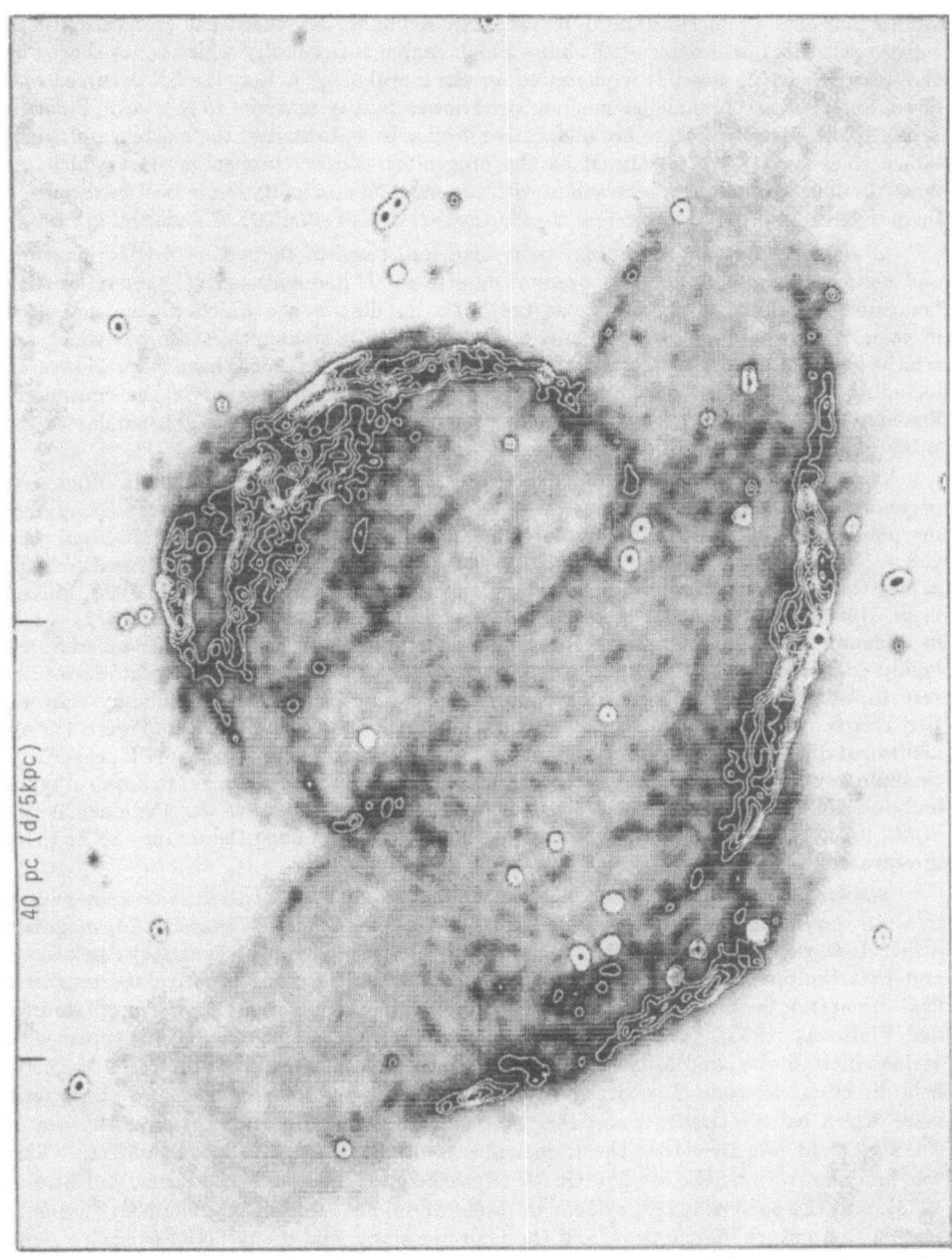

40 pc (d/5kpc)

**Figure 1.** A λ49 cm map of the SNR VRO42.05.01 made with the Westerbork Telescope.

almost looks like a superposition of two separate remnants (as indeed has on occasion been suggested). The northeastern half shows a high degree of circularity which is not shared by the western part. A possible explanation for the morphology is that the SN occurred at a discontinuity in the interstellar medium, with lower density material to the west (Pineault *et al.*, 1985). However, there are alternative mechanisms, including the existence of wind-blown shell structure(s) produced by the progenitor and/or companion stars which are now "illuminated" through interaction with the shock, a possibility which will be discussed further when we come to IC443 (see also, in the case of VRO42.05.01, Pineault *et al.*, 1987).

In addition to their continuum radio emission, some of the well-evolved or mature, and hence presumably old, SNR produce detectable HI line emission. (I prefer the term "mature" to "old" in this context as the latter implies an age which can be measured in years, while there may be remnants well-advanced in a dynamical sense but which are relatively young in a strictly temporal one.) Braun and Strom (1986a) have mapped several, including VRO42.05.01 and IC443. Much of the HI is concentrated in the continuum filaments and the velocity spread observed, up to several hundred km/s, is similar to the values measured in the optical line emission in these objects.

The optical emission observed in mature remnants differs from what is often seen in young ones in several important respects. One of the most striking of these is purely morphological: the optical filaments in young objects like Tycho, Kepler, SN1006 and even Cas A (see *e.g.* Van den Bergh *et al.*, 1973; Van den Bergh, 1978) bear a crude resemblance at best to their radio continuum emission (Duin and Strom, 1975; Henbest, 1980; Matsui *et al.*, 1984; Reynolds and Gilmore, 1986; Braun *et al.*, 1987; Dickel *et al.*, 1982), while in remnants like IC443 it has long been apparent that the optical and radio emission are highly correlated (*e.g.* Duin and Van der Laan, 1975). This behavior can also be readily seen in VRO42.05.01, as illustrated by Figure 2 where a red exposure (mainly showing Hα) clearly reveals most of the ridges which dominate the radio structure (Figure 1). An additional difference is the faintness of the optical emission in many young SNRs, especially the Balmer-dominated ones like Tycho and SN1006. The whole question of the optical types and possible classification schemes has been discussed in a qualitative way (Van den Bergh *et al.*, 1973; Van den Bergh, 1978), and it is clear that even among the mature SNRs there are several morphological types.

The synchrotron radiation through which SNRs reveal themselves at radio wavelengths is by its very nature dependent upon, and hence evidence for the presence of, magnetic fields. It is well-known that the radio emission will, in general, be (partially) polarized, and that the direction of the plane of polarization can be used to derive the magnetic field orientation in the emission region projected onto the plane of the sky (*e.g.* Gardner and Whiteoak, 1966). The radio polarization in many remnants has been mapped with various instruments, and Milne (1987) has recently published a compendium of magnetic field directions for some 27 objects. It is an oft-stated view that SN shells have magnetic fields which can be classified as either predominantly radial or tangential in the rim of the shell, and moreover that the former always occurs among the young objects. This was first shown to be the case for Cas A (Rosenberg, 1970) and Tycho (Duin and Strom, 1975), and the pattern is still evident in recent additions. Tangential alignments are only seen in the mature remnants well-off the Galactic plane, and many SNRs exhibit a more complex behavior. A tangential configuration might simply arise from compression of the ambient field by the shock. The cause of the radial field has been a topic of much well-

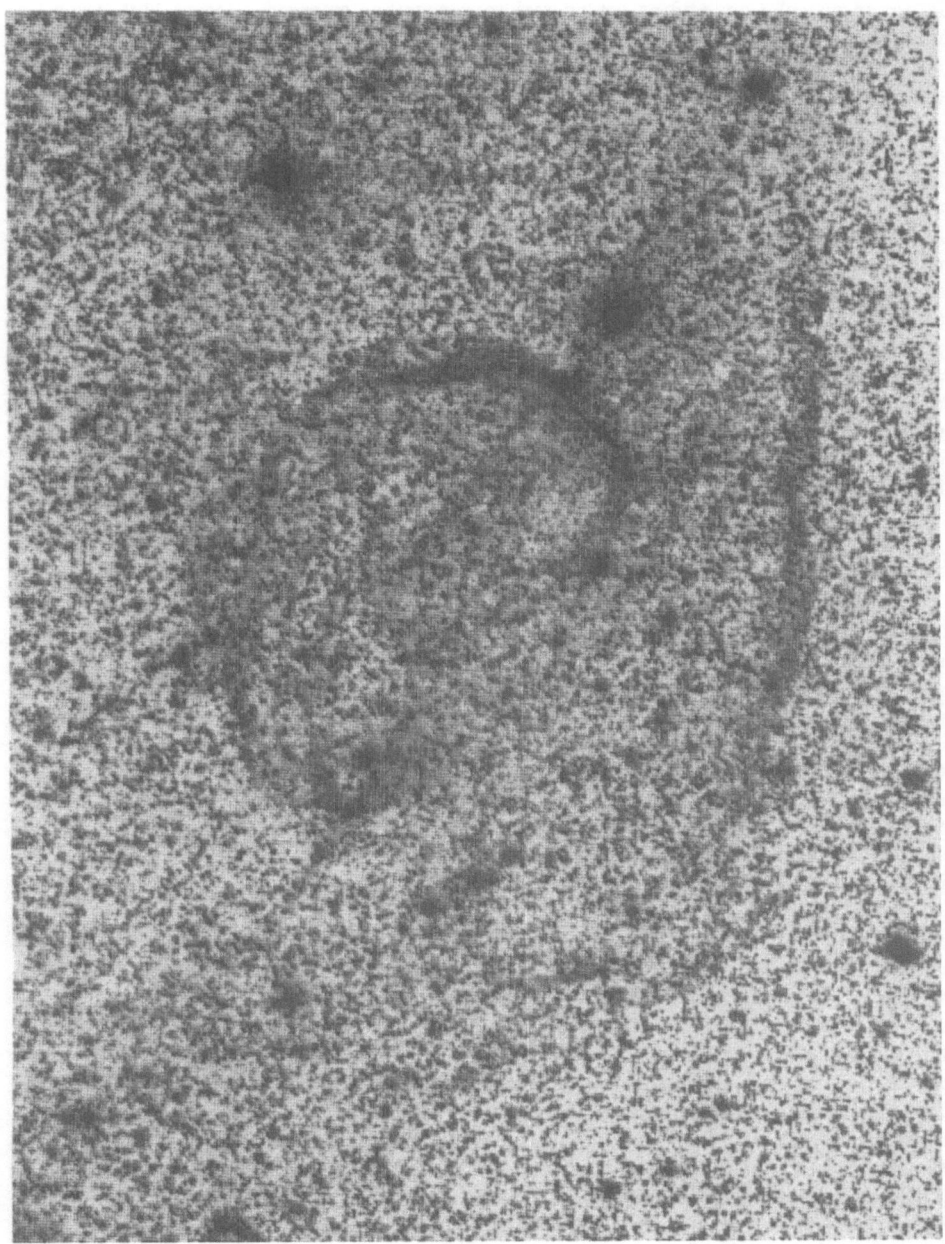

**Figure 2.** A red (mainly Hα) image of VRO42.05.01 to the same scale as Figure 1.

informed speculation, and as yet no real consensus has emerged. Recent high resolution maps indicate that whatever its origin, it must be intimately related to the interaction between the shock wave and the interstellar medium, for the radial configuration is present at the very edge of young remnants.

Turning to the general problem of shell morphology, seen in any wavelength region (but especially radio or X-ray, which seem to give the most complete picture), much attention has recently been paid to the idea that SN shells have a three-dimensional structure akin to an old-fashioned wooden barrel, with the emission concentrated in the staves, and very little coming from the end caps (Kesteven and Caswell, 1987). Although VRO42.05.01 does not conform to this picture, its bow-shaped western half is quite similar to *half* a barrel sliced along its major axis. Kesteven and Caswell argue that most shells fit in their scheme (a ring would correspond to a barrel viewed along its major axis). They suggest several possible mechanisms for the origin of this structure, including anisotropic emission and a possible role for magnetic fields, but prefer an explanation in which the SN outburst itself was largely toroidal. While there are some striking examples of the barrel morphology, a degree of caution is probably warrented when it comes to claims that all shells are members of the class. It is all too easy to try to force each SNR into a scheme whose main rule is the requirement that there be an axis of (approximate) mirror symmetry.

## 3. Consideration of Some Individual Shells

The Cygnus Loop is an impressive and well-studied SNR, largely because its proximity makes it uniquely accessible for investigation throughout the electromagnetic spectrum, and at the same time puts its projected location off the most crowded part of the Galactic plane, reducing the confusion by background objects. Braun and Strom (1986b) have used infrared maps from IRAS and other information to investigate shock-heated dust in the Cygnus Loop and its ambient medium. They find that standard models for shock heating of dust particles (Draine, 1981) can explain the emission seen, and they also present evidence for adjacent gas containing cooler dust. The observations strongly suggest that a dense shell predating the SN outburst has been heated by the passage of the blast wave, and this is the main component now so prominent at optical, radio, X-ray and infrared wavelengths.

Cox (1988) identifies at least four distinct density regimes in the Cygnus Loop, ranging from just over 0.1 cm$^{-3}$ to more than 10 cm$^{-3}$. They are found to vary widely in temperature ($\simeq 10^3 - > 10^6$ K) with corresponding shock speeds reaching up to some 400 km/s. Optically, the most striking of these (and the dominant component in spectacular and often publicized photographs) consists of the high density and rather slow-moving forbidden line filaments at some $10^4$ K. These have been widely studied, and their relative brightness means that they are quite amenable to spectroscopic investigations (*e.g.* Fesen *et al.*, 1982). Recently, through the use of imaging Fabry-Perot techniques such as TAURUS, it has become feasible to do detailed two-dimensional kinematic work on bright filaments such as those in the Cygnus Loop (Greidanus and Strom, 1988a; 1988b). A pattern is found in the [OIII] emission in particular which is qualitatively consistent with many of the filaments being sheets seen nearly edge-on. Modelling which has been done by Hester (1987) is able to produce features bearing an uncanny resemblance to filaments in the Cygnus Loop and other SN shells both in terms of their morphology and kinematic properties.

Let us turn now to IC443, another impressive shell SNR seen in the Galactic anticenter region. Although its distance is somewhat greater than that of the Cygnus Loop, the fact that it does lie in the anticenter region and is somewhat displaced from the plane makes it nonetheless a suitable object for detailed study. IC443 lies on the edge of a cloud complex, some of which can be seen stretching to the northeast in optical photographs (Figure 3).

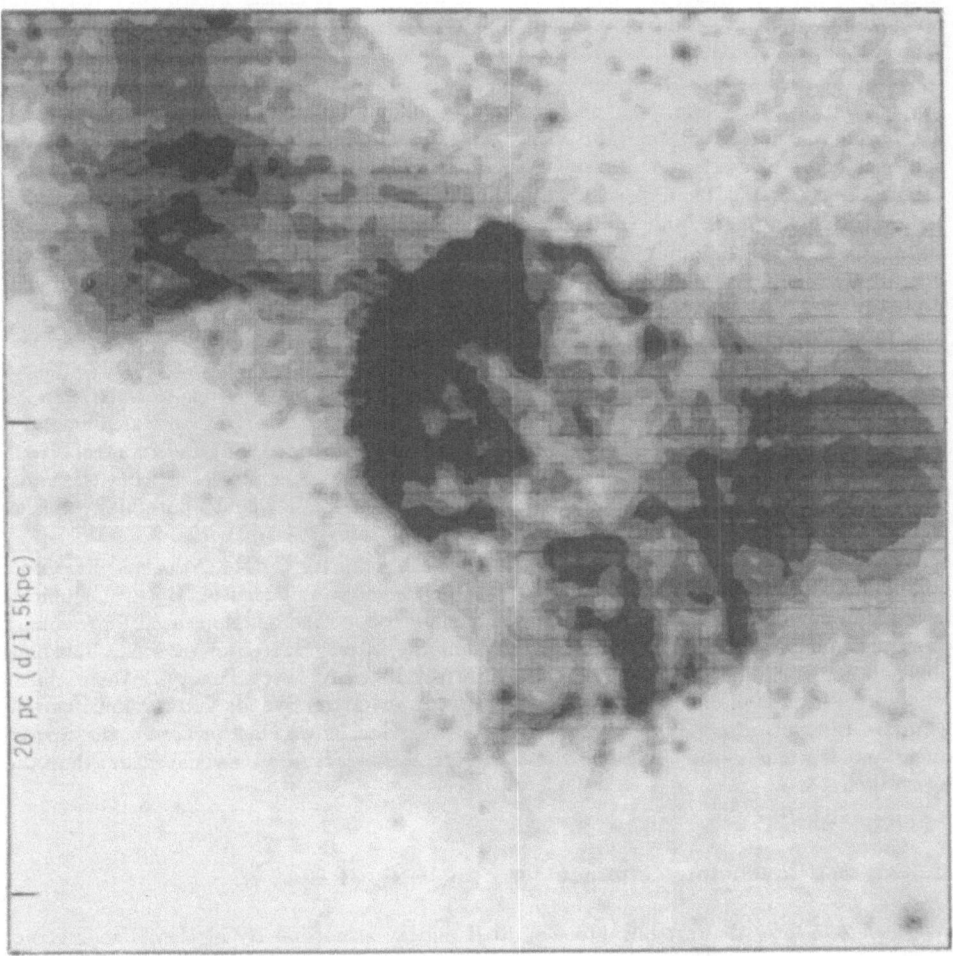

**Figure 3.** A red (mainly Hα) exposure of IC443, also showing some of the adjacent emission from the complex to the northeast. The image has been spatially filtered and smoothed to remove stars and enhance the more extended structures. The gray patch on the western edge is the remnant of a bright star, not true extended emission.

The shell itself is dominated by a bright rim on its northeastern edge (suggesting that it might be interacting with the adjacent cloud), while what looks like a larger and fainter shell can be discerned to the southwest. Braun and Strom (1986c) have studied the entire region at infrared and radio wavelengths. IRAS survey data were used to investigate the distribution of dust with different characteristics throughout the region. In the infrared, the entire complex clearly stands out from other Galactic components. By using a spectral decomposition·technique, it proves possible to separate the emission into cool and warm components, the latter clearly showing the IC443 shell and hence being (as expected) mainly the result of shock-heated dust.

Radio observations with the Westerbork Telescope produced something of a surprise, for in addition to the two shells described above, they reveal a third large shell stretching east from the northeast rim (and visible in the optical emission), and south from the southern edge. These features can be seen in the radio map (Figure 4). Braun and Strom (1986c) suggest that as in the Cygnus Loop, we are seeing large bubbles which existed before the SN outburst, and which are now illuminated by the interaction with the shock front. They are able to pinpoint more than 10 members of the Gem OB1 complex which are candidate sources of the gas and energy needed to evacuate such large cavities, and calculate that the required energy is in principle available.

In addition to the components already mentioned, the southern edge of the bright rim in IC443 contains much neutral gas. This has been mapped in great detail in the 21 cm neutral hydrogen line (Braun and Strom, 1986a), and there is an excellent correlation with the ridges of radio continuum emission (which are also seen as bright optical filaments). Moreover, molecular hydrogen has also been observed (Burton, 1988) and its structure is very similar to the HI. A number of other molecules have been detected from the same area, including OH, CO, $HCO^+$, HCN and CS. Further studies could potentially tell us much about the kinematics of and molecule formation near strong shocks.

In view of the multiple, interconnecting shells comprising IC443, the peculiar morphology of VRO42.05.01 (Figures 1 and 2) is no longer so outlandish. Both phenomena could well have a common origin, whatever it might be, so there is no reason for invoking multiple, independent SN explosions to explain the latter. Moreover, it seems morphologically unlikely that IC443 could be explained by a density discontinuity in the ambient medium (Pineault *et al.*, 1985), making this appear less attractive for VRO42.05.01 as well (but see Pineault *et al.*, 1987). Whether or not the wind-blown shell picture is the correct one, there is accumulating evidence that a hierarchical structure is the rule rather than the exception.

## 4. General Concluding Comments

We have seen how the recognizable shell SNR display a multiplicity of partial shell structures, at least among the mature remnants. The young shell SNR all show clean, single shells, although the remnant of SN185 (the oldest of the group) may display a secondary appendage to the southwest. Why should a rapidly evolving object like a SNR retain a high degree of circularity? Once it has become adiabatic the shock radius depends upon density as, $\rho^{-1/5}$ (*e.g.* Woltjer, 1972). Although the similarity solution from which this is derived is strictly speaking only valid for a constant density environment, we can imagine a shock front developing in two adjacent sectors of different density, and it is clear that the radius will be very weakly dependent upon $\rho$. It is the nature of an adiabatic shock to propagate in a manner relatively insensitive to its environment, so it has a high degree of built-in sphericity.

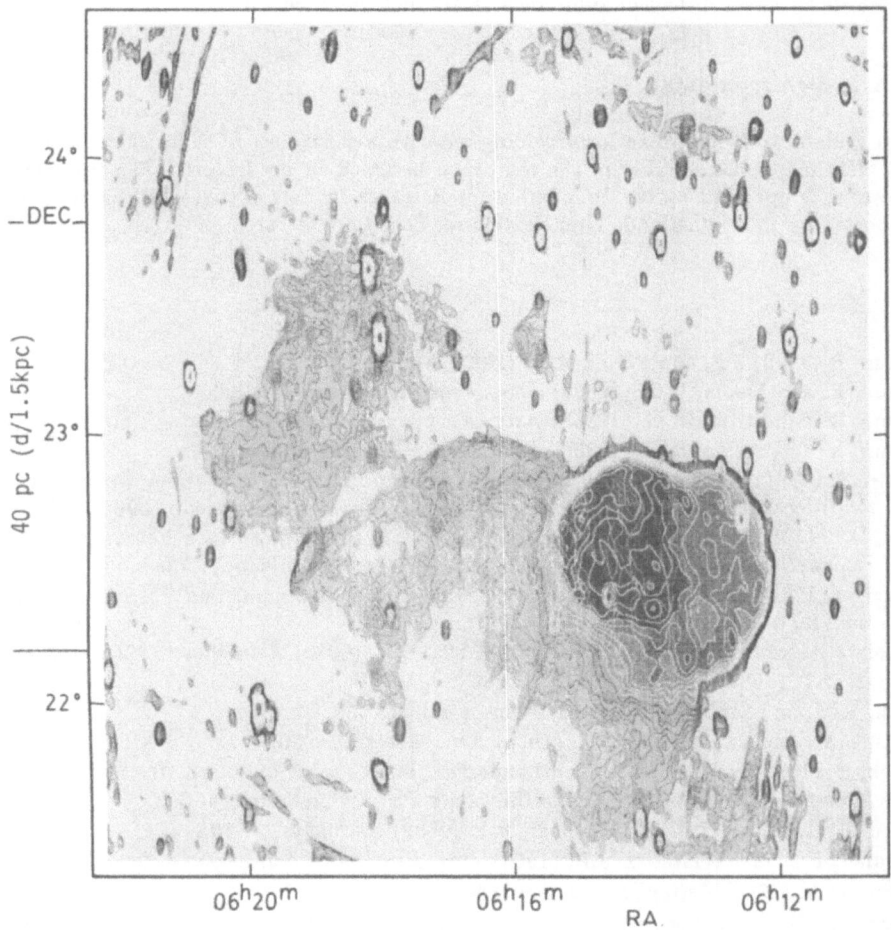

**Figure 4.** A $\lambda92$ cm map of the IC443 complex made with the Westerbork Telescope.

Another point worth considering is how the shock front will propagate through the wind-blown bubbles postulated above to explain what is seen in IC443. If a SN explodes in such a cavity, the ejecta will initially expand through the low density interior with little deceleration. Upon encountering the cavity wall, a relatively slow shock will be excited and the high density present will ensure that a long time is required for it to traverse the shell, and presumably also be responsible for the high brightness emission found there. This is the region of the dense forbidden line filaments. So if wind-blown cavities are the explanation for the structures seen in mature remnants, then the reason that we see so many SN shells in this stage is a combination of their brightness and the fact that the shocks spend most of their lifetimes passing through the shell walls.

## 5. Acknowledgements

I am grateful to Doug Milne for providing me with a superb set of slides showing MOST 843 MHz maps of a number of the remnants discussed in my lecture. The Westerbork Telescope is operated by the Foundation for Research in Astronomy, which is financially supported by the Netherlands Organisation for Scientific Research (NWO).

## 6. References

Braun, R., Gull, S.F. and Perley, R.A. (1987) *Nature* **327**, 395.

Braun, R. and Strom, R.G. (1986a) *Astr. Astrophys. Suppl.* **63**, 345.

Braun, R. and Strom, R.G. (1986b) *Astr. Astrophys.* **164**, 208.

Braun, R. and Strom, R.G. (1986c) *Astr. Astrophys.* **164**, 193.

Burton, M.G. (1988) in Roger, R.S. and Landecker, T.L. (eds.) *Supernova Remnants and the Interstellar Medium*, Cambridge University Press, Cambridge, p. 399.

Cox, D.P. (1988) in Roger, R.S. and Landecker, T.L. (eds.) *Supernova Remnants and the Interstellar Medium*, Cambridge University Press, Cambridge, p. 73.

Danziger, J. and Gorenstein, P. (eds.) (1983) *Supernova Remnants and Their X-ray Emission*, Reidel, Dordrecht.

Dickel, J.R., Murray, S.S., Morris, J. and Wells, D.C. (1982) *Astrophys. J.* **257**, 145.

Draine, B.T. (1981) *Astrophys. J.* **245**, 880.

Duin, R.M. and Strom, R.G. (1975) *Astr. Astrophys.* **39**, 33.

Duin, R.M. and Van der Laan, H. (1975) *Astr. Astrophys.* **40**, 111.

Dwek, E. (1988) in Roger, R.S. and Landecker, T.L. (eds.), *Supernova Remnants and the Interstellar Medium*, Cambridge University Press, Cambridge, p. 363.

Fesen, R.A., Blair, W.P. and Kirshner, R.P. (1982) *Astrophys. J.* **262**, 171.

Gardner, F.F. and Whiteoak, J.B. (1966) *Ann. Rev. Astr. Astrophys.*, 4, 245.

Green, D.A. (1988) *Astrophys. Sp. Sci.* **148**, 3.

Greidanus, H. and Strom, R.G. (1988a) in Roger, R.S. and Landecker, T.L. (eds.) *Supernova Remnants and the Interstellar Medium*, Cambridge University Press, Cambridge, p. 433.

Greidanus, H. and Strom, R.G. (1988b) in Kundt, W. (ed.) *Supernova Shells and Their Birth Events*, Springer-Verlag, Berlin, p. 140.

Henbest, S.N. (1980) *Mon. Not. Roy. Astr. Soc.* **190**, 833.

Hester, J.J. (1987) *Astrophys. J.* **314**, 187.

Kesteven, M.J. and Caswell, J.L. (1987) *Astr. Astrophys* **183**, 118.

Kundt, W. (ed.) (1988) *Supernova Shells and Their Birth Events*, Springer-Verlag, Berlin.

273

Mathewson, D.S., Ford, V.L., Dopita, M.A., Tuohy, I.R., Long, K.S. and Helfand, D.J. (1983) *Astrophys. J. Suppl.* **51**, 345.

Mathewson, D.S., Ford, V.L., Dopita, M.A., Tuohy, I.R., Mills, B.Y. and Turtle, A.J. (1984) *Astrophys. J. Suppl.* **55**, 189.

Matsui, Y., Long, K.S., Dickel, J.R. and Greisen, E.W. (1984) *Astrophys. J.* **287**, 295.

Mills, B.Y., Little, A.G., Durdin, J.M. and Kesteven, M.J. (1982) *Mon. Not. Roy. Astr. Soc.* **200**, 1007.

Milne, D.K. (1987) *Aust. J. Phys.* **40**, 771.

Pineault, S., Pritchet, C.J., Landecker, T.L., Routledge, D. and Vaneldik, J.F. (1985) *Astr. Astrophys.* **151**, 52.

Pineault, S., Landecker, T.L. and Routledge, D. (1987) *Astrophys. J.* **315**, 580.

Raymond, J.C. (1984) *Ann. Rev. Astr. Astrophys.* **22**, 75.

Reynolds, S.P. and Gilmore, D.M. (1986) *Astr. J.* **92**, 1138.

Roger, R.S. and Landecker, T.L. (eds.) (1988) *Supernova Remnants and the Interstellar Medium*, Cambridge University Press, Cambridge.

Rosenberg, I. (1970) *Mon. Not. Roy. Astr. Soc.* **151**, 109.

Srinivasan, G. and Radhakrishnan, V. (1985) *Supernovae, their Progenitors and Remnants*, Indian Academy of Sciences, Bangalore.

Van den Bergh, S. (1978) *Astrophys. J. Suppl.* **38**, 119.

Van den Bergh, S., Marscher, A.P. and Terzian, Y. (1973) *Astrophys. J. Suppl.* **26**, 19.

Weiler, K.W. and Sramek, R.A. (1988) *Ann. Rev. Astr. Astrophys.* **26**, 295.

Woltjer, L. (1972) *Ann. Rev. Astr. Astrophys.* **10**, 129.

# SUPERNOVA REMNANTS III: EXOTICS

RICHARD G. STROM
*Netherlands Foundation for Research in Astronomy*
*Radiosterrenwacht*
*P.O. Box 2*
*7990 AA Dwingeloo*
*Netherlands*

ABSTRACT. Taking exotic supernova remnants to be principally those objects which are morphologically unusual, the class is shown to mainly consist of remnants which, like the Crab Nebula, have radio and X-ray emission peaking near the center, indicating that they are powered by a central energy source. The group, which includes neutron stars seen as pulsars, is briefly considered. A related category, which appears to combine a Crab-like object with a shell or less well-defined surrounding emission, is also discussed, with particular reference to CTB80. The so-called axisymmetric sources are considered, and it is concluded that there is little evidence to indicate that they should be treated as a separate category. Finally, the linear features variously referred to as jets, hoses and chimneys are discussed. The best studied one, in the Crab Nebula, exhibits outward motion in all directions, and may represent discharge of relativistic fluid from the remnant's center, although our limited understanding of the phenomenon is emphasized.

## 1. Introduction

ĕxŏ´ tĭc ...strange, bizarre; attractively strange or unusual; of kind newly brought into use (*The Concise Oxford Dictionary*).

If we are to discuss exotic supernova remnants (SNRs), as I now propose to do, then we have to agree upon the objects which we will include. The definition quoted above is not a bad starting point: exotic SNR are indeed unusual. Moreover, the observant will note that they are uncommonly attractive to astronomers, and it is not even too far-fetched to conclude that the most recently discovered specimens are quite likely to be proclaimed singular, unique, a modern Rosetta Stone (and much more). No, the problem is not so much one of definition as of inclusion: where do we draw the line between truly bizarre and merely somewhat out of the ordinary? Astronomers would probably universally agree that the Crab Nebula is exotic, but its unique position has been undermined by discoveries of similar objects going back to the first high resolution study of 3C58 (Weiler and Seielstad, 1971). Cas A is unusual in many respects (*e.g.* Van den Bergh and Kamper, 1983; Tuffs,

*W. Kundt (ed.), Neutron Stars and Their Birth Events, 274–280.*
© *1990 Kluwer Academic Publishers.*

1986; Braun *et al.*, 1987), but it is really just a young (hence extreme) version of a shell SNR. Objects like VRO42.05.01 and IC443 have, as we saw in my last lecture, their abnormalities, but they are nevertheless shells. Indeed, each SNR, if examined in sufficient detail, is likely to display some unique quality. So we must take care not to be too *inclusive*.

But how *exclusive* should we be? As noted above, the Crab Nebula has become somewhat less unique, so perhaps we should discard it and its brethren. Shells, of course, don't get a look in, and even extreme examples of morphological irregularity are to be ignored. CTB80 (Strom *et al.*, 1980; Angerhofer *et al.*, 1981) can be dropped; like the Crab, it has a pulsar (Strom, 1987; Kulkarni *et al.*, 1988). Vela (Harnden *et al.*, 1985) must suffer the same fate, and so will RCW89 (Seward *et al.*, 1984), MSH15-5*6* (Kesteven and Caswell, 1987), ... And before we know it, we will be left with practically nothing. Like the Cheshire cat which disappeared before Alice's eyes (Dodgson, 1865), we will (if we are lucky) end up staring at an ironical grin (the mystical smile of W50?).

No, we must return to the beginning. Our fundamental starting point will be one based on morphology: exotic remnants *look* unusual. As we saw in the second lecture, some 80% of SNRs can be classified as shells, so now we will, loosely speaking, be considering the rest. If we categorize the observed properties of SNR in general to be:

> extended radio source
> —nonthermal
> —generally polarized
> thermal line emitting optical nebulosity
> infrared from shock-heated dust
> coextensive X-ray emission

then we might typify the class of exotics as:

> radio emission peaks near center
> —flat radio spectrum ($\alpha \geq -0.3$)
> nonthermal optical continuum
> —polarized
> centrally peaked nonthermal X-ray emission
> —follows on from radio and optical

In terms of their astrophysics, exotics are usually found to have a central energy source powering them.

## 1.1. GENERAL LITERATURE

For general background, the references cited at the beginning of the first lecture (Kundt, 1988; Roger and Landecker, 1988; Green, 1988; Srinivisan and Radhakrishnan, 1985; Raymond, 1984; and Danziger and Gorenstein, 1983) should again prove useful. A recent review (Weiler and Sramek, 1988), although devoted to all kinds of SNRs (and their initiating supernovae as well), pays special attention to the nonshell morphologies. In addition a workshop several years ago was devoted to the Crab Nebula and its relatives, the proceedings of which (Kafatos and Henry, 1985) provide an overview of the subject at the time. An appendix catalogues the then known Crab-like SNR (Weiler, 1985).

## 2. The Crab Nebula and Its Relatives

The Crab Nebula is, as we saw in the first lecture, without doubt the result of a supernova which exploded in the middle of the eleventh century. It is consequently a supernova remnant, but if we knew nothing about SN1054 we might have some doubts, for the expansion speed of the most prominent nebulosity is substantially less than that observed in other contemporary SNR (SN1006, for example; see Long *et al.*, 1988). The filaments have, moreover, undergone significant acceleration (while those in shell SNR like SN1006 have been substantially decelerated), and the energy required to power the entire nebula is reasonably balanced by the spindown of the central neutron star, PSR0531+21 (see Manchester and Taylor, 1977, and references therein). In other words, energetically, there appears to be no need for the supernova explosion itself. In the Crab this can be attributed to the unusually high rate at which the neutron star is losing kinetic energy, which may simply swamp the remains of the original explosion, although it may be something of an embarrassement in other less luminous but otherwise similar objects. The class of composite SNR (*e.g.* Weiler, 1985) offers at least a partial solution to the problem, as they may represent those objects in which both the initial ejecta and the neutron-star-excited filaments can be seen.

The term "plerion" has been devised to denote those objects which share the Crab Nebula's center-filled morphology (Weiler and Panagia, 1978), and there are perhaps 10 pure plerions known, including the Crab. Since such objects represent emitting material which derives all its power from a neutron star near the center, any pulsar may be surrounded by similar emission (and at some level, they probably all are). Seward and Wang (1988) have made an investigation along these lines and find an empirical relationship between the X-ray luminosity of the surrounding nebula and the rate at which the powering neutron star loses energy (see also Strom, 1988). They then go on to predict the likely properties of undetected pulsars in similar objects not known to have neutron stars, remnants like 3C58 (*e.g.* Reynolds and Aller, 1988).

### 2.1. COMPOSITE REMNANTS: THE CASE OF CTB80

There are several examples of plerions with surrounding emission of a steeper spectrum, in some cases in the form of a shell which is plausibly the result of a blast wave excited by the original supernova (for some examples, see Weiler, 1985). One which does not look very shell-like in its radio structure is CTB80, whose structure was first studied in detail by Strom *et al.* (1980) and Angerhofer *et al.* (1981). Its $\lambda$49 cm radio structure is shown in Figure 1, where we apparently see a bundle of arcs and filaments emanating from a central, amorphous plateau (the latter is shown as light gray in Figure 1, surrounded by darker shading because a double-valued mapping of radio brightness onto the gray shading has been used). At the western end of this plateau, there is a high-brightness component, itself rather amorphous with a diameter of about 30" arc and a very flat radio spectrum ($\alpha \simeq 0$). It has been known for some time that this component contains a compact X-ray source (Becker *et al.*, 1982), *prima facie* evidence for the existence of a neutron star.

Strom *et al.* (1984) mapped the flat spectrum component with sufficient resolution to

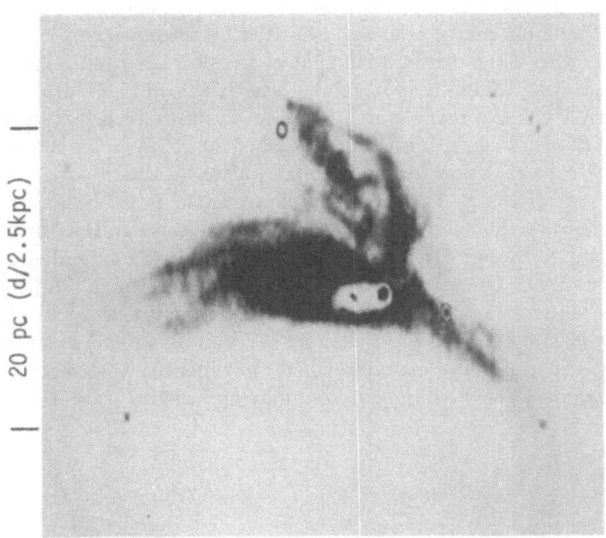

**Figure 1.** A λ49 cm map of CTB80 made with the Westerbork Telescope. The outer emission stretches over nearly 1°.

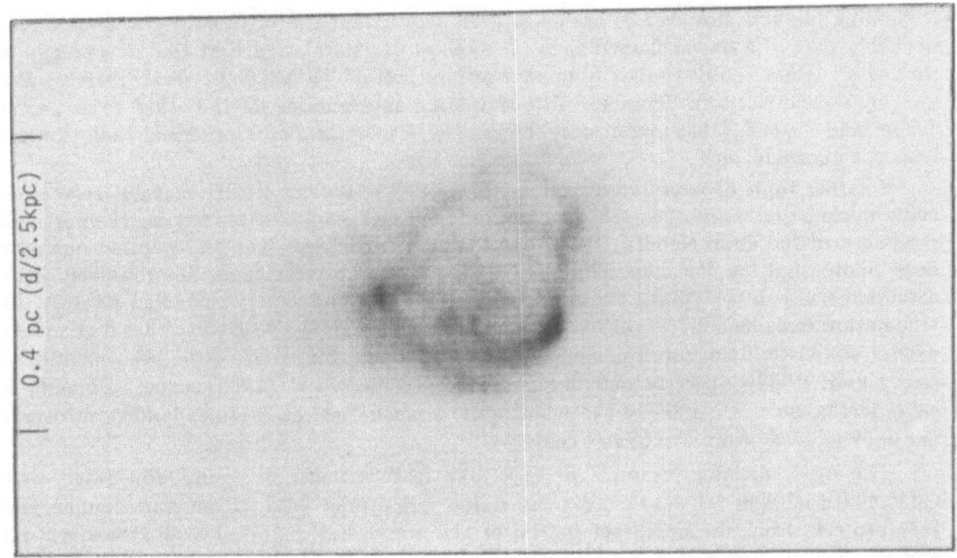

**Figure 2.** A λ6 cm map of the central, flat spectrum component of CTB80 made with the Very Large Array.

show that it has a dominant hot spot near the southwestern edge, and speculated that this might represent some local center of activity excited by a compact object. The position of this feature agrees, within the errors, with that of the point X-ray source found by Becker *et al.* A higher resolution λ6 cm map obtained later is shown in Figure 2. For comparison an observation at λ20 cm was made which revealed an unresolved feature to the northeast of the hot spot which also falls within the error box of the X-ray point source (Strom, 1987), and this turns out to be emission from a 39.5 ms pulsar (Kulkarni *et al.*, 1988). As in the Crab Nebula, the rate of energy loss from the neutron star in CTB80 agrees well with the energy required to power the nebula. This would appear to solve the "mystery" of the flat spectrum component, although its possible relationship to the extended complex of arms and arcs remains unclear. One explanation is along the lines of the outer structure being the remains of the supernova which produced the neutron star, since its characteristic age is quite large, notwithstanding the short period. Fesen *et al.* (1988) point out the existence of a shell-like structure in infrared emission which they identify with a SNR shell now being re-energized by its interaction with the outward moving neutron star (see also Strom, 1988).

## 3. Discussion

With the discovery of two somewhat unusual funnel-shaped nonthermal sources near the Galactic center (Becker and Helfand, 1985; Shaver *et al.*, 1985), Helfand and Becker (1985) suggested the possible existence of a new class of "axisymmetric Galactic radio sources" produced by energy released from a fast-moving object (presumably a neutron star, or something like it). However, it has since been shown that one of the objects, G5.4-1.2, is probably part of a barrel-shaped shell (Caswell *et al.*, 1987), and that the morphology of its "wing" is not too dissimilar from the western half of VRO42.05.01 (see Figures 1 and 2 in my second lecture). There are also competing explanations for the other object, so as Weiler and Sramek (1988) point out, the case for a new class of nonthermal radio sources is as yet unconvincing.

Another topic of some recent interest concerns the hose or jet-like features seen especially in the radio maps of certain remnants. The best-known of these adjoins the northern periphery of the Crab Nebula. It is ironical that its existence was first pointed out on a deep photograph by Van den Bergh (1970) nearly twenty years ago, but relatively little attention was paid to it until Gull and Fesen's (1982) deep imagery rekindled interest. Its true nature remains a matter of some speculation, although the possibility that it is a place where relativistic fluid from the nebula's interior is being vented has often been mentioned (*e.g.* Kundt, 1983). Spectroscopic observations by Wilson *et al.* (1985) suggest a kinematical model in which the walls of the cylindrically-shaped "jet" are both expanding outwards and moving away from the Crab's center.

The other striking example of a jet-like feature is to be found associated with G332.4+0.1 (Roger *et al.*, 1985). At a low brightness level, a narrow sinuous feature projects from the northeastern rim of this somewhat distorted shell, broadens out and appears to merge with an amorphous "plume" of low brightness emission. Neither the SNR nor its appendage have been detected in other wavelength bands, mainly because of extinction. One possible explanation is that, as suggested for the Crab Nebula, the "jet" marks a point where material is venting from the shell's interior. It must however be said that this hypothesis has not really been subjected to the critical scrutiny required to fully test its physical plausibility. The radio observations of G332.4+0.1 are fully consistent with the "jet" being thermal and in view of the crowded nature of the field the chance superposition of a background object cannot be entirely discounted. A final judgment must await additional observational material.

## 4. Concluding Comments

To the exotic SNR we can thus assign both morphological (center-filled) and operational (still significant energy input) definitions. Although the Crab Nebula remains the prototype, its prodigious energy output makes it rather atypical. With the discovery of less luminous examples, there has been progress in extending our investigation of their parameter space. Our understanding of how a neutron star deposits the energy which it releases in the surrounding environment is still far from complete, however, and it should be mentioned that similar processes must be taking place near binary X-ray sources (see some of the other lectures presented during this ASI).

Finally, I have refrained from discussing W50, the extended nonthermal source which harbors the unusual binary system SS433, surely one of the most exotic objects in the Galaxy. There can be no doubt that the binary and outer shell go together, for the influence of the former can be seen in the morphology of the latter, which appears to have been pushed outward along the directions of the twin jets (e.g. Downes et al., 1986). What remains unclear is whether anything we see has been the result of shock waves initiated by a supernova explosion. The question is posed by Green (1988): "Is this a supernova remnant?"

## 5. Acknowledgements

I am grateful to Doug Milne for several beautiful slides of remnants with which I was able to illustrate this lecture in Erice. The Westerbork Telescope is operated by the Foundation for Research in Astronomy, which is financially supported by the Netherlands Organisation for Scientific Research (NWO). The Very Large Array is a facility of the National Radio Astronomy Observatory, the latter being operated by Associated Universities, Inc., under contract with the U.S. National Science Foundation.

## 6. References

Angerhofer, P.E., Strom, R.G., Velusamy, T. and Kundu, M.R. (1981) *Astr. Astrophys.* **94**, 313.
Becker, R.H., Helfand, D.J. and Szymkowiak, A.E. (1982) *Astrophys. J.* **255**, 557.
Becker, R.H. and Helfand, D.J. (1985) *Nature*, **313**, 115.
Braun, R., Gull, S.F. and Perley, R.A. (1987) *Nature* **327**, 395.
Caswell, J.L., Kesteven, M.J., Komesaroff, M.M., Haynes, R.F., Milne, D.K., Stewart, R.T. and Wilson, S.G. (1987) *Mon. Not. Roy. Astr. Soc.* **225**, 329.

Danziger, J. and Gorenstein, P. (eds.) (1983) *Supernova Remnants and Their X-ray Emission*, Reidel, Dordrecht.
Dodgson, C.L. (Lewis Carroll) (1865) *Alice's Adventures in Wonderland*, Macmillan, London.
Downes, A.J.B., Pauls, T. and Salter, C.J. (1986) *Mon. Not. Roy. Astr. Soc.* **218**, 393.
Fesen, R.A., Shull, J.M. and Saken, J.M. (1988) *Nature* **334**, 229.
Green, D.A. (1988) *Astrophys. Sp. Sci.* **148**, 3.
Gull, T.R. and Fesen, R.A. (1982) *Astrophys. J.* **260**, L75.
Harnden, F.R., Grant, P.D., Seward, F.D. and Kahn, S.M. (1985) *Astrophys. J.* **299**, 828.
Helfand, D.J. and Becker, R.H. (1985) *Nature* **131**, 118.
Kafatos, M.C. and Henry, R.B.C. (eds.) (1985) *The Crab Nebula and Related Supernova Remnants*, Cambridge University Press, Cambridge.
Kesteven, M.J. and Caswell, J.L. (1987) *Astr. Astrophys* **183**, 118.
Kulkarni, S.R., Clifton, T.C., Backer, D.C., Foster, R.S., Fruchter, A.S. and Taylor, J.H. (1988) *Nature* **331**, 50.
Kundt, W. (1983) *Astr. Astrophys.* **121**, L15.
Kundt, W. (ed.) (1988) *Supernova Shells and Their Birth Events*, Springer-Verlag, Berlin.
Long, K.S., Blair, W.P. and Van den Bergh, S. (1988) *Astrophys. J.* **333**, 749.
Manchester, R.N. and Taylor, J.H. (1977) *Pulsars*, W.H. Freeman, San Francisco.
Milne, D.K. (1987) *Aust. J. Phys.* **40**, 771.
Raymond, J.C. (1984) *Ann. Rev. Astr. Astrophys.* **22**, 75.
Reynolds, S.P. and Aller, H.D. (1988) *Astrophys. J.* **327**, 845.
Roger, R.S., Milne, D.K., Kesteven, M.J., Haynes, R.F. and Wellington, K.J. (1985) *Nature* **316**, 44.
Roger, R.S. and Landecker, T.L. (eds.) (1988) *Supernova Remnants and the Interstellar Medium*, Cambridge University Press, Cambridge.
Seward, F.D., Harnden, F.R., Szymkowiak, A. and Swank, J. (1984) *Astrophys. J.* **281**, 650.
Seward, F.D. and Wang, Z.-R. (1988) *Astrophys. J.* **332**, 199.
Shaver, P.A., Salter, C.J., Patnaik, A.R., Van Gorkom, J. and Hunt, G.C. (1985) *Nature* **313**, 113.
Srinivasan, G. and Radhakrishnan, V. (1985) *Supernovae, their Progenitors and Remnants*, Indian Academy of Sciences, Bangalore.
Strom, R.G. (1987) *Astrophys. J.* **319**, L103.
Strom, R.G. (1988) in Kundt, W. (ed.) *Supernova Shells and Their Birth Events*, Springer-Verlag, Berlin, p. 91.
Strom, R.G., Angerhofer, P.E. and Velusamy, T. (1980) *Nature* **284**, 38.
Strom, R.G., Angerhofer, P.E. and Dickel, J.R. (1984) *Astr. Astrophys.* **139**, 43.
Tuffs, R.J. (1986) *Mon. Not. Roy. Astr. Soc.* **219**, 13.
Van den Bergh, S. (1970) *Astrophys. J.* **160**, L27.
Van den Bergh, S. and Kamper, K.W. (1983) *Astrophys. J.* **268**, 129.
Van den Bergh, S., Marscher, A.P. and Terzian, Y. (1973) *Astrophys. J. Suppl.* **26**, 19.
Weiler, K.W. (1985) in Kafatos, M.C. and Henry, R.B.C. (eds.) *The Crab Nebula and Related Supernova Remnants*, Cambridge University Press, Cambridge, p. 265.
Weiler, K.W. and Seielstad, G.A. (1971) *Astrophys. J.* **164**, 455.
Weiler, K.W. and Panagia, N. (1978) *Astr. Astrophys.* **70**, 419.
Weiler, K.W. and Sramek, R.A. (1988) *Ann. Rev. Astr. Astrophys.* **26**, 295.
Wilson, A.S., Samarasinha, N.H. and Hogg, D.E. (1985) *Astrophys. J.* **294**, L121.

# WHICH TYPES OF SUPERNOVAE LEAVE NEUTRON STARS?

David Branch
Department of Physics and Astronomy
University of Oklahoma
Norman, OK 73019, U.S.A.

ABSTRACT. The question of which types of supernovae are associated with neutron stars is considered from the perspective of supernova observations and current ideas about their progenitors and explosion mechanisms. SNe Ia do not leave neutron stars; some but not necessarily all SNe II do; SNe Ib may or may not. These conclusions lead to rough limits on the fraction of Galactic supernovae that leave neutron stars and on the rate at which neutron stars are born via Galactic supernovae. The question of whether these limits are in conflict with evidence from supernova remnants and pulsars is discussed briefly.

## 1. Introduction

Which of the three observed supernova types - Type II (having hydrogen), Type Ia (lacking hydrogen and helium), and Type Ib (having helium but not hydrogen) - leave neutron stars? The direct way to address this question would be to associate supernova types with Galactic supernova remnants that do and do not contain neutron stars. But it is difficult to be sure that a remnant does *not* contain a neutron star, and for most remnants it is difficult to be sure of the type of the supernova progenitor. If it is clear that the remnant includes hydrogen that was ejected during the explosion, then the supernova was of Type II. So we know that the progenitor of the Crab was a Type II (but an unusual one, if there was no high-velocity hydrogen), and that it left a neutron star. Cas A, on the other hand, shows no signs of containing a neutron star, yet it also appears to have ejected hydrogen (Fesen, Becker, and Blair 1987). Neutron stars are not seen in the remnants of Tycho's and Kepler's supernovae either, but the shapes of their light curves, as reconstructed from contemporary reports, are consistent with those of either Type Ia, Type Ib, or the Type II supernovae with "linear" light curves, Type II-L.

In this contribution, the connection between supernovae and neutron stars is considered in a less direct way. Given the observations of extragalactic supernovae of each type, and the current ideas about their stellar progenitors and explosion mechanisms - is associated neutron-star formation expected, or not?

A good place to begin is to consider what can be inferred about the initial (main sequence) masses of supernova progenitor stars from the observed supernova *sites*. Nearly all observed SNe II, and most SNe Ia, are found in spiral galaxies, but in places that are almost mutually exclusive. Huang (1987) found that among a

281

W. Kundt (ed.), Neutron Stars and Their Birth Events, 281–288.

sample of 29 SNe II for which good site photographs were available, 25 were within 5 arcsec of giant H II regions; three of the others were at least in spiral arms, and one was in the star-forming nucleus of M83. Thus the association between SNe II and star formation is practically complete. SNe Ia in spirals are associated neither with giant H II regions (Huang 1987) nor with spiral arms (Maza and van den Bergh 1976). They do, however, come from the disk population rather than from the halo: the SNe Ia rate per unit galaxian mass increases along the Hubble sequence from ellipticals to late-type spirals, and many of the ellipticals that have produced SNe Ia show signs of star formation in the relatively recent past ($\sim 10^8$ yr) or are otherwise peculiar (Kochhar 1989).

Because the sites, and therefore the initial-mass ranges, of the SNe II and SNe Ia progenitors appear to be distinct, it is commonly supposed that SNe II and SNe Ia come from stars whose initial masses are *above* and *below* respectively, the critical mass ($M_c \simeq 8\ M_\odot$) for nondegenerate carbon ignition. Hence the general expectation that SNe II involve core collapse and neutron-star formation, and that SNe Ia are thermonuclear explosions. The inference that the two mass ranges lie entirely on opposite sides of the critical mass is reasonable, but good quantitative limits to the mass ranges (lower for SNe II, upper for SNe Ia), based on the statistics of the sites, are needed.

A few years ago, when SNe Ib were first emphasized, an apparently strong association with star formation was stressed. The association was consistent with several very high spectroscopic estimates of ejected mass, and with the idea that the SNe Ib progenitors are Wolf-Rayet stars (e.g., van den Bergh 1988). Panagia and Laidler (1988) find, however, that the association between SNe Ib and giant H II regions is not as strong as that found by Huang for SNe II. The initial-mass range of SNe Ib progenitors now appears to be in the gray region between the mass ranges of SNe II and SNe Ia, possibly overlapping either or both. It is not clear whether SNe Ib progenitors are above or below the critical mass $M_c$.

## 2. Do Supernovae of Type Ia Leave Neutron Stars?

No, they do not.

*Single* stars having $M_i < M_c$ are not expected to produce neutron stars, but they may not make supernovae, either. Some of the most massive of them ($8\ M_\odot \gtrsim M_i \gtrsim 5\ M_\odot$) conceivably could ignite degenerate carbon and explode as supernovae (without neutron stars), but the paucity of AGB stars having bolometric magnitude brighter than -6, in Galactic and LMC clusters that contain Cepheids, suggests that all stars having $M_i < M_c$ lose enough mass to avoid carbon ignition and become white dwarfs (e.g., Iben 1985). Even if some of these intermediate-mass stars do explode, they should have hydrogen or helium envelopes, and therefore would not be SNe Ia. To get SNe Ia from stars having $M_i < M_c$, one must appeal to accreting white dwarfs in binary systems.

The response of a white dwarf to matter accretion depends (in numerical simulations that assume spherical symmetry) on the initial white-dwarf composition, the composition of the accreted matter, and the accretion rate. One can find a combination of parameters that produces an appealing model for SNe Ia. A carbon-oxygen white dwarf accretes at a rate on the order of $10^{-7}$ $M_\odot$/yr, converts the accreted hydrogen and/or helium to carbon and oxygen by means of thermal flashes (unless the accreted matter is carbon and oxygen in the first place [Iben and Tutukov 1984, Webbink 1984]), ignites central degenerate carbon upon approaching the Chandraskehar mass, and disrupts completely via a subsonic nuclear burning front called a carbon deflagration (Nomoto, Thielemann, and Yokoi 1984, Woosley, Axelrod, and Weaver 1984). With a final kinetic energy of $10^{51}$ ergs, an ejected mass of 1.4 $M_\odot$, a mass of radioactive $^{56}$Ni, $M_{Ni}$, of about 0.6 $M_\odot$, and a comparable mass of intermediate-mass elements from oxygen to calcium, the carbon deflagration model predicts spectra (Branch et al. 1985, Wheeler et al. 1986, Axelrod 1980a,b) and light curves (e.g., Sutherland and Wheeler 1984, Graham 1987) that are in good agreement with observation.

Exploding white dwarfs near the Chandrasekhar mass might be expected to produce a rather homogenous set of supernovae. SNe Ia *are* remarkably similar in their observational properties, but they are not strictly identical (Branch, Drucker, and Jeffery 1988). One possibility for the cause of the differences is that some of the white dwarfs partially solidify before they begin to accrete, and that their subsequent explosions don't disrupt the solid cores (Canal et al. 1988). Such events would leave white dwarf remnants, but not neutron stars.

It also is possible to find parameters such that an accreting white dwarf collapses to form a neutron star. For example, accrete carbon and oxygen onto a C-O or O-Ne-Mg white dwarf at a rate in excess of $3\times10^{-6}$ $M_\odot$/yr (Saio and Nomoto 1985). These "silent" neutron-star births would not, however, simultaneously produce SNe Ia. The accretion-induced collapse might eject a small amount of matter carrying a significant kinetic energy, but the event (if seen) would not be called a Type Ia. The ejected masses of intermediate-mass elements and radioactive nickel both would be much less than the amounts inferred from the spectra, and the peak luminosity would be much too faint. For $H_o$=60 km/s/Mpc, the brightest SNe Ia need $M_{Ni}$=0.6 $M_\odot$ (Arnett, Branch, and Wheeler 1985); other SNe Ia may be somewhat less luminous and eject less nickel, but not orders of magnitudes less.

The physics of the carbon-deflagration is complicated, and the details of the model are subject to change, but the basic picture of SNe Ia as nickel-powered explosions of accreting CO white dwarfs is probably correct. And the simultaneous production of a Type Ia supernova *and* a neutron star would be asking too much of an accreting white dwarf.

## 3. Do Type II Supernovae Leave Neutron Stars?

Yes. At least some of them.

The general expectation is that stars having $M_c < M_i < M_{BH}$ undergo core collapse and produce supernovae with neutron stars. The collapse of stars having $M_{BH} < M_i \lesssim 100$ M$_\odot$ makes black holes, with or without supernovae, and $M_i \gtrsim 100$ M$_\odot$ leads to pair instabilities and ensuing thermonuclear burning that disrupts the star completely. Since stars having $M_i \gtrsim M_c$ are much more numerous than those having $M_i \gg M_c$, most SNe II should leave neutron stars, provided that $M_{BH}$ is not too close to $M_c$. The SN 1987A neutrinos testified to a core collapse, and the duration of the neutrino signal suggests the formation of a neutron star rather than a black hole (Burrows and Lattimer 1987). It will be reassuring, nevertheless, to see some direct signs of the neutron star. If it's there, then $M_{BH} > 20$M$_\odot$, comfortably above $M_c$. Many of the observed SNe II have "plateaus" on their light curves, signifying substantial ejected mass (typically 6M$_\odot$, according to Litvinova and Nadyozhin 1985); if $M_{BH} > 20$M$_\odot$, then most of these SNe II-P will probably fall into the neutron-star mass range.

Not *all* SNe II necessarily leave neutron stars, of course. The pair-instability explosions won't, but they surely are rare (SN 1961V is the favorite candidate). Stars whose cores collapse to black holes might contribute some SNe II without neutron stars, but only if they a) do not lose all of their hydrogen before collapsing and b) eject their outer layers rather than collapse completely.

There also is the possibility that, in spite of the evidence cited above, some stars having $M_i < M_c$ do ignite degenerate carbon and undergo thermonuclear disruptions. If they still have some hydrogen, they will be SNe II. SNe II-L, which account for about one quarter of the observed SNe II (Barbon, Ciatti, and Rosino 1979), have light-curve shapes that are not so different from the characteristic SN I shape (Doggett and Branch 1986); in particular, the light curve of the SN II-L 1979C was very much like that of the SN Ia 1972E both in shape and peak luminosity (Bartunov and Tsvetkov 1986). So one wonders whether some SNe II-L are thermonuclear disruptions of stars having $M_i < M_c$. However, I am not aware of any evidence, from the observed sites, for a systematic difference in progenitor mass for SNe II-P and II-L. SN 1979C itself appeared in or near an H II region (de Vaucouleurs et al. 1981).

The detailed published models of stars having $M_i$ just a little above $M_c$ predict that they evolve to core collapse, but the internal evolution of these stars is complicated and the possibility that some of them explode via oxygen deflagrations upon ignition of central oxygen cannot yet be ruled out (Canal 1989). If stars do explode in this way, with their hydrogen envelopes intact, they would be SNe II without neutron stars. The Crab is not necessarily in conflict with this possibility, because the high neon abundance and the low nitrogen abundance in the Crab point to $M_i > 9.5$ M$_\odot$ (Henry 1986).

## 4. Do SNe Ib Leave Neutron Stars?

Perhaps.

The association between SNe Ib and star formation, the presence of helium in the early spectra, and some very high estimates from the later spectra of the mass of ejected oxygen, led to suggestions that SNe Ib are the explosions of stars that are massive enough to lose their hydrogen envelopes via winds, i.e., Wolf-Rayet stars. However, the association with star formation now appears to be looser than it is for SNe II (Panagia and Laidler 1988) and recent spectroscopic estimates of the oxygen mass are low, $\lesssim 1$ M$_\odot$ (Fransson and Chevalier 1988, Axelrod, 1988). This is just as well, because Wolf Rayet stars, having $M_i \gtrsim 30$ M$_\odot$, might have trouble accounting for the SN Ib frequency, radio emission (Panagia and Laidler 1988), and narrow light curve peak (Ensman and Woosley 1988). (One SN Ib, 1985F, does not necessarily have any of these problems.)

If SNe Ib are from stars that are massive enough to experience core collapse, they must be from stars having $M_i \gtrsim M_c$, i.e., stars that would explode as SNe II if they did not lose their hydrogen, presumably to binary companions. SN 1987K had conspicuous hydrogen lines near maximum light but lost them a few weeks later and began to look like an SN Ib (Filippenko 1988). This may have been from the explosion of a star that lost almost but not quite all of its hydrogen, narrowly avoiding the Type Ib label (Filippenko proposes to call SN 1987K a Type IIb). The discovery of SN 1987K provides evidence for a close link between SNe Ib and SNe II. The peak luminosity of SNe Ib may be a problem for the core-collapse model, however. SNe Ib appear to be less luminous than SNe Ia at maximum light by a factor of about 4, in which case they eject $M_{Ni} \simeq 0.15$ M$_\odot$. But SN 1987A, with a 20 M$_\odot$ progenitor, managed only $M_{Ni} = 0.07$ M$_\odot$ (Catchpole et al. 1988, Woosley 1988). If the typical SN Ib progenitor is less massive than that of SN 1987A, it should eject less nickel, not more.

An alternative idea for SNe Ib, involving off-center ignition of helium in accreting white dwarfs (Branch and Nomoto 1986, Iben et al. 1987) rather than core collapse and neutron stars, may have severe problems putting enough energy into the late, nebular spectra (Fransson and Chevalier 1988). In any case, the white-dwarf speculation involves a completely different composition (helium and nickel, initially) in the outer layers of the ejected matter, so forthcoming quantitative abundance estimates for SNe Ib, with non-LTE effects taken into account, will decide its fate.

Other possibilities for SNe Ib without neutron stars include the carbon-deflagration disruptions of stars having $M_i \lesssim M_c$ or oxygen-deflagrations in those having $M_i \gtrsim M_c$, as mentioned above in connection with SNe II, but now in stars that have lost their hydrogen. Either of these hypothetical events would provide a way to get sufficient nickel for the light curve without resorting to stars that are too massive. The radio emission could still be a problem.

## 5. Constraints From Supernova Remnants and Pulsars

What do the foregoing considerations imply for the fraction of Galactic supernovae that leave neutron stars, and for the rate at which neutron stars are produced by supernovae in the Galaxy? For the purposes of the following excercise, suppose that the relative proportions of Galactic SNe Ia, II, and Ib are 3:11:4, as found by van den Bergh, McClure, and Evans (1987) for "a typical Shapley Ames galaxy", and that the mean interval between Galactic supernovae is 50 years (e.g., Capellaro and Turatto 1988, van den Bergh et al. 1987). For an optimistic yield of neutron stars, assume that although SNe Ia don't leave neutron stars, all SNe II and SNe Ib do. Then the fraction, f, of supernovae that leave neutron stars is 0.8, and the interval, T, between supernovae that leave neutron stars is 60 years. For a pessimistic yield, suppose that SNe Ia, SNe Ib, and half of the SNe II fail to yield neutron stars; now f=0.3 and T=170 years. Obviously the range between the optimistic and pessimistic limits, although already wide, is not as wide as it should be, because we have not allowed for the considerable uncertainties in the relative and absolute supernova rates.

Are these limits inconsistent with what we know about neutron stars in supernova remnants, and the Galactic population of neutron stars? On the optimistic side, T=60 yr is not inconsistent with estimates of the pulsar birthrate (remember the uncertainties); whether the optimistic estimate f=0.8 is inconsistent with the evidence from Galactic SNRs depends on whom you ask. The pessimistic f=0.3 surely is not inconsistent with the SNRs, but T=170 yr *is* too long to be consistent with the pulsar birthrate. This discrepancy can be explained away, however, by saying that if most of the pulsars are not made by the kinds of supernovae that we *see*, maybe they are made by the kinds we *don't see*. If SNe Ib are not from core collapse in massive stars that have lost their hydrogen, this may only mean that such explosions are too faint to be discovered, not that they don't occur. Another way to get a significant number of neutron stars from supernovae we don't see is from subluminous SNe II such as 1987A. Now that subluminous SNe II have been forced on our attention by SN 1987A, we realize that they may be common in space. SN 1987A probably would not have been noticed if it had occurred much beyond the Local Group (Schmitz and Gaskell 1988), yet neither the relative nor absolute supernova rates adopted here have allowed for such faint SNe II.

A different way to look at the pulsar constraint is to compare the local pulsar birthrate with the local population of massive stars. Taking this approach, Blaauw (1985) estimates that stars having masses as low as 6 $M_\odot$ must make pulsars. If taken literally, this result would have important implications for our ideas about which supernovae leave neutron stars. SNe II probably come from stars more massive than 6 $M_\odot$, so in spite of what has been said above they all would have to leave neutron stars. For SNe Ib it would become a close call. From the available evidence Panagia and Laidler (1988) estimate that SNe Ib have $M_i \gtrsim 6$ $M_\odot$, but considering the uncertainties we could not quite conclude that SNe Ib must leave neutron stars, too. We need to build up a bigger observational sample of SNe Ib, and use the statistics of their sites to put more accurate limits on the initial masses of their progenitor stars.

Acknowledgements

I would like to thank Wolfgang Kundt for calling my attention to the import of the paper by Blaauw, David Jeffery and Hans Ritter for their careful readings of the manuscript, and Rolf Kudritzki for the hospitality of the Institut für Astronomie und Astrophysik der Universitäts München, where the written version of this contribution was prepared. This work has been supported by the Humboldt Foundation and the National Science Foundation.

## REFERENCES

Arnett, W. D., Branch, D., and Wheeler, J. C. 1985, Nature, 314, 337.

Axelrod, T. S. 1980a, in Type I Supernovae, ed. J. C. Wheeler (Austin, University of Texas), p. 80.

Axelrod, T. S. 1980b, Ph.D Thesis, University of California at Santa Cruz.

Axelrod, T. S. 1988, Lecture Notes in Physics, 305, 3754.

Barbon, R., Ciatti, F., and Rosino, L. 1979, Astr. Ap., 72, 287.

Bartunov, O. S. and Tsvetkov, D. Yu. 1986, Ap. Sp. Sci., 122, 343.

Blaauw, A. 1985, in Birth and Evolution of Massive Stars and Stellar Groups, ed. W. Boland and H. van Woerden (Dordrecht, Reidel), p. 211.

Branch, D., Doggett, J. B., Nomoto, K., and Thielemann, F.-K. 1985, Ap. J., 294, 619.

Branch, D., Drucker, W., and Jeffery, D. J. 1988, Ap. J. (Letters), 330, L117.

Branch, D. and Nomoto, K. 1986, Astr. Ap., 164, L13.

Burrows, A. and Lattimer, J. M. 1987, Ap. J. (Letters), 318, L63.

Canal, R. 1989, in Evolutionary Phenomena in Galaxies, ed. B. E. J. Pagel and J. Beckman (Cambridge, Cambridge University), in press.

Canal, R., Isern, J., and Lopez, R. 1988, Ap. J. (Letters), 330, L113.

Cappellaro, E. and Turatto, M. 1988, Astr. Ap., 190, 10.

Catchpole, R. M. and 23 others, 1988, M.N.R.A.S., 231, 75P.

de Vaucouleurs, G., de Vaucouleurs, A., Buta, R., Ables, H.D., and Hewitt, A.V. 1981, P.A.S.P., 93, 36.

Doggett, J. B. and Branch, D. 1986, A.J., 90, 2303.

Ensman, L. and Woosley, S. E. 1988, Ap. J., 333, 754.

Fesen, R. A., Becker, R. H., and Blair, W. P. 1987, Ap. J., 313, 378.

Filippenko, A. V. 1988, A.J., in press.

Fransson, C. and Chevalier, R. A. 1988, Ap. J., in press.

Graham, J. R., 1987, Ap. J., 315, 588.

Henry, R. B. C. 1986, P.A.S.P., 98, 1044.

Huang, Y.-L. 1987, P.A.S.P., 99, 461.

Iben, I. I., Jr. 1985, in Mass Loss from Red Giants, ed. M. Morris and B. Zuckerman (Dordrecht, Reidel), p. 1.

Iben, I. I., Jr., Nomoto, K., Tornambe, A., and Tutukov. A. V. 1987, Ap. J., 317, 717.

288

Iben, I. I., Jr. and Tutukov, A. V. 1984, Ap. J. Suppl., 54, 335.
Kochhar, R. K. 1989, Ap. Sp. Sci., in press.
Litvinova, I. Yu. and Nadyozhin, D. K. 1985, Sov. Astr. Lett., 11, 145.
Maza, J. and van den Bergh, S. 1976, Ap. J., 204, 519.
Nomoto, K., Thielemann, F.-K., and Yokoi, K. 1984, Ap. J., 286, 644.
Panagia, N. and Laidler, V. G. 1988, in Supernova Shells and Their Birth Events,
    ed. W. Kundt (Springer-Verlag, Berlin); Lecture Notes in Physics, in press.
Saio, H. and Nomoto, K. 1985, Astr. Ap., 150, L21.
Schmitz, M.F. and Gaskell, C. M. 1988, in Supernova 1987A in the Large
    Magellanic Cloud, ed. M. Kafatos and A. Michelitsianos (Cambridge,
    University of Cambridge).
Sutherland, P. G. and Wheeler, J. C. 1984, Ap. J., 280, 282.
van den Bergh, S. 1988, Ap. J., 327, 156.
van den Bergh, S., McClure, R. D., and Evans, R. 1987, Ap. J., 323, 844.
Webbink, R. 1984, Ap. J., 277, 355.
Wheeler, J. C., Harkness, R. P., Barkat, Z., and Schwartz, D. 1986, P.A.S.P., 98,
    1018.
Woosley, S. E. 1988, Ap. J., 330, 218.
Woosley, S. E., Axelrod, T. S., and Weaver, T. A. 1984, in Stellar Nucleosynthesis,
    ed. C. Chiosi and A. Renzini (Dordrecht, Reidel), p. 263.

# The circumstellar environment of SN 1987A

Peter Lundqvist
Lund Observatory
Box 43
S-221 00 Lund
Sweden

ABSTRACT: Observations of SN 1987A, with particular emphasis on the evidence for a circumstellar medium around the supernova are reviewed. Models of the observations are discussed. In particular, it is shown that narrow emission lines in the UV and optical can be used to put constraints on the presupernova evolution, as well as the unobserved soft X-ray burst occurring a few hours after the core collapse.

## 1. Introduction

During the last decade, radio observations of Type II and Type Ib supernovae (SNe) have shown that progenitors of these types of SNe suffer from extensive mass loss (cf. reviews by Chevalier, 1984, and Fransson, 1986). In particular, the mass loss rates for the progenitors of the radio luminous Type II SNe SN 1979C and SN 1980K have been estimated to be $1.2 \times 10^{-4}$ $(v/10 \text{ km s}^{-1}) \text{ M}_\odot \text{ yr}^{-1}$ and $3 \times 10^{-5}$ $(v/10 \text{ km s}^{-1}) \text{ M}_\odot \text{ yr}^{-1}$ (Lundqvist and Fransson, 1988). Here v is the wind speed of the red supergiant wind. A significant fraction of the presupernova envelope is thus lost to the circumstellar medium (CSM) before the explosion. This modifies both the supernova environment and the surface composition of the progenitor. In an analysis of UV emission lines from SN 1979C, Fransson et al. (1984) found N/C~8. This N/C-ratio is roughly 50 times the solar value, and was interpreted by Fransson et al. to be attributed to a high mass loss rate, in combination with surface mixing during the progenitor evolution. Both these effects result in an exposure of CNO-processed material at the surface of the progenitor, and hence a high N/C-ratio.

Progenitor mass loss also affects the dynamics of the blast wave, and the interaction of the CSM with the supernova envelope gives rise to radio, EUV and X-ray emission (Chevalier, 1984). With the emission from the SN seen as a background source, the unshocked CSM may show up as narrow absorption lines, mainly in the UV (Lundqvist and Fransson, 1988). In the infrared, both SN 1979C and SN 1980K had a strong continuum emission. This emission was interpreted by Dwek (1983) to come from circumstellar dust, heated by the UV-visual radiation from the SN.

W. Kundt (ed.), Neutron Stars and Their Birth Events, 289–302.
© 1990 Kluwer Academic Publishers.

Prior to February 23 1987, Type II SN precursors were all thought to be red supergiants. With the advent of SN 1987A, this picture has changed since the progenitor of SN 1987A was the blue supergiant Sanduleak (Sk) -69°202 (Sonneborn et al., 1987; West et al., 1987; White and Malin, 1987), classified by Rousseau et al. (1984) as a B3 Ia star. Because red and blue supergiants have very different mass loss characteristics (see e.g. de Jager et al., 1988), we expect the CSM of SN 1987A to be different from that around 'normal' Type II SNe. Here we concentrate on recent attempts to understand the CSM of SN 1987A.

## 2. Progenitor evolution and explosion

The first signal from the core collapse of Sk -69°202 was the neutrino burst observed on February 23.316 (UT) (Hirata et al., 1987; Bionta et al., 1987). Three hours later, the first optical observations showed that the visual brightness had increased from V~12 (Rousseau et al., 1984) to V~6.4 (Mc Naught, 1987). During a few days the brightness was roughly constant at this level. It then started to increase, and the peak of the visual luminosity occurred ~90 days after the explosion with V~2.9. Since the effective temperature was roughly constant during the building up of the peak (Catchpole et al., 1987), the expanding photosphere must have been powered by some internal energy source. The heat source was most likely radioactively decaying $^{56}Ni$ (e.g. Shigeyama et al., 1988; Woosley, 1988). Due to the expansion, the stored internal energy eventually diffused away fast enough for the light curve to experience a turn-over. Between ~90 and ~130 days after the explosion, the visual magnitude increased from ~2.9 to ~4.3. Then the bolometric luminosity drop off settled to the decay rate of radioactive $^{56}Co$ ($\propto exp(-t/111$ days$)$). Three hundred days after the explosion, the *visual* luminosity started to decline slightly faster (Hamuy and Phillips, 1988; Catchpole and Whitelock, 1988). This is interpreted as due to an increasing fraction of the gamma rays produced during the cobalt decay starting to escape freely through the supernova envelope (Matz et al., 1988). The *bolometric* luminosity, including the gamma rays, still decayed at the same rate as the decay rate of $^{56}Co$.

To explain the light curve, model calculations of the presupernova evolution and the subsequent explosion have been made by several groups (e.g. Arnett, 1988; Shigeyama et al., 1988; Woosley, 1988). (For a general discussion of SN light curves, see the paper by Branch in this volume.) Shigeyama et al. get a best fit to the light curve if they explode a blue supergiant with a 6 $M_\odot$ He core and a 6.7 $M_\odot$ hydrogen envelope. The radius of the presupernova is $3 \times 10^{12}$ cm and the explosion energy (=E) $1 \times 10^{51}$ erg, giving $E/M_{env}=1.5 \times 10^{50}$ erg $M_\odot^{-1}$, where $M_{env}$ is the mass of the hydrogen envelope. In the 10 H-models by Woosley, $M_{env}=10$ $M_\odot$ and the mass of the He core is 6 $M_\odot$. The explosion energy is $1.4 \times 10^{51}$ erg, corresponding to $E/M_{env}=1.4 \times 10^{50}$ erg $M_\odot^{-1}$. The mass of the He core is adjusted to fit the bolometric luminosity of Sk -69°202, while the other parameters are chosen to model the early part of the light curve. The hydrodynamic calculations of the explosion all agree on the fact that it is necessary to power the peak of the light curve by a radioactive decay of

~0.07 $M_\odot$ $^{56}Ni$. The nickel is created at the explosion, and the width of the peak indicates that part of the nickel is mixed into the He core. The calculations show that there is a density inversion in the expanding core matter because of the radioactive heat input. The density inversion results in Rayleigh-Taylor instabilities, which can explain the nickel mixing. The early (150-200 days after the explosion) emission of hard X-rays and gamma rays from SN 1987A (Dotani et al., 1987; Sunyaev et al., 1987; Matz et al., 1988) also suggests that part of the nickel is mixed into the He core during the explosion (e.g. Itoh et al., 1987; Pinto and Woosley, 1988).

All hydrodynamic calculations of the explosion also agree that when the shock wave breaks through the surface of the envelope, there is a burst of EUV and soft X-ray emission. The number of ionizing photons is of order ~$10^{57}$ and the peak effective temperature, $T_{peak}$, is $(2-5) \times 10^5$ K. The models, however, have a very simple treatment of the radiative transfer during the burst which is likely to underestimate the effective temperature (Sections 4 and 5; Fransson and Lundqvist, 1989).

The calculations of Shigeyama et al. indicate a ~50% higher helium fraction in the envelope than for the sun, which points at surface mixing and mass loss from the presupernova. Both these effects, together with the low metallicity in the Large Magellanic Cloud (LMC), are important for the evolutionary track of the progenitor in the HR-diagram. We now believe that Sk -69°202 underwent a blue-red-blue supergiant evolution rather than a blue only. The most evident reason for this is the large number of observed red supergiants in the LMC (Humphreys and Davidson, 1979). A blue-red-blue evolution can be a result of either mass loss (Maeder, 1987) or low metallicity (Woosley et al., 1988). Models which only include mass loss can not explain the red-blue evolution, if not the hydrogen envelope is lost completely (Maeder, 1987). However, from the simulations of the explosion, we know that several solar masses of hydrogen-rich material must have been left at the explosion (see earlier in this Section). Barkat and Wheeler (1988) show that for a constant luminosity and envelope composition, self-consistent solutions for the envelope fall into two possible mass regimes. For models with solar abundances, the excluded mass interval for the envelope of Sk -69°202 ranges between ~$(0.17-12.4)$ $M_\odot$. If the helium mass fraction in the envelope, Y, is increased, the excluded mass interval becomes smaller. If $Y > 0.5$, virtually all envelope masses are possible. This means that the preferred values for $M_{env}$ from the hydrodynamic calculations indicate that part of the helium rich layer was dredged up and mixed into envelope during the evolution. Saio et al. (1988) have made similar calculations as Barkat and Wheeler and show that a massive (~7-10 $M_\odot$) hydrogen envelope may be present at the explosion if mass loss is combined with mixing of the hydrogen envelope during the He-shell burning. The blueward evolution from the red supergiant stage to the position of Sk -69°202 in the HR-diagram occurs more easily if the metallicity is low. As a result of the mixing, the stellar surface at the moment of the explosion should contain N-rich material because CNO-processed gas has been brought up from underlying layers. The photospheric gas should thus show high N/C and N/O ratios, just as was observed for SN 1979C (Fransson et al., 1984). In the calculations of Saio et al., a He-core of 6 $M_\odot$ corresponds to a total mass on the main sequence of 21 $M_\odot$, and it takes ~$6\times10^3$ years for the star to evolve from a red supergiant to become a SN. For a hydrogen envelope of ~$(7-10)$ $M_\odot$ left at the explosion, ~$(5-8)$ $M_\odot$ should have been lost by the presupernova in the form of stellar winds and/or transient mass expels.

## 3. Circumstellar environment from the evolution of the progenitor

If Sk -69°202 underwent a blue-red-blue evolution, the stellar wind during the first blue supergiant stage creates a cavity in the interstellar medium (ISM). Around this cavity there is an interaction region of shocked gas (Weaver et al., 1977). The radius of this shell increases as $R_{blue}=80$ $(\dot{M}/10^{-7}$ $M_\odot$ $yr^{-1})^{0.2}(n_0/1$ $cm^{-3})^{-0.2}(v_{blue}/2000$ km s$^{-1})^{0.4}(t_{blue}/10^7$ yrs$)^{0.6}$ pc. Here $n_0$ is the gas density of the ISM, $\dot{M}$ the stellar mass loss rate, $v_{blue}$ the stellar wind velocity and $t_{blue}$ the age. The evolutionary models of Sk -69°202 favour an initial stellar mass of ~20 $M_\odot$, and that the duration of the first blue supergiant stage ~$(1-1.5)\times10^7$ yrs (e.g. Maeder, 1987; Woosley et al., 1987). On the main sequence a 20 $M_\odot$ star corresponds to an O9 V star, and in a recent compilation by de Jager et al. (1988) for *galactic* stars, an O9 V star has a mass loss rate of ~$10^{-7}$ $M_\odot$ yr$^{-1}$. Observations of hot stars in the LMC (Garmany and Conti, 1985), indicate that mass loss rates for these types of stars are roughly the same as in the Galaxy. The expression for the expansion of the shell was derived under the assumption of no radiative losses in the interaction region. Since the expression is sensitive to radiative cooling (Weaver et al., 1977), the radius should be regarded as an upper limit, but shows that a large cavity is created by the stellar wind. The interior of the bubble is in pressure equilibrium and has a typical temperature of a few x $10^6$ K (Chevalier, 1987). The simple picture of a smooth wind-blown bubble may be distorted by interactions with bubbles from other blue stars. Projected distances to the two closest stars, classified as B0 V and B1.5 V stars (Gilmozzi et al., 1987), are only ~0.7 and ~0.4 pc, respectively (Walborn et al., 1987).

The duration of the subsequent red supergiant phase is ~$10^6$ yrs (e.g. Maeder, 1987; Woosley et al., 1987). During this period, the wind speed is expected to decrease to ~$(10-30)$ km s$^{-1}$, and the mass loss rate to increase to ~$(10^{-6}$- few x$10^{-5})$ $M_\odot$ yr$^{-1}$ (see Sect.5). The maximum radius of the red supergiant wind then is $R_{red}=10$ $(v_{red}/10$ km s$^{-1})$ $(t_{red}/10^6$ yrs) pc, using the same nomenclature as above for the blue wind. The red supergiant wind is expected to be contaminated with dust particles (e.g. Dwek, 1988; Emmering and Chevalier, 1989), like the winds around SN 1979C and SN 1980K (Dwek, 1983). As the star evolves back towards the blue again, the wind speed increases, and shock waves will form when the fast blue supergiant wind runs into the slow red supergiant wind. This has been discussed in detail by Chevalier (1987), who finds that the interaction region between the two winds consists of shocked red and blue supergiant winds, where the red supergiant wind is compressed to a thin shell due to radiative cooling. The radius of the shell is for plausible parameters ~$10^{18}$ cm, and the density of the compressed shell $10^3$-$10^4$ cm$^{-3}$ (Chevalier, 1987). The shell is most likely fragmented because of Rayleigh-Taylor instabilities. Shells due to transient mass losses of the presupernova may also be present. In particular, this is expected if the star performs rapid transitions from the red to the blue supergiant stage. Models by Woosley (1988) show such a behaviour during and immediately after the helium burning.

## 4. Circumstellar environment from the observations

The observational mapping of the CSM of SN 1987A range from radio wavelengths to X-rays. In the radio, SN 1987A showed a wavelength dependent turn-on and subsequent decay (Turtle et al., 1987). At 0.843 GHz and 1.4 GHz, the peak emission occurred ~3.6 and ~2.6 days after the core collapse, respectively, while for 2.3 GHz and 8.4 GHz, the emission peaked before the first observations. Chevalier and Fransson (1987) have used the same model as for 'normal' Type II SNe in which the synchrotron emission comes from a region close to the SN blast wave when it interacts with the stellar wind. The delayed radio emission is due to free-free absorption in the wind outside the blast wave, and the time it takes for the free-free optical depth to fall below unity scales as $t_\nu \propto (x_e \dot{M}/v)^{2/3} \nu^{-2/3} T^{-1/2} V_b^{-1}$. Here M and v are for the presupernova mass loss rate and wind speed just before the explosion, respectively, $x_e$ the fraction of free electrons in the wind, $\nu$ the radio frequency, T the wind temperature and $V_b$ the velocity of the blast wave. The frequency dependence compares well with the observations, and an updated analysis by Chevalier (1987) of the results by Chevalier and Fransson (1987) gives $\dot{M} \sim 7.5 \times 10^{-6}$ (T/$10^5$ K)$^{3/4}$ (v/550 km s$^{-1}$) M$_\odot$ yr$^{-1}$, assuming a complete ionization of the free-free absorbing stellar wind. From this expression we see that M is sensitive to the wind temperature, and in a time dependent study of the ionization of the wind, Lundqvist and Fransson (1987) find that the wind temperature was ~(0.65-2)x$10^5$ K. The main uncertainty in this result is the spectral shape and duration of the EUV soft X-ray burst which ionizes and heats the CSM.

Also important for the ionization of the wind is the radiation from the interaction region between the supernova envelope and the blue supergiant wind (Lundqvist and Fransson, 1987). This region consists of the shocked SN envelope by a reverse shock and the shocked presupernova wind by the blast wave (Chevalier, 1982). For 'normal' Type II SNe, the main ionizing component is radiation from the outer shock (Fransson, 1984; Lundqvist and Fransson, 1988). In the shocked stellar wind, photospheric photons gain energy when inversely Compton scattered by hot (~$10^9$ K) electrons. The scattered photons thereby add a power law tail in the EUV and soft X-ray to the photospheric spectrum. For SNe like SN 1979C and SN 1980K, this radiation is probably more important for the ionization of the CSM than the radiation from the outburst (Lundqvist and Fransson, 1988). The wind around SN 1987A had too low a density for the inverse Compton scattering to be effective (Chevalier and Fransson, 1987), and instead the X-rays from the reverse shock dominated the ionizing emission from the interaction region. Chevalier (1987) estimates the luminosity to ~($10^{37}$-$10^{38}$)x $t_{days}^{-1}$ erg s$^{-1}$ and the temperature to ~(1-5)x$10^8$ x $t_{days}^{-2/5}$ K. Lundqvist and Fransson (1987) argue that an ionizing component from the reverse shock with these characteristics is needed to add to the burst to prevent the unshocked wind from recombining to ions such as C IV, N V and O VI.

These ions would otherwise have given rise to broad (~500 km s$^{-1}$) absorption lines in the UV, which were not observed (de Boer et al., 1987; Dupree et al., 1987).

There is an alternative model for the radio emission from SN 1987A where the free-free absorption occurs *within* the radio emitting region (Storey and Manchester, 1987). This model, however, only tolerates significantly lower mass loss rates of the progenitor than found by Chevalier and Fransson, or that the wind should suffer from very effective clumping. Another problem with the internal absorption model is that it gives good fits to the observations only for shock velocities of ~2x10$^4$ km s$^{-1}$. This velocity is, however, considerably less than the velocities inferred from the maximum blueshift of the Hα line (~3.1x10$^4$ km s$^{-1}$, Hanuschik and Dachs, 1987).

Since the first ~50 days after the radio burst, SN 1987A has been radio quiet. In the infrared, Chalabaev et al. (1989) claim that speckle interferometry in June and August 1987 at λλ2-5μm reveal a light echo, most likely from pre-existing dust heated by the SN outburst. The observations may indicate a dust sphere with inner radius <1.5x10$^{17}$ cm and outer radius >2.6x10$^{17}$ cm. The detected signals are, however, very faint, close to that of systematic errors. Rank et al. (1988) have reported an IR excess in their observations between 5.3-12.6 μm. They attribute this excess to dust emission, while Aitken et al. (1988) conclude that a similar excess in their observations between 8-13.5 μm is more likely to be associated with emission from gas in the SN envelope than from dust. Models of the dust emission by Emmering and Chevalier (1989) support the latter conclusion.

UV observations with the IUE (International Ultraviolet Explorer) have since May 1987 shown a number of narrow emission lines (Fransson et al., 1989). During the first ~400 days after the explosion, the luminosity of most of the lines (C III] λ1909, N III] λ1750, O III] λ1664 and N V λ1240) increased nearly linearly with time, while He II λ1640 and N IV] λ1486 were nearly constant. After ~400 days, the line fluxes started to decline (Sonneborn et al., 1989). Observations in the optical, ~ 300 days after the explosion by Wampler and Richichi (1989), have shown narrow [O III] and Balmer emission lines, as well as He II λ4686. The UV and optical observations provide important constraints on the parameters of the emitting gas. From the relative intensities in the C III] λλ1906.7-1908.7 doublet, Fransson et al. (1989) derive ~2.6x10$^4$ cm$^{-3}$ for the electron density, and a simple abundance analysis gives N/C and N/O ratios corresponding to ~37 and ~12 times solar, respectively. The electron temperature of the emitting gas may be estimated from the [O III] lines, and Wampler and Richichi (1989) find ~5x10$^4$ K, 300 days after the explosion.

The most likely interpretation for the line emission is that gas in a shell-like structure was heated and ionized by the EUV and soft X-ray burst, and that the gas now reemits the radiation at longer wavelengths (Lundqvist and Fransson, 1987; Fransson and Lundqvist, 1989; Fransson et al., 1989). If that is the case, the temporal behaviour of the line emission depends on a combination of the initial state of the gas after the ionization, how fast the subsequent recombination and cooling rates proceed, and on the importance of light travel time effects. The initial degree of ionization depends on the ionizing spectrum and is closely coupled to $T_{peak}$, while the recombination and the cooling rates are mainly governed by the gas density.

For a temperature in the shell of $10^5$ K and an electron density of $2.6 \times 10^4$ cm$^{-3}$, the recombination times are ~182 days for N VI to N V, ~11 days for N V to N IV, ~10 days for N IV to N III, and ~13 days for N III to N II. These time scales are short, or comparable to the evolutionary time scale of the lines, which is of order ~100 days. Recombination is thus indeed important for the evolution of the emitted line luminosities, $L_e(t_e)$. The *received* (= observed) luminosity, $L_r$, is modified by the finite speed of light according to

$$L_r = \frac{c}{2 R_s} \int_{t_{min}}^{t} L_e(t_e) \, dt_e.$$

Here $t_{min} = \max(0, t-2 R_s/c)$ and $R_s$ the shell radius. If $L_e(t_e)$ drops rapidly after the initial ionization, i.e. if recombination and cooling effects are important, $L_r$ is constant during the first $t_s = 2 R_s/c \sim 386$ ($R_s/5 \times 10^{17}$ cm) days, and then falls rapidly. The opposite case is if $L_e(t_e)$ is constant. Then $L_r = c \, t \, L_e/2 R_s$ for $t < t_s$ and is for later times constant until recombination sets in. Since we have already noticed from the above comparison between the time scales that recombination is important, the first of these two cases probably best explains the observations. The time between the explosion and the turn-over for the UV-lines ($t_s \sim 400$ days, Sonneborn et al., 1989) then directly translates into a shell radius of ~$5 \times 10^{17}$ cm.

To model the line emission from the shell, Fransson and Lundqvist (1989) and Lundqvist and Fransson (1989) have made photoionization calculations, taking the ionizing radiation from the hydrodynamic calculations of the explosion (see Sect. 2). (To check against the observed line intensities, see Fig. 3 in Fransson et al., 1989.) The effective temperature in the hydrodynamic simulations peaks at $(2-5) \times 10^5$ K and then falls to ~$3 \times 10^4$ K in ~5 hours. Since this is much less than the recombination and cooling time scales, the calculations for the evolution of the gas in the shell must be done time dependently. The shell radius is assumed to be $5 \times 10^{17}$ cm, as inferred from the turn-over time of the UV lines, while the electron density is $2.6 \times 10^4$ cm$^{-3}$ (Fransson et al., 1989). The elements included are H, He, C, N, O, Ne and S with abundances He/H=0.1, N/C=5, N/O=2. The overall metallicity, Z, is $0.3 \times Z_{\odot}$. This metallicity agrees with the results for LMC H II-regions found by Dufour (1984). In Fig. 1 results using the ionizing radiation from Shigeyama et al.'s 11E1Y6 model are shown. The upper panel shows that the temperature in the middle of the shell initially becomes ~$1.35 \times 10^5$ K, and that the metals attain their He-like stages, like N VI, while hydrogen and helium are fully ionized. The gas then recombines to C III, N III etc. in ~(200-300) days. However, since collisional ionization from C II, N II etc. is faster than recombination from C III, N III etc. for temperatures higher than ~$(3-4) \times 10^4$ K, the gas does not recombine further until after ~500 days. This explains why the observed luminosities of the C III, N III and O III lines increase almost linearly with time (lower panel). N IV, on the other hand, is less affected by collisional ionization, and quickly recombines to N III. Therefore, the *emitted* N IV] $\lambda$1486 luminosity quickly decays (middle panel), while the *received* luminosity becomes constant (lower panel). The model explains well the general evolution of the observed line

296

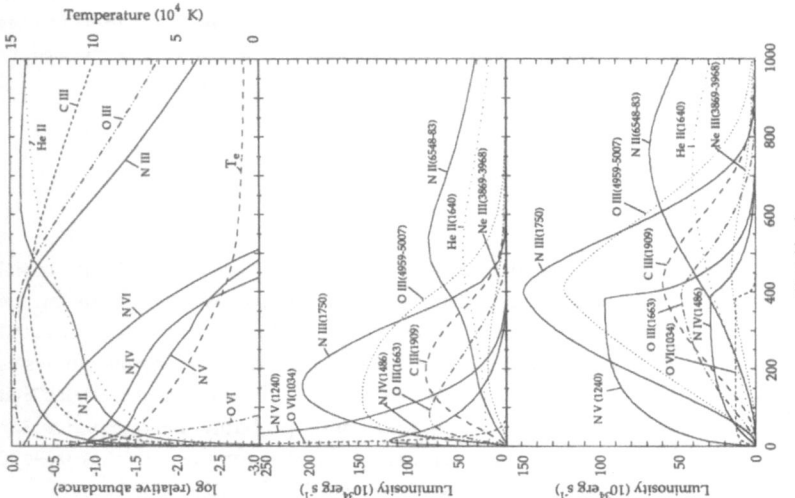

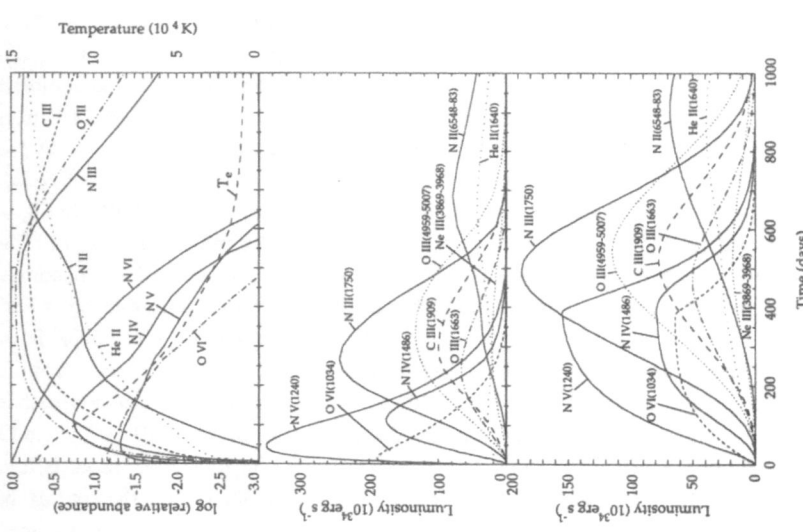

**Fig. 1 and 2.** Evolution of a shell ionized by the EUV burst from SN 1987A (see text for details). The upper panel shows the temperature and relative abundances in the shell as a function of time. Note the successive recombination of the nitrogen ions. The middle shows the *instantaneous* luminosities of the different lines and the lower the *received* (= observed) luminosities, integrated over the observable parts of the shell. In the left figure (Fig. 1) the ionizing flux has been taken from Shigeyama et al.'s 11E1Y6 model and in the right from Woosley's 10L model.

luminosities by Fransson et al. (1989), and the temperature at ~300 days (~5.5x10$^4$ K) is close to that deduced from the [O III] lines.

In Fig. 2, results are shown for an analogous calculation using the ionizing radiation from Woosley's 10L model. Because of the lower $T_{peak}$ (2x10$^5$ K), the initial ionization and temperature in the shell are lower, giving a more rapid recombination. This results in a "flatter" light curve as well as a more rapid decay after $t_s$=2 $R_s$/c for N V. The low wind temperature is also reflected in the relatively high [O III] λλ4959-5007 emission and a high N III] λ 1750 / N IV] λ 1486 ratio.

To model the burst spectrum of SN 1987A, Fransson and Lundqvist (1989) vary $T_{peak}$ between 10$^5$-10$^6$ K. For the range of electron densities between (1-4)x10$^4$ cm$^{-3}$, constrained by the C III] observations (Fransson et al., 1989), the calculations show that $T_{peak}$ must have been (4-8)x10$^5$ K. The basic constraints for this result are the N III] λ 1750 / N IV] λ 1486 ratio 200-400 days after the explosion, and the temperature at ~300 days deduced from the O [III] lines. Another constraint is that the N V line flux must not be too strong compared to the N III] and the N IV] line fluxes. This excludes $T_{peak}$ higher than ~8x10$^5$ K. A further constraint for the lower limit of $T_{peak}$ is that the observationally "flat" He II λ1640 line flux indicates that collisional excitation is important for this line. This requires a value of $T_{peak}$ higher than ~3x10$^5$ K.

The total mass in the shell is sensitive to the ionizing spectrum. For the range of densities and values of $T_{peak}$ discussed above, the shell mass is within the interval 0.02-0.05 M$_\theta$. The determinations of the absolute metallicity, as well as the He/H ratio depends on the observed Hβ and He II λ4686 fluxes. Unfortunately, the errors in these observations are quite large, so the results should be taken with some caution. Preliminary results (Lundqvist and Fransson, 1989) suggest that the total CNO abundance is 0.2-0.3 times solar, while the He/H ratio is in the range 0.1-0.2, by number. The N/C and N/O ratios are for these models ~5 and ~2, respectively, confirming that the assumptions in Fransson et al. (1989) are reasonable approximations.

Except for the observed UV, and optical [O III], He II and Balmer lines, there are also other lines that may be observable. In the far UV the C III λ990 and N III λ970 resonance lines are both strong. In the same way as the C IV λ1550 line they may, however, be unobservable due to interstellar absorption. The O VI λ1034 line is especially sensitive to $T_{peak}$ (Figs. 1 and 2), and observations of this line are therefore particularly important. In the optical, the [N II] λλ6548-83, 5755 lines are expected to become strong after ~300 days. This coincides with the recombination from N III to N II. Also the [Ne III] λλ3869-3968 lines can be fairly strong, as well as [O II] λλ3726-29.

The values for the gas density and the shell radius are close to those expected for the compressed part of the red supergiant wind (cf Sect. 3). If indeed the narrow lines originates in such a region, the emitting gas should contain gas of different densities. This is because gas in this region is expected to suffer from clumping because of Rayleigh-Taylor instabilities (Chevalier, 1987). The steady increase of the N V line luminosity between 250-400 days could be attributed to a low density component, whereas other lines may come from density enhanced regions. The density of the low density component should be a factor ~(2-3) lower than found by Fransson et al. (1989) for the C III] emitting region. Any low density component lowers the allowed maximum value of $T_{peak}$ in the calculations by Fransson and Lundqvist (1989).

After $t_x \sim 9.0\ (R_s/5 \times 10^{17}\ cm)^{1.15}\ (\dot{M}/6 \times 10^{-6}\ M_\odot\ yr^{-1})^{0.15}\ (v_{blue}/550\ km\ s^{-1})^{-0.15}$ yrs, where $\dot{M}$ is the mass loss rate during the last blue supergiant phase of Sk $-69°202$, the supernova blast wave is expected to hit the shell (Chevalier, 1987), resulting in a sudden increase of the radio and soft X-ray emission (Itoh et al., 1987; Chevalier, 1987). The expression for $t_x$ is mainly sensitive to the slope of the density profile in the expanding supernova envelope, and was derived for the density profile $\rho \propto r^{-9.58}$. For a steeper density profile the shock reaches the shell earlier. The time $t_x$ may thus provide useful information about the structure in the expanding envelope. Short term mass expels during the final blue supergiant stage of the progenitor (see Sect. 3) may give rise to rapid X-ray brightenings prior to $t_x$. Masai et al. (1988) model the transient increase in luminosity of the soft X-ray component observed by Ginga ~330 days after the explosion (Dotani et al., 1987) in terms of interaction between the blast wave and circumstellar matter at a distance of $(1-3.2) \times 10^{16}$ cm from the supernova. In their model, the circumstellar matter only resides at the far side of the SN and covers less than ~30 % of the total spherical area. The central region of the interaction region is assumed to be observationally blocked off by the SN ejecta. The geometry is crucial for modelling the X-ray emission and may explain why no radio emission was observed. If the absorbing gas is fully ionized hydrogen, the free-free optical depth is $\tau_{ff} \sim 2.7 \times 10^{-27}\ (g_{ff}/10)\ (v/1.4\ GHz)^{-2}$ $(T/10^5\ K)^{-3/2}$ EM, where $g_{ff}$ is the temperature averaged Gaunt factor, and T and EM the temperature and the emission measure of the absorbing gas, respectively. The radial emission measure through a shell of mass $M_s$ at distance $R_s$ is $\sim 1.3 \times 10^{23}\ (R_s/5 \times 10^{17}\ cm)^{-4}\ (M_s/0.03\ M_\odot)^2$ $cm^{-5}$. Using parameter values for the UV-emitting shell given earlier in this Section, we see that this shell is transparent to radio emission. By contrast, the SN ejecta which contains several solar masses at a radius ~few x $10^{16}$ cm, is not. In the model by Masai et al., the radio radiation has to penetrate the envelope, and will not be observable unless the gas in the envelope is almost completely neutral. Although the model by Masai et al. may explain the observations, and that a circumstellar shell close to the presupernova has also been proposed for SN 1984E (Gaskell, 1984), the circumstellar gas distribution in Masai et al.'s model seems difficult to understand. There is also the problem that the Ginga observations show intrinsic strong time variabilities (see the discussion in Chevalier, 1988). If real and if the SN envelope is very filamented, the soft X-rays could instead come from a pulsar synchrotron nebula (Bandiera et al., 1988) or accretion from a binary companion (Fabian and Rees, 1988).

A circumstellar environment similar to that in the model of Masai et al. (1988) has been used by Hillebrandt et al. (1987) to explain the 'mystery spot' inferred from speckle interferometry observations ~(30-50) days after explosion (Meikle et al., 1987; Nisenson et al., 1987). The source was only ~3 magnitudes fainter than the SN in the R band, and at ~50 days the projected distance between the SN and the companion was ~$6 \times 10^{16}$ cm. During the observations, the source appeared to move away from the SN with a projected velocity of

~0.4c. Hillebrandt et al. argue that the emission could be Hα radiation coming from a rapidly moving recombination region in a cloud ionized by the EUV and soft X-ray outburst. In their model, the cloud radius is ~1.5x10$^{17}$ cm, and the distance between the SN and the cloud is 1x10$^{17}$ cm. For a gas density of 10$^6$ cm$^{-3}$, this means a cloud mass of ~12 M$_\odot$. Furthermore, Hillebrandt et al. assume T$_{peak}$~10$^6$ K and a total number of ionizing photons of ~10$^{58}$, corresponding to a total energy of ~5x10$^{48}$ erg. This energy, however, is roughly an order of magnitude higher than found in the hydrodynamic simulations (Shigeyama et al., 1988; Woosley, 1988). There is also the problem that Hillebrandt et al. need a very high efficiency in converting the ionizing energy to Hα emission. Lundqvist and Fransson (1987) show that only ~10 % of the ionizing energy is converted to Hα emission, and that [O I] λλ 6300-64, C IV λ1550 and C III] λ1909 are roughly as strong as Hα. Because of the low conversion efficiency to Hα emission and since neither Hα, nor the other lines were seen spectroscopically at the time of the speckle observations, the photoionization model seems not a very likely model to explain the 'mystery spot'. Several other models for the 'mystery spot' have been proposed (see e.g. the discussion by Phinney, 1988).

## 5. Discussion

From the discussion in the previous Sections, we have seen that the model of the CSM of SN 1987A from the presupernova evolution agrees well with the observational picture from the radio and narrow line observations. In the current picture there is a blue supergiant wind (~6x10$^{-6}$ M$_\odot$ yr$^{-1}$) close to the supernova, extending out to R$_s$~5x10$^{17}$ cm. Inside this wind there may be regions of higher densities due to transient mass ejections of the presupernova, and outside the blue supergiant wind there is most likely a dusty red supergiant wind. Between the two winds there is an interaction region, in which the shocked red supergiant wind has had time to collapse because of cooling. For a mass loss rate during the red supergiant stage of $\dot{M}_{red}$, the swept up red supergiant wind contains ~0.08 (R$_s$/5x10$^{17}$ cm) (v$_{red}$/20 km s$^{-1}$)$^{-1}$ ($\dot{M}_{red}$/10$^{-5}$ M$_\odot$ yr$^{-1}$) M$_\odot$, whereas the mass in the unshocked red supergiant wind is ~10 ($\dot{M}_{red}$/10$^{-5}$ M$_\odot$ yr$^{-1}$) (t$_{red}$/10$^6$ yrs) M$_\odot$. For $\dot{M}_{red}$ of the order 10$^{-5}$ M$_\odot$ yr$^{-1}$ these masses are roughly the same as found for the UV-emitting shell, and expected from models of the presupernova evolution, respectively. The mass of the red supergiant wind is, however, not yet known from the observations. In Sect. 4 we found that the UV-emitting shell was transparent to the early radio emission. Since the ratio of the radial emission measure of the red supergiant wind to that of the UV shell is ~0.018 (R$_s$/5x10$^{17}$ cm) ($\dot{M}_{red}$/10$^{-5}$ M$_\odot$ yr$^{-1}$)$^2$ (M$_s$/0.03 M$_\odot$)$^{-2}$, also the wind is transparent to radio emission. The density of the wind, ~1.2x10$^2$ (R$_s$/5x10$^{17}$ cm)$^{-2}$ ($\dot{M}_{red}$/10$^{-5}$ M$_\odot$ yr$^{-1}$) cm$^{-3}$, is also too low to contribute to the narrow line emission. So far, the most promising method for the mass determination is radiation from dust in the red supergiant wind (Chevalier, 1987; Dwek, 1988), although a fairly uncertain assumption must be made of the gas to dust ratio. From the present lack of such radiation, Emmering and Chevalier (1989) estimate an upper limit of ~10 M$_\odot$ if the shell radius is 5x10$^{17}$ cm. This is consistent with the presupernova models.

Outside the red supergiant wind, the outer blue supergiant wind and the ISM probably have created a shell-like interaction region. In principle, gas in this region could give rise to narrow emission lines similar to those from the UV-emitting shell. There are, however, several problems to overcome. The ionizing radiation from the soft X-ray burst only ionizes ~0.8 $(S/10^{57})$ $M_\odot$, where S is the number of ionizing photons. Since we expect S to be ~$10^{57}$ (see Sect. 2), no ionizing photons will reach the outer blue supergiant wind if the *red* supergiant wind is not preionized by the progenitor. Hydrogen in the red supergiant wind may be preionized if $\dot{M}_{red}$ is less than ~$10^{-5}$ $(R_s/5 \times 10^{17}$ cm$)^{1/2}$ $M_\odot$ yr$^{-1}$, whereas helium may still absorb a significant fraction of the radiation from the burst. There is also the problem that the photoionization time scales for the outer interaction region may be long compared to the duration of the burst and to recombination time scales even for low ionization stages.

One of the most important results from the studies of the CSM of SN 1987A is that we may put constraints on the ionizing radiation from the soft X-ray burst. In Sect. 4 we found that the 11E1Y6 model by Shigeyama et al. (1988) did better than the 10L model by Woosley (1988) in the models of the narrow UV and optical emission lines. The reason for this was the higher peak effective temperature in the 11E1Y6 model. However, all hydrodynamic models of SN 1987A, treat the supernova atmosphere as an LTE blackbody. From previous studies of more extended progenitors, which take the effect of electron scattering into account (Klein and Chevalier, 1978; Falk, 1978), we know that a blackbody may underestimate the radiation temperature by a factor ~(2-3). If this is true also for compact presupernovae like Sk -69°202, this would, for example, move the 10L model into the allowed temperature range. Better calculations for the spectrum of the outburst in the hydrodynamic models are strongly encouraged.

SN 1987A was the brightest supernova since 1604, and has, of course, given us an unusually good opportunity to test current SN theory. Clear evidence for a core collapse came from the neutrino pulse, the first neutrinos ever associated with a specific object outside the solar system. The inflection in the optical light curve ~600 days after the explosion (Hamuy et al., 1988) indicated that the collapse resulted in a rotating neutron star. This was directly confirmed by the recent detection of a 0.5 ms pulsar in SN 1987A (Middleditch et al., 1989). Future observations of this rapidly spinning object will, hopefully, teach us more about supernova explosions. Finally, we expect an increased X-ray and radio emission from SN 1987A in the 1990's when the blast wave hits the UV-emitting shell.

**Acknowledgements:** I am indebted to Claes Fransson for stimulating discussions. This research was supported by the Royal Swedish Academy of Sciences and the Swedish Natural Science Research Council.

**REFERENCES**

Aitken, D.K., Smith, C.H., James, S.D., Roche, P.F., Hyland, A.R., and McGregor, P.J. 1988, M.N.R.A.S. **231**, 7P.

Arnett, W.D. 1988, in *Supernova 1987A in the Large Magellanic Cloud*, eds. M. Kafatos and A.G. Michalitsianos, (Cambridge: C.U.P.), p. 301.

Bandiera, R., Pacini, F., and Salvati, M. 1988, Nature **332**, 418.

Barkat, Z., and Wheeler, J.C. 1988, Ap.J. **332**, 247.

Bionta, R.M. et al. 1987, Phys. Rev. Letters **58**, 1494.

Catchpole, R.M., and Whitelock, P.A. 1988, IAU Circ. No. 4544.

Catchpole, R.M. et al. 1987, M.N.R.A.S. **229**, 15P.

Chalabaev , A.A., Perrier, C., and Mariotti,J.-M. 1989, Astr. Ap. **210**, L1.

Chevalier, R.A. 1982, Ap. J. **259**, 302.

Chevalier, R.A. 1984, Ann. N.Y. Acad. Sci. **422**, 215.

Chevalier, R.A. 1987, in *Proc. of the ESO Workshop on SN 1987A*, ed. I.J. Danziger, p. 481.

Chevalier, R.A. 1988, Nature **332**, 514.

Chevalier, R.A., and Fransson, C. 1987, Nature **328**, 44.

de Boer, K.S., Grewing, M., Richtler, T., Wamsteker, W., Gry, C., and Panagia, N. 1987, Astr. Ap. **177**, L37.

de Jager, C., Nieuwenhuijzen, H., and van der Hucht, K.A. 1988, Astr. Ap. Suppl. **72**, 259.

Dufour, R.J. 1984, in *Strucure and Evolution of the Magellanic Clouds* (IAU Symp. 108), eds. S. van den Bergh and K.S. de Boer, (Reidel: Dordrecht, Boston, Lancaster), p. 353.

Dupree, A.K., Kirshner, R.P., Nassiopoulos, G.E., Raymond, J.C., and Sonneborn, G. 1987, Ap. J. **320**, 597.

Dwek, E. 1988, in *Supernova 1987A in the Large Magellanic Cloud*, eds. M. Kafatos and A.G. Michalitsianos, (Cambridge: C.U.P.), p. 240.

Emmering, R.T., and Chevalier, R.A. 1989, submitted to Ap. J.

Fabian, A.C., and Rees, M.J. 1988, Nature **335**, 50.

Falk, S.W. 1978, Ap. J. (Letters), **225**, L133.

Fransson, C. 1984, Astr. Ap. **133**, 264.

Fransson, C. 1986, in *Radiation Hydrodynamics in Stars and Compact Objects*, eds. D. Mihalas and K.H.A. Winkler, (Springer; Berlin, Heidelberg, New York), p. 141.

Fransson, C., and Lundqvist, P. 1989, Ap. J. (Letters), in press.

Fransson, C., Benvenuti, P., Gordon, C., Hempe, K., Palumbo, G.G.C., Panagia, N., Reimers, D., and Wamsteker, W. 1984, Astr. Ap. **132**, 1.

Fransson, C., Cassatella, A., Gilmozzi, R., Kirshner, R.P., Panagia, N., Sonneborn, G., and Wamsteker, W. 1989, Ap. J. **336**, 429.

Garmany, C.D., and Conti, P.S. 1985, Ap.J. **293**, 407.

Gaskell, C.M. 1984, P.A.S.P. **96**, 789.

Gilmozzi, R., Cassatella, A., Clavel, J., Fransson, C., Gonzalez, R., Gry, C., Panagia, N., Talavera, A., and Wamsteker, W. 1987, Nature **328**, 318.

Hamuy, M., and Phillips, M. 1988, IAU Circ. No. 4534.

Hamuy, M., Suntzeff, N., Martin, G., and Gonzalez, R. 1988, IAU Circ. No. 4680.

Hanuschik, R.W., and Dachs, J. 1987, Astr.Ap. 177, L4.

Hillebrandt, W., Hoflich, P., Schmidt, H.U., and Truran, J.W. 1987, Astr. Ap. 186, L9.

Hirata, K. et al. 1987, Phys. Rev. Letters 58, 1490.

Humphreys, R.A., and Davidson, K. 1979, Ap. J. 232, 409.

Itoh, H., Hayakawa, S, Masai, K., and Nomoto, K. 1987, Publ. Astron. Soc. Japan 39, 529.

Itoh, M., Kumagai, S., Shigeyama, T., Nomoto, K., and Nishimura, J. 1987, Nature 330, 233.

Klein, R.I., and Chevalier, R.A. 1978, Ap. J. (Letters) 223, L109.

Lundqvist, P., and Fransson, C. 1987, in *Proc. of the ESO Workshop on SN 1987A*, ed. I.J. Danziger, p. 495.

Lundqvist, P., and Fransson, C. 1988, Astr. Ap. 192, 221.

Lundqvist, P., and Fransson, C. 1989, in preparation.

Maeder, A. 1987, in *Proc. of the ESO Workshop on SN 1987A*, ed. I.J. Danziger, p. 251.

Masai, K., Hayakawa, S., Inoue, H., Itoh, H., and Nomoto, K. 1988, Nature 335, 804.

Matz, S.M., Share, G.H., Leising, M.D., Chupp, E.L., Vestrand, W.T., Purcell, W.R., Strickman, M.S., and Reppin, C. 1988, Nature 331, 416.

Mc Naught, R.H. 1987, IAU Circ. No. 4316.

Meikle, W.P.S., Matcher, S.J., and Morgan, B.L. 1987, Nature 329, 608.

Middleditch, J. et al. 1989, IAU Circ. No. 4735.

Nisenson, P., Papaliolios, C., Karovska, M., and Noyes, R. 1987, Ap. J. (Letters) 320, L15.

Phinney, E.S. 1988, Nature 331, 566.

Pinto, A.P., and Woosley, S.E. 1988, Nature 333, 534.

Rank, D.M., Bregman, J., Witteborn, F.C., Cohen, M., Lynch, D.K., and Russell, R.W. 1988, Ap. J. (Letters) 325, L1

Rousseau, J., Martin, N., Prevot, L., Rebeirot, E., Robin, A., and Brunet, J.P. 1978, Astr. Ap. Suppl. 31, 243.

Saio, H., Nomoto, K., and Kato, M. 1988, Nature 334, 508.

Sonneborn, G., Altner, B, and Kirshner, R.P. 1987, Ap. J. (Letters) 323, L35.

Sonneborn, G. et al. 1989, IAU Circ.

Shigeyama, T., Nomoto, K., and Hashimoto, M. 1988, Astr. Ap. 196, 141.

Storey, M.C., and Manchester, R.N. 1987, Nature 329, 421.

Turtle, A.J., Campbell-Wilson, D., Bunton, J.D., Jauncey, D.L., Kesteven, M.J., Manchester, R.N., Norris, R.P., Storey, M.C., and Reynolds, J.E. 1987, Nature 327, 38.

Walborn, N.R., Lasker, B.M., Laidler, V.G., and Chu, Y.-H. 1987, Ap. J. (Letters) 321, L41.

Wampler, E.J., and Richichi, A. 1989, submitted to Astr. Ap.

Weaver, R., McCray, R., Castor, J., Shapiro, P., and Moore, R. 1977, Ap. J. 218, 377.

West, R.M., Lauberts, A., Jorgensen, H.E., and Schuster, H.-E. 1987, Astr. Ap. 177, L1.

White, G.L., and Malin, D.F. 1987, Nature 327, 36.

Woosley, S.E. 1988, Ap. J. 330, 218.

Woosley, S.E, Pinto, P.A., and Ensman, L. 1987, Ap. J. 324, 466.

# DIFFUSIVE SHOCK ACCELERATION OF RELATIVISTIC PARTICLES

S.A.E.G. FALLE,
Department of Applied Mathematical Studies,
The University,
Leeds LS2 9JT, U.K.

ABSTRACT. In this lecture I am going to look at the mechanism of diffusive shock acceleration of energetic particles which has been proposed by Krymsky (1977), Axford, Leer and Skadron (1977) and Bell (1978) as the means by which both relativistic electrons in supernova remnants and the galactic cosmic rays are produced. Most of the work on this subject has neglected the effect that the relativistic particles have on the dynamics of the thermal gas. We shall see that current theories of the interaction between the relativistic particles and the thermal gas are not sensible once such effects are included and that we really need to look much more carefully at the fundamental physics of the problem.

## 1. Introduction

The basic idea of diffusive shock acceleration is extremely simple; a population of relativistic particles is supposed to interact with a thermal gas by scattering off magnetic irregularities which more or less comove with the gas. A shock in the thermal gas is assumed to be transparent to the relativistic particles and they can pick up energy from the mean flow by being scattered back and forth across such a shock. Clearly the probability that a particle will be accelerated to a given energy is simply the probability that it will be scattered enough times before it finally escapes downstream.

W. Kundt (ed.), Neutron Stars and Their Birth Events, 303–318.

The attraction of this mechanism is that in simple situations it predicts a power law spectrum in energy for the relativistic particles whose slope agrees very well with that observed for both galactic cosmic rays and the synchrotron emitting electrons in supernova remnants. Unfortunately the simple theory, which neglects the dynamical effects of the relativistic particles on the thermal gas, predicts that the process is very efficient, so much so that most of the energy in the thermal gas should be transferred to the relativistic particles. In that case one can clearly not neglect the dynamical effects of the energetic particles.

## 2. Fokker-Planck Equation

Let $f(\mathbf{x}, p, t)$ be the distribution function for the relativistic particles so that $4\pi p^2 f dp$ is the number of particles per unit volume with momenta between $p$ and $p + dp$ at $\mathbf{x}$ at time $t$. We have assumed that the distribution is isotropic in momentum space and that there is only one type of particle. Skilling (1975) has shown that provided the fluid speed is much smaller than the particle speed, $f$ satisfies the Fokker-Planck equation

$$\frac{\partial f}{\partial t} + \mathbf{u}.\nabla f = -\frac{1}{3}p\frac{\partial f}{\partial p}\nabla.\mathbf{u} + \nabla.(\kappa\nabla f) \qquad 2.1$$

where $\mathbf{u}$ is the velocity of the thermal gas. In 2.1 the second term on the left is due to advection by the thermal gas, the first term on the right is adiabatic compression and the final term represents space diffusion through the thermal gas. Here $\kappa$ is a diffusion coefficient which in principle can depend on both the state of the thermal gas and on the momentum of the particles.

If we now assume that we have a steady gas shock and that the particles have no effect on the gas, then it is easy to show (see e.g. Drury 1983) that

$$f_{-\infty}(p) = \alpha p^{-\alpha}\int_0^p y^{\alpha-1}f_\infty(y)\,dy + bp^{-\alpha}, \qquad 2.2$$

where $f_{-\infty}(p)$ is the downstream distribution and $f_\infty(p)$ is the upstream one. $\alpha$ is given by

$$\alpha = \frac{3r}{(r-1)} \qquad 2.3$$

where r is the shock compression ratio.

The first term in 2.2 represents particles advected through the shock from upstream whilst the second comes from particles injected at low energies at the shock. The latter always have a power law spectrum with slope $-\alpha$ while the downstream spectrum of the advected particles has slope $-\alpha$ as $p \rightarrow \infty$ if $f_\infty$ has a steeper slope than $\alpha$. Since for a strong shock we have $r = 4$, the theory predicts $\alpha = 4$ for such shocks irrespective of whether the particles are injected at the shock or are accelerated from some softer upstream distribution. This agrees very nicely with the galactic cosmic ray spectrum which has a slope of 4.3 at the relevant energies.

Another advantage of this simple theory is that nothing depends on the diffusion coefficient $\kappa$ as long as we have a steady state and the energetic particles have no dynamical effect on the gas. $\kappa$ only affects the scale over which f changes from $f_\infty$ to $f_{-\infty}$. This is just as well since it is very hard to determine $\kappa$ either theoretically or from observation.

Unfortunately the theory contains the seeds of its own destruction. The relativistic particle pressure is

$$p_c = \frac{4\pi}{3}\int_0^\infty p^3 vf\,dp \qquad 2.4$$

where v is the particle velocity. In the relativistic limit $v = c$ and so if $f \sim p^{-\alpha}$ for large p the integral diverges for large p if $\alpha \leq 4$. This means that the energetic particle pressure becomes important for strong shocks if we have enough time to accelerate particles to very high energies.

## 3. Two-Fluid Model

In order to shed some light on this problem Drury and Völk (1981) constructed a simple one dimensional two fluid model which included the effect of the energetic particles on the thermal gas. They did this by integrating equation 2.1 over momentum to get an equation for the relativistic particle energy density $E_c$ ,

$$\frac{\partial E}{\partial t}c + \frac{\partial}{\partial x}(uE_c) = -p\frac{\partial u}{\partial x} + \frac{\partial}{\partial x}\bar{\kappa}\frac{-\partial E}{\partial x}c \ , \qquad 3.1$$

where

$$\bar{\kappa} = \frac{\int_0^\infty \{\kappa(p)p^2 T \partial f/\partial x\}dx}{\int_0^\infty (p^2 T \partial f/\partial x)dx} \ , \qquad 3.2$$

is a mean difusion coefficient and $T(p)$ is the particle kinetic energy.

The gas is described by the usual equations with an additional pressure due to the energetic particles.

$$\frac{\partial \rho}{\partial t} + \frac{\partial \rho u}{\partial x} = 0 \ , \qquad 3.3a$$

$$\frac{\partial \rho u}{\partial t} + \frac{\partial}{\partial x}(p_c + p_g + \rho u^2) = 0 \ , \qquad 3.3b$$

$$\frac{\partial}{\partial t}\left[\frac{1}{2}\rho u^2 + \frac{p_g}{(\gamma_g - 1)}\right] + \frac{\partial}{\partial x}\left[u\left[\frac{\gamma_g p_g}{(\gamma_g - 1)} + \frac{1}{2}\rho u^2\right]\right] = -u\frac{\partial p}{\partial x}c \ . \qquad 3.3c$$

In these equations $p_g$ is the gas pressure and $\gamma_g$ is the ratio of specific heats for the gas. The relativistic particle pressure $p_c$ is related to the relativistic particle energy density by

$$P_c = (\gamma_c - 1)E_c .$$  3.4

$\gamma_c$ should really be determined from the energetic particle distribution function, but in this simple model one has to assume a value of $\gamma_c$ in the range $4/3 \leq \gamma_c \leq 5/3$ to close the system of equations.

We want to look at the steady shock structures admitted by these equations. For steady flow we have

$$\rho u = \text{const.,}$$  3.5a

$$p_c + p_g + \rho u^2 = \text{const.,}$$  3.5b

$$\rho u \left( \frac{1}{2} u^2 + \frac{\gamma_g p_g}{(\gamma_g - 1)\rho} + \frac{\gamma_c p_c}{(\gamma_c - 1)\rho} \right) - \frac{\bar{\kappa}}{(\gamma_c - 1)} \frac{\partial p_c}{\partial x} = \text{const.}$$  3.5c

If there is no gas shock then

$$\frac{p}{\rho^\gamma} = \text{const.,}$$  3.5d

otherwise we have to apply the usual jump conditions at the gas shock since the diffusive term in 3.1 ensures that $p_c$ is continuous.

It can be shown that equations 3.5 have two different types of solution depending on the shock Mach number and the upstream particle pressure. For weak shocks and low upstream values of $p_c$ we get the kind of structure shown in figure 1a. There is a gas shock which is preceded by a precursor whose width scales like $\kappa/u$ where $u$ is a typical fluid speed. The gas behaves isentropically in the precursor, but there is the usual jump in entropy at the gas shock with the result that the gas pressure dominates behind the shock. For strong shocks or those with very large particle pressures upstream the solution looks like that in figure 1b. There is no gas shock and the gas behaves isentropically throughout. The energetic particle pressure dominates downstream of the shock and for very strong shocks the compression ratio is

$$r = \frac{(\gamma_c + 1)}{(\gamma_c - 1)} \cdot$$ 
3.6

Since $\gamma_c \leq 5/3$ this means that the compression ratio exceeds 4 and it would appear from equation 2.4 that the particle pressure is infinite. However, the solution 2.2 is only valid for a discontinuous shock structure. If we assume that the diffusion coefficient is constant, then the two fluid model becomes exact as far as the dynamics is concerned. It is then possible to solve for the distribution function for such a shock structure and show that the particle pressure obtained from equation 2.4 agrees with that required by equations 3.5 (Drury, Axford and Summers 1982). So this two fluid model is at least internally consistent.

Unfortunately we do not expect the diffusion coefficient to be constant, but to increase with particle momentum. There will therefore always be high energy particles which see the shock as a discontinuous structure and they will want to adopt the distribution given by equation 2.2. If the shock compression exceeds 4 then these particles will exert a large pressure which must modify the shock structure. So either there is no steady state or something is missing from the basic equations.

## 4. Time Dependent Numerical Calculations

On the whole steady state solutions are not a very good idea, partly because of the above difficulties, but also because this is a non-linear system and we are not guaranteed that steady states either exist or are unique. In fact there is a non-uniqueness problem for the two-fluid shock structures. It turns out that for strong shocks there are at least three different solutions for a given upstream state and downsteam pressure. Although one of these solutions can be shown to be unstable, that still leaves two entirely plausible solutions and which one is chosen clearly depends on how the steady state is set up.

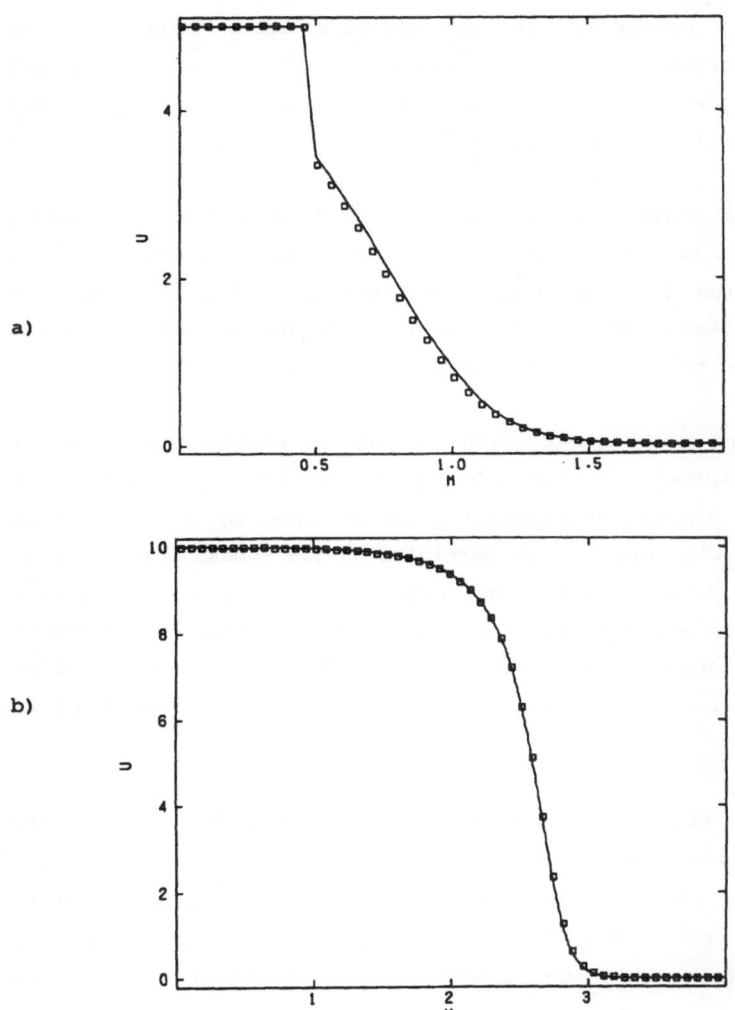

a)

b)

Figure 1.  Gas velocity in the two-fluid shock structure with a gas sub-shock (figure 1a) and without a gas sub-shock (figure 1b).  The curves are obtained from equations 3.5 and the squares from a numerical integration of equns. 3.1 and 3.3 .  $M := \int \varrho \, dx$ = Lagrangian coordinate.

Unfortunately it is not possible to make much progress with time dependent solutions by analytic means and so we have to resort to numerical calculations. For this purpose we used a first order Lagrangian Godunov scheme for the gas dynamics and a Crank-Nicholson form for the diffusive term in equation 2.1. The details of the numerical scheme and various test calculations are given in Drury and Falle (1986) and Falle and Giddings (1987).

Figure 1 shows a fairly typical test calculation for the two-fluid model. The curves are the analytic results obtained from equations 3.5 whilst the squares are the numerical results. The agreement is excellent except that at this resolution the numerical dissipation leads to the gas entropy being slightly greater than it should be.

Clearly diffusive shock acceleration is only a viable mechanism for accelerating particles if it can inject particles into the acceleration process from the thermal distribution, so we have to introduce some process which injects relativistic particles at gas shocks. It is known from studies of interplanetary shocks that this occurs, but it is hard to extrapolate from such obervations to the rather different conditions in interstellar shocks. Fortunately the details of the injection mechanism do not matter too much as long as they make some kind of physical sense.

We simply assume that the shock injects a small fraction $\varepsilon$ of the incoming thermal particles at an energy $\lambda$ (typically $\lambda = 2$) times the post-shock thermal energy. Numerically this was effected by injecting the appropriate number of particles with a Gaussian distribution in space of a few mesh points and removing the energy of these particles from the gas energy. Since weak shocks are not expected to inject particles we switch off injection for shocks with Mach numbers less than 1.3.

It is shown in Falle and Giddings that the results obtained with this scheme agree both with the test particle and two-fluid solutions under the appropriate conditions and so should be reliable for the full problem. The scheme is not particularly accurate, but for one dimensional problems we can use very high resolution and so ensure respectable accuracy without consuming excessive quantities of computer time.

To study the questions of existence and uniqueness of the steady solutions we set up a one dimensional piston problem in which a constant speed piston drives a shock into undisturbed gas which contains no relativistic particles. The diffusion coefficient is allowed to depend on momentum according to

$$\kappa = \kappa_i \left(\frac{p}{p_i}\right)^{1/4} , \qquad\qquad 4.1$$

where $\kappa_i$ is the diffusion coefficient at the injection momentum $p_i$. This dependence is chosen entirely for numerical convenience. We expect $\kappa$ to be an increasing function of particle momentum, but since the length scale is determined by $\kappa/u$, we would end up with a very large range of length scales if we allowed $\kappa$ to increase rapidly with particle momentum. Although the form of $\kappa$ affects the results quantitatively, the qualitative results are the same provided $\kappa$ increases with particle momentum.

We looked at three cases which illustrate the various aspects of the theory. The first is a low Mach shock whose speed is very subrelativistic. In that case the particles are not accelerated to relativistic energies, $\gamma_c$ is close to 5/3 and the shock compression is less than 4. We would therefore expect a steady solution to exist and the post-shock particle distribution should resemble that given by 2.2 if the relativistic particles do not have much dynamical effect on the thermal gas.

As can be seen from figure 2, this is indeed the case. The solution becomes steady on a timescale which is about that calculated from test

312

particle theory and the post-shock particle pressure is proportional to
the injection rate ε for low values of ε and then saturates as the
dynamical effects of the relativistic particles begins to broaden the
shock structure. These results are very similar to those obtained by
Ellison and Eichler (1984) from Monte-Carlo simulations of such shocks.

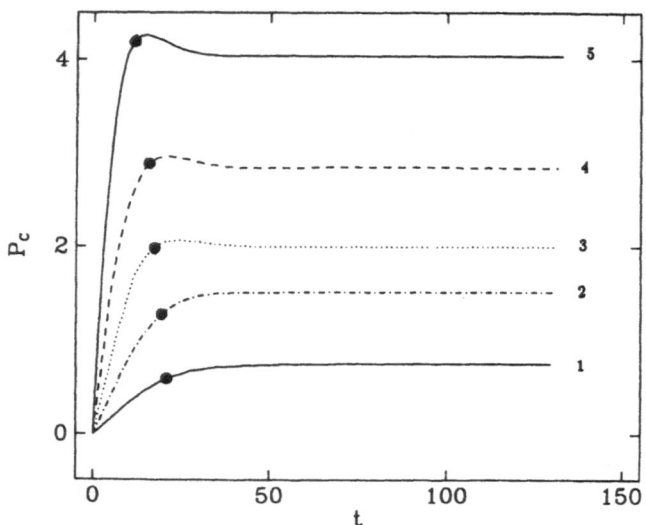

Figure 2. Postshock particle pressure as a function of time for a low
Mach shock (piston Mach no = 2.0). The injection rates ε are
(1) 0.0009, (2) 0.00225, (3) 0.0045, (4) 0.009, (5) 0.0225. The solid
circles mark the acceleration time according to test particle theory.

Next we consider a shock with an intermediate Mach number (piston Mach
number 3.5) and let the shock speed be such that the particles are
injected at a momentum $0.1 m_p c$ where $m_p$ is the particle mass. We then
expect the particles to be accelerated to relativistic energies and
hence $\gamma_c$ should be substantially less than 5/3. Then if the particle
pressure becomes important the compression ratio should exceed 4 and
there should be no steady solution.

In fact what we get is a quasi-steady state. Although the distribution
of relativistic particles with energy changes continously, the

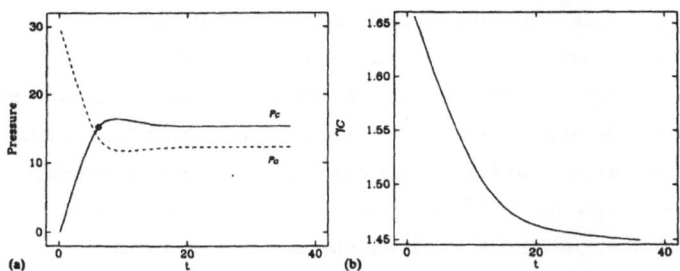

Figure 3. Time evolution of the post-shock state for a piston Mach number of 3.5. a) Particle pressure $P_c$ and gas pressure $P_G$, b) Post-shock value of $\gamma_c$. The solid circle represents the test particle acceleration time.

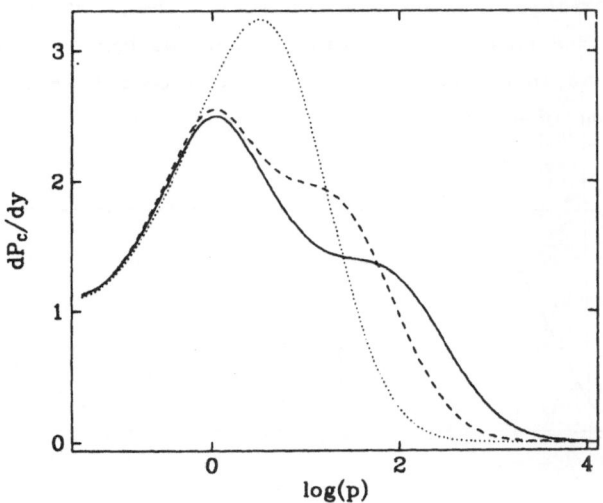

Figure 4. This shows how the particle pressure is distributed over momentum for a piston Mach number of 3.5. The area under the curves is proportional to $P_c$. The times are t = 18.1 (dotted line), t = 27.1 (dashed line), t = 36.1 (solid line).

314

post-shock particle pressure does not. What happens is that the particles are continually being accelerated to higher energies and this leads to an increase in the effective value of the diffusion coefficient so that the shock structure gets progressively broader. However, this happens on the acceleration timescale of the most energetic particles and this is always much larger than the flow time through the shock structure. Hence the shock evolves through a succession of quasi-steady states.

This can be seen in figures 3 and 4. Figure 3a shows how the post-shock pressures become constant in about the time it takes a typical particle to be accelerated. $\gamma_c$, however, changes continually as particles are accelerated to more and more relativistic energies. Eventually $\gamma_c$ will reach 4/3 when the particle pressure becomes dominated by highly relativistic particles. The way in which the post-shock particle distribution shifts to higher and higher energies can be seen in figure 4 which shows the integrand in the pressure integral as a function of particle momentum (y = log p).

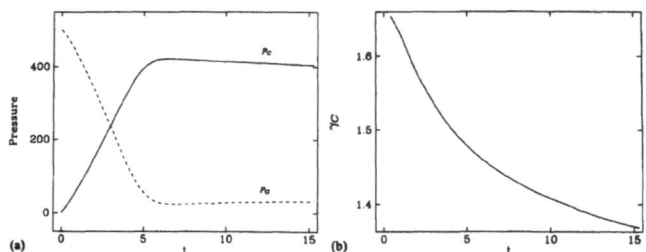

Figure 5. Time evolution of the post-shock state for a piston Mach number of 15. a) Particle pressure $P_c$ and gas pressure $P_G$, b) post-shock value of $\gamma_c$.

Finally we looked at a high Mach number shock again with an injection momentum of $0.1 m_p c$. In this case we expect the particle pressure to become large behind the shock and for the gas shock to disappear. If the only source of particles is injection at the gas shock, then the total number of particles in the shock structure will decrease since they are being continually advected downstream. Hence either the particle pressure will decrease and the gas shock reappear or an ever decreasing number of particles at higher and higher energies must supply the downstream particle presssure.

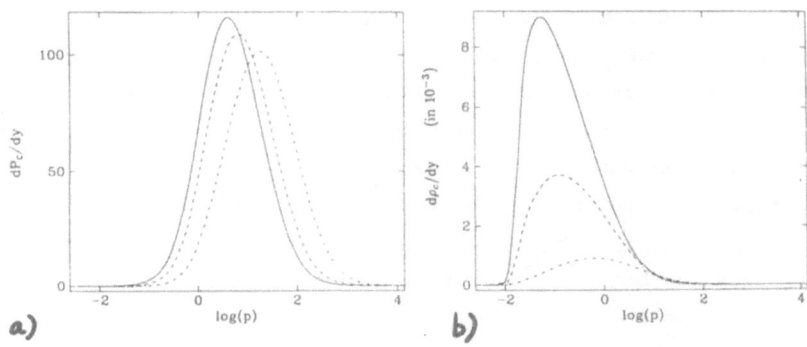

Figure 6. Post-shock particle distribution for a piston Mach number of 15. a) The integrand in the pressure integral, b) the particle number as a function of momentum.

Figure 5 shows that the particle pressure continues to dominate over the gas pressure after the gas shock disappears at $t = 5.1$, although it does decrease very slowly. $\gamma_c$ decreases towards 4/3 as the particles are accelerated to more and more relativistic energies. This shift to higher and higher energies can be seen in figure 6a which again shows the integrand in the pressure integral. The decrease in the number of particles is shown in figure 6b. What we have here is a shock structure in which there are no upstream particles and no injection at the shock, but the particle pressure nevertheless dominates downstream.

This seems ridiculous, but the two-fluid equations 3.5 do in fact admit a steady solution with these properties, although Drury and Völk rejected it as unphysical. Our solution is not actually steady since there is a secular change in both the post-shock pressure and density, but it does look very much like a steady solution and we believe that it will become steady once $\gamma_c$ stops changing.


## 5. Conclusions

I pointed out at the beginning of this article that diffusive shock acceleration seemed a very promising mechanism for producing both the galactic cosmic rays and the synchrotron emitting electrons in radio sources such as supernova remnants and radio jets. This is because it naturally produces the observed distribution with energy and it is an efficient process provided the diffusion coefficient is not too large.

However, once one begins to look more closely at it, a number of serious flaws appear. In the first place it is much too efficient since for any but the weakest shocks it leads to most of the energy dissipated at the shock being transferred to the relativistic particles. This conflicts with observations of supernova remnants which show that both young and old remnants contain substantial amounts of hot gas. Apart from that nice simple spectra are only obtained if the relativistic particles do not exert any dynamical effect on the gas.

What is wrong with the standard theory is that there is an irreversible mechanism for transferring energy from the thermal gas to the energetic particles, but there is only a reversible mechanism, adiabatic compression, for transferring it the other way. Since physical systems try to increase their entropy, the inevitable result is that most of the energy ends up in the relativistic particles. A sensible theory should clearly include some mechanism for irreversibly transferring energy from the energetic particles to the gas.

Several such mechanisms have been suggested. Völk, Drury and McKenzie (1984) showed that the energetic particles can excite Alvén waves which are then dissipated and so heat the gas. However, this is not a very effective process since Alvén waves are very slowly damped. There also exists an instability which leads to the amplification of sound waves in the shock structure (Drury and Falle 1986), but again it does not seem to be an effective way of transferring energy to the thermal gas.

It seems to me that the real problem lies with equation 2.1. A lot of assumptions and approximations have gone into its derivation and these have not really been looked at much since Skilling's original paper. One of the consequences of this is that nobody has much idea about how the diffusion coefficient depends on particle momentum and the state of the thermal gas.

Nevertheless, diffusive shock acceleration does have some things going for it. Observations tell us that there are relativistic electrons present whenever we have strong long-lived shocks and the electron spectra are about what we would expect from this mechanism. In extragalactic radio jets the shock speeds and timescales are such that we would expect the relativistic particles to dominate the pressure totally and indeed one never sees any thermal gas actually inside such jets. Finally there does appear to be a power law tail in the electron distribution downstream of the Earth's bow shock although the observations certainly do not provide conclusive proof that this is due to diffusive shock acceleration.

## 6. References

Axford, W.I., Leer, E. & Skadron, G.,1977.Proc. 15th Int. Cosmic Ray Conf.,Plovdiv,**11**,132.

Bell, A.R.,1978.Mon. Not. R. astr. Soc.,**182**,147.

Drury, L.O'C.,1983.Rep. Prog. Phys.,**46**,973.

Drury, L.O'C.,Axford, W.I. & Summers, D.,1982.Mon. Not. R. astr. Soc.,**198**,833.

Drury, L.O'C. & Falle, S.A.E.G.,1986.Mon. Not. R. astr. Soc.,**223**,353.

Drury, L.O'C. & Völk, H.J.,1981.Astrophys. J.,**248**,344.

318

Ellison, D.C. & Eichler, D.,1984.Astrophys. J.,**277**,429.

Falle, S.A.E.G. & Giddings, J.R.,1987.Mon. Not. R. astr. Soc.,**225**,399.

Krymsky, G.F.,1977.Dokl. Akad. Nauk. SSSR,**234**,1306. (Engl. Trans. Sov.

Phys. Dokl.,**23**,327).

Skilling, J.,1975.Mon. Not. R. astr. Soc.,**172**,557.

Völk, H.J., Drury, L.O'C. & McKenzie, J.F.,1984.Astr. Astrophys,**130**,19.

**Acknowledgements**

I would like to thank NATO for financial support during the course of
this meeting and also Wolgang Kundt for organising such an enjoyable and
stimulating summer school.

# SYMMETRY PROPERTIES OF THE BIFURCATED RADIO STRUCTURE ASSOCIATED WITH ACTIVE GALACTIC NUCLEI

D.G.BANHATTI
School of Physics
Madurai-Kamaraj University
Madurai 625 021
INDIA

ABSTRACT Symmetry parameters of active galaxies are reviewed. These observational properties are considered in the light of currently popular theories of such objects.

## 1 Introduction

In the 1920s, Hubble classified the regular optical forms of galaxies into ellipticals, lenticulars, normal spirals and barred spirals. Active nuclei are found to reside in all of these galaxies: LINERs (Low Ionization Nuclear Emission-line Regions) and Seyfert nuclei in spirals, quasars in spirals and ellipticals, and radio galaxies and BL Lac objects in ellipticals. Nuclear activity modifies their appearance in all wavebands: X-rays, ultraviolet, optical, infrared and radio; near the nucleus and all the way to distances upto megaparsecs from it. Radio emission originates in the nuclei (subparsec), the jets (parsec to Mpc), the hotspots (subkpc), and the diffuse structures surrounding all of these. The radio emission is synchrotron radiation from relativistic electrons with energies of a few 100 GeV gyrating in magnetic fields of a few to hundreds of microGauss. This is modified by various energy loss mechanisms: synchrotron loss, inverse-Compton scattering, adiabatic expansion, etc. Often, the particles have a power-law distribution of energies, $N_E dE \propto E^{-\gamma} dE$, rather flat when they are young, evolving with time to a power-law index of $\gamma \approx -3$ or steeper since the higher-energy particles lose energy faster. Repeated acceleration through shocks naturally gives rise to a power-law energy spectrum (Achterberg 1987; see however Kundt 1984, Falle 1987 and this volume), as inferred from the observed spectra (Impey 1987). Cygnus A was the first object shown to possess the double radio structure (Jennison & DasGupta 1953), since found to be characteristic of a majority of extragalactic radio sources. Examples are the low-resolution (few arcminutes) map of Fornax A (Moffet 1966), $Cyg_u$ A observed with high resolution (subarcsec) and dynamic range ($\sim 10^4$)(Perley et al 1984) and 1313-099, a giant radio galaxy similar in appearance to Cyg A (A.K.Singal & C.J.Salter, private communication). NGC6251 exemplifies a radio jet that extends from pc to Mpc scales (Readhead et al 1978).

## 2 Extended Structure

The extended structures of extragalactic radio sources have bilateral symmetry and fall into three partially overlapping sequences (Miley 1980): the edge-brightening sequence (Virgo A, 3C449, 3C236 and

319

*W. Kundt (ed.), Neutron Stars and Their Birth Events, 319–325.*

3C390.3), the rotational symmetry (or twisting) sequence (Cyg A, 3C47, NGC315, 3C315) and the bending sequence (3C449, Virgo A, 1610-608, 3C465, IC708, 3C83.1B and IC310). Radio luminosity increases along the edge-brightening sequence (Fanaroff & Riley 1974), dividing the straight doubles into classes FR1 (low luminosity, edge-darkened sources) and FR2 (high-luminosity, edge-brightened sources). The luminosity at the dividing line (~$10^{25}$ WHz$^{-1}$ sr$^{-1}$ at 178MHz, with $H_0$ =50 kms$^{-1}$ Mpc$^{-1}$ and $q_0$ =1/2) is very close to that above which the space density of radio sources shows a steep increase at earlier epochs (i.e., at larger redshifts) (Auriemma et al 1977). The same luminosity also sharply divides the bending sequence into more bent doubles of low radio and optical luminosity and less bent ones of high radio and optical luminosity. This trend does not extend to straight doubles which may have low or high radio luminosities so that they belong to classes FR1 or FR2 respectively (Valentijn 1979). The strong 3CR and weaker B2 radio galaxies are about equally divided between twisted (S-shape) and bent (C-shape) doubles, with the distortion angle similarly distributed for both. However, radio galaxies in Abell clusters have much larger bending angles, with only C-shape structures being present (Ekers 1982). The trends for FR1 doubles are supplemented by the increase of radio luminosity with the mean of the distances to the first brightness peaks on the two sides (Birkinshaw et al 1978). For FR2 doubles, the fraction of emission in the hotspots increases with the total radio luminosity (shown with coarse - resolution observations by Jenkins and McEllin 1977 and confirmed, e.g., by Swarup et al 1982,1984).

### 3 Symmetry Parameters

The bilateral symmetry of the extended radio structure is related to the optical and radio structure at the nucleus. It can be usefully quantified by (a) the ratios of armlengths, hotspot strengths and total component strengths on the two sides, (b) the relative strengths of the radio core, the hotspots, and the total lobe emission, (c) the misalignment of the two arms from a straight line, and (d) the misalignment between the axes of the bilateral radio structure on nuclear and larger scales; the angles giving the orientation of the optical image, the dust lane, the rotation axis and the optical and radio polarization axes.

### 4 Active Galactic Nuclei

An active galactic nucleus may show broad and (possibly variable) narrow nonstellar emission lines. There are broad permitted lines, and the narrow lines may either be permitted ones or forbidden ones. An AGN may also have a (possibly variable) nonstellar infrared-to-X-ray continuum and may be either radio-loud or -quiet. It may be a LINER, Seyfert nucleus, quasar, BL Lac object or radio galaxy. In general, there is a central engine of size <0.01 pc emitting energy at the rate of $10^{40}$ to $10^{48}$ ergs$^{-1}$ first into the broad-line region of size ≤1 pc with gas clouds of electron density >$10^{10}$ cm$^{-3}$ at $10^4$ K electron temperature filling ≪ 0.01 of the volume, moving at ~5000 to 30,000 kms$^{-1}$ in a plasma having 0.01 photons per ionizable nucleon. The radiation then travels further into the narrow-line region of size ~1 kpc with gas clouds of electron density ~$10^2$ to $10^8$ cm$^{-3}$ at similar electron temperature filling <<0.01 of the volume moving at ~200 to 1000 kms$^{-1}$ in a plasma having ~0.001 (for a LINER) or ~0.01 to 0.1 (for a Seyfert 2) photons per ionizable nucleon (Lawrence 1987). Bilaterally symmetric narrow-line regions coextensive with the extended radio structure ( ~10 kpc to few Mpc) have recently been found whenever looked for deeply enough (e.g., van Breugel & Heckman 1982, Baum & Heckman 1989 and references therein). The velocity structure in these extended narrow-line regions is indicative of systematic bulk motions (e.g., Baum 1989). Active galactic nuclei can be classified in a 5-parameter space: narrow vs broad lines, continuum vs lines, optical luminosity, radio luminosity (or power), and excitation level of the ions (Lawrence 1987). There is also a definite order in a 'radio colour-magnitude diagram' plotting log($z^2 \cdot$ flux density(5GHz)) vs spectral index at 5GHz. Spirals and Seyferts are less radio luminous than radio galaxies (ellipticals), all of which have steep spectra. There is a sequence of spectral flattening and increasing radio luminosity from spirals and Seyferts through normal ellipticals to BL Lac objects (which may have inverted spectra and are at least as luminous as radio galaxies). Quasars are the most luminous sources and may have steep (for straight or wide angled bent doubles) or flat spectra (for asymmetric radio structures)(Usher 1978).

## 5 Jets in Galaxies and Stars

The narrowest jets are seen in high-luminosity radio galaxies and quasars, and emanate from the most luminous cores. Low-luminosity radio galaxies, Seyferts and normal spirals form a continuation of the sequence to wider jets from less luminous cores (Sofue 1986), continuing down to stellar (radio) jets. Thus, bilateral (radio) structure is also seen in symbiotic stars (e.g., CH Cygnii: Seaquist & Taylor 1986, R Aquarii:Michalitsianos et al 1986), bipolar flows from young stars exciting Herbig-Haro objects and maser knots (Lada 1985, Schwartz 1986, Buehrke et al 1988) and radio-emitting X-ray binaries (neutron star sources)(ScoX-1, Cyg X-3, SS433: Feigelson 1986). On formation, every star loses mass in a wide bipolar flow sometimes embedding narrow jets (seen in optical and radio) which may excite Herbig-Haro objects en route to a bow shock far from the exciting young star. Bipolar flows around young stellar objects consist of blueshifted and redshifted (~25 kms$^{-1}$) lobes of cold (10 to 90 K), dense (300 to 3000 cm$^{-3}$) molecular gas (1 to 100 M(sun)) symmetrically situated about an embedded infrared source or young star (Lada 1985). Close binary stars form bilateral structure associated with the accretion of matter from the distended companion late in its life onto the primary which may be a neutron star left over possibly after a supernova explosion at the end of its life. It is intriguing that the young stellar object S106 shows emission spectrum (with neutral hydrogen lines removed) very similar to the active galactic nucleus Markarian 402 (Persson 1986). (See Kundt 1987a for a detailed comparison of galaxian and stellar sources showing jets. See also Koupelis & Van Horn 1988 for a model of acceleration of jets applied to AGNi, neutron star sources like SS433 and bipolar flows from protostellar objects.)

## 6 Models

Many models have been proposed to account for the observed properties of active galactic nuclei. For example, Morrison(1969) gives a model for extragalactic radio sources based on pulsar models; a giant $10^8$ to $10^{10}$ M(sun) sphere of magnetized plasma of size $\sim 10^{13}$ to $10^{15}$ cm (of density about that of water) rotating at about $10^{-7}$ rads$^{-1}$ (about once in 10 days). The magnetic field falls as $1/r^2$ or $1/r^3$ (for a dipole) or steeper (for higher multipoles) upto the light cylinder (at which the linear rotation speed = c) and is mainly toroidal thereafter, falling as $1/r$. A more modern variant taking account of recent observations is the supermassive magnetized core model of an accretion disk (Kundt 1987b). Another model is the slingshot theory in which the jets are illuminated plasma trails left by orbiting supermassive black holes, and the hotspots turning points in their trajectory (Valtonen 1984). Currently most popular is the beam theory in which the spin of a supermassive black hole-acretion disk system generates two opposing beams which first plough through the interstellar medium and then into the intergalactic medium forming subpc to kpc and Mpc-scale radio structure. For a double source aligned towards the line of sight, the beam pointing towards us is brighter due to Doppler boosting of its radiation and is seen as an intense jet, the counterjet being Doppler diminished, very often below the limit of detection.

## 7 Bilateral Structure and Beaming

The earliest indications of general bilateral (i.e. two-sided) structure in extragalactic radio sources came with interferometric observations by Fomalont(1969) and aperture synthesis observations by Mackay(1971). With ~arcminute resolution, these show that about 70% of the sources selected at metrewavelengths have double structures with the brighter, more compact component closer to the nucleus. The misalignment of the two arms at the nucleus is a few degrees. With the appearance of surveys at cm-wavelengths, more compact sources were found; later to be separated into cores, jets and other structures by subarcsec resolutions. When the resolution reached ~milliarcsec, apparent proper motions of upto tens of c (called superluminal motions) were discovered. Relativistic beams oriented close to the line of sight constitute the simplest model which accounts for both of these observations as well as the fact that most quasars are radio-quiet (Scheuer & Readhead 1979, Orr & Browne 1982). One-sided and core-halo structures may be considered intrinsically bilaterally symmetric ones oriented close to the line of sight. There is other more indirect evidence which implies highly relativistic bulk velocities (Lorentz factors >7) close to the nucleus and mildly relativistic speeds (~few tenths of c) for the advance of hotspots (Scheuer 1987). Superluminal motion has been found in all structural types except compact steep spectrum doubles (Pearson et al 1987). Many studies designed to test the relativistic beaming model lead to the conclusions that (a) all known superluminals have extended structure, (b)core asymmetries are related to asymmetries in the extended structure, and (c)there are no obvious radio features which distinguish superluminals from the general population of radio sources (Rusk & Rusk 1986, Browne 1987). However, there may be other indicators like high optical polarization, which point to the presence of superluminal motion (Impey 1987). The 10 to 1000 pc compact steep-spectrum doubles, which apparently do not fit into the relativistic beaming model, may be the earliest stages in the life of a source near the sky plane where the beams have not entrained enough material yet to be illuminated. They are characterized by >90% emission from two nearly equal components with separations >3 times their diameter, with substructures within the components indicative of tails toward the nucleus. The two components in these sources have very similar spectra with self-absorption peaks and steep optically thin sections at high radio frequencies. Component sizes are comparable to the sizes derived from synchrotron self-absorption calculations. They show both low (<2%) radio polarization and variability (<10%). No larger scale structure (upto arcsec) is seen with >3000:1 dynamic range. The nucleus is a galaxy showing narrow emission lines (Hodges & Mutel 1987).

## 8 Arm Ratios, Flux Ratios and Misalignment Angles

One of the earliest clear results on arm ratios and arm misalignments showed that for the strong 3C sample, extended straight doubles have an average arm ratio of about 1.4 and that they are misaligned by about 7° (Ingham & Morrison 1975). The arm ratios for doubles assumed straight can be used to derive an upper limit on the speed of advance of the hotspots by attributing all the asymmetry to light-travel time effects from a population of randomly oriented intrinsically symmetric doubles. Banhatti(1979,1980) found that this gives ~0.3c. Swarup & Banhatti(1981) further used this and flux density ratios for hotspots to show that hotspot luminosity must decline as the inverse cube of age. Macklin(1981) found that the lobe flux density ratio behaves differently from that for hotspots, which show no tendency for the brighter hotspot to be the nearer to the nucleus. Earlier results to the contrary were due to coarse observing resolution. Macklin also confirmed earlier results on the misalignment angle (~10°). Teerikorpi(1984,1986) further confirmed that the closer hotspot is not always the brighter. Considering a sample of double-lobed quasars, he showed that doubles with the closer hotspot fainter are equally common, and are more luminous on an average. Banhatti(1988) finds that the linear sizes of the latter type of source conform better than the former to the predictions of an intrinsically symmetric model.

## 9 Magnetic Fields

Examining the difference in the radio polarization position angle and the large scale radio axis, Clarke et al(1980) find that the more luminous edge-brightened doubles (FR2) have their magnetic field parallel to the radio axis, while the less luminous edge-darkened doubles (FR1) have it perpendicular. Using a much better observed sample of radio sources with jets, Bridle & Perley (1984) found that the field is parallel near the core and perpendicular farther away. The earlier result is due to the coarser resolution then available. Studies of hydrodynamic laboratory jets and numerical simulations show that FR1 jets have lower Mach numbers (<2) than FR2 jets, both being light jets in a heavy ambient medium (e.g., Norman 1986). Even detailed structures in the brightness distributions have been successfully modelled. A full understanding, taking magnetic fields into account is, however, lacking. (See Koupelis & Van Horn 1988 for a simple model in which both rotation and magnetic fields are important.)

## 10 Recent Results

With observations of better resolution, higher dynamic range and sensitivity becoming available, reexamination of the statistics of symmetry parameters of extragalactic double radio sources should be rewarding. One recent result is that for the luminous edge-brightened (FR2) radio doubles, there is more depolarization on the side opposite the jet, implying more magnetoionic material on the counterjet side (Garrington et al 1988, Laing 1988). This would seem to be one more piece of evidence in favour of the relativistic beaming model. Structures of higher - redshift sources are now being increasingly mapped and show that galaxies and their halos were smaller in the past (i.e., at higher redshifts)(Wiita & Gopal-Krishna 1987). The structures are also found to be more distorted at earlier epochs, indicative of a denser ambient medium around radio sources of higher redshifts (Lonsdale 1986). Detection of extended narrow-emission-line regions coextensive with extended radio lobes in double radio sources (e.g., Durret 1989) is another important recent result.

## Acknowledgments and apologies

I thank the University Grants Commission, New Delhi for partial financial assistance, D.J.Saikia for exchange of views and Chris Salter and T.P.Srinivasan for many suggestions. I have made no attempt to include all aspects of the topic, nor to give priority-based credit, and apologize to those whose results or names may have been omitted.

## References

Achterberg,A:1987:in Kundt,W(ed)**ApJets & Their Engines** 223-36
Auriemma,C, Perola,G C, Ekers,R, Fanti,R, Lari,C, Jaffe,W J, Ulrich,
  M -H:1977:**AA 57** 41-50
Banhatti,D G:1979:**Bull A S India** 7 116
Banhatti,D G:1980:**AA 84** 112-4
Banhatti,D G:1988:**ApSpSc** 140 291-300
Baum,S:1989:in **ESO W/s on Extranuclear Activity in Galaxies**, to appear
Baum,S,Heckman,T:1989:**ApJ 336** 702-21
Birkinshaw,M,Laing,R,Scheuer,P,Simon,A:1978:**MN 185** 39P-43P
Bridle,A H,Perley,R A:1984:**Ann Rev AA 22** 319-58
Browne,I W A:1987:in Zensus,J A,Pearson,T J(eds)**Superlum
        Rad Srcs** 129-47
Buehrke,T,Mundt,R,Ray,T P:1988:**AA 200** 99-119
Clarke,J N,Kronberg,P P,Simard-Normandin,M:1980:**MN 190** 205-15
Durret,F:1989:in **ESO W/s on Extranuclear Activity in Galaxies**, to appear
Ekers,R D:1982:**IAU Symp 97** 465-74
Falle,S A E G:1987:in Kundt,W(ed)**ibid** 163-70

Fanaroff,B L,Riley,J M:1974:**MN 167** 31P-35P
Feigelson,E D:1986:**Can J Phys 64** 421-5
Fomalont,E B:1969:**ApJ 157** 1027-45
Garrington,S T,Leahy,J P,Conway,R G,Laing,R A:1988:**Nat 331** 147-9
Hodges,M W,Mutel,R L:1987:in Zensus,J A,Pearson,T J(eds)**Superlum Rad Srcs** 168-73
Impey,C:1987:**ibid** 233-50
Ingham,W,Morrison,P:1975:**MN 173** 569-77
Jenkins C,McEllin,M:1977:**MN 180** 219-25
Jennison,R C,DasGupta,M K:1953:**Nat 172** 996-7
Koupelis,T,Van Horn,H M:1988:**ApJ 324** 93-111
Kundt,W:1984:**J AA 5** 277-83
Kundt,W:1987a:in Kundt,W(ed)**ibid** 1-13
Kundt,W:1987b:in Kundt,W(ed)**ibid** 13-20
Lada,C:1985:**Ann Rev AA 23** 267-317
Laing,R A:1988:**Nat 331** 149-51
Lawrence,A:1987:**PASP 99** 309-34
Lonsdale,C J:1986:**Can J Phys 64** 445-8
Mackay,C D:1971:**MN 154** 209-27
Macklin,J T:1981:**MN 196** 967-86
Michalitsianos,A G,Hollis,J M,Kafatos,M:1986:**Can J Phys 64** 523-6
Miley,G K:1980:**Ann Rev AA 18** 165-218
Moffet,A T:1966:**Ann Rev AA 4** 145-70
Morrison,P:1969:**ApJ 157** L73-6
Norman,M L:1986:**IAU Colloq 89** 425-37
Orr,M J L,Browne,I W A:1982:**MN 200** 1067-80
Pearson,T J,Readhead,A C S,Barthel,P D:1987:in Zensus,J A, Pearson,T J(eds),**ibid** 94-103
Perley,R A,Dreher,J W,Cowan,J:1984:**ApJ 285** L35-8+plates
Persson,S E:1986:**Can J Phys 64** 421-5
Readhead,A C S,Cohen,M H,Blandford,R D:1978:**Nat 272** 131-4
Rusk,R,Rusk,A C M:1986:**Can J Phys 64** 440-4
Scheuer,P A G:1987:in Kundt W(ed) **ApJets and Their Engines** 129-36
Scheuer,P A G,Readhead,A C S:1979:**Nat 277** 182-5
Schwartz,R D:1986:**Can J Phys 64** 414-20
Seaquist,E R,Taylor,A R:1986:**Can J Phys 64** 520-2
Sofue,Y:1986:**Can J Phys 64** 527-30
Swarup,G,Banhatti,D G:1981:**MN 194** 1025-32
Swarup,G,Sinha,R P,Hilldrup,K:1984:**MN 208** 813-43
Swarup,G,Sinha,R P,Salter,C J:1982:**IAU Symp 97** 411-2
Teerikorpi,P:1984:**AA 132** 179-86
Teerikorpi,P:1986:**AA 164** L11-2
Usher,P D:1978:in Wolfe,A M(ed) **Pittsburgh Confce on BL Lac Objs** 197-203+209-10
Valentijn,E A:1979:**AA 78** 367-72
Valtonen,M J:1984:**QJRaS 25** 28-52
van Breugel,W,Heckman,T:1982:**IAU Symp 97** 61-4
Wiita,P,Gopal-Krishna:1987:in Ulmer,M P(ed) **13th Texas Symp in Relativistic Ap** 355-6

digesting neutron stars

330